Lecture Notes in Computer Scie

T0238230

Commenced Publication in 1973
Founding and Former Series Editors:
Gerhard Goos, Juris Hartmanis, and Jan van Leeuwen

Bernd Mohr   Jesper Larsson Träff
Joachim Worringen   Jack Dongarra (Eds.)

Recent Advances in Parallel Virtual Machine and Message Passing Interface

13th European PVM/MPI User's Group Meeting
Bonn, Germany, September 17-20, 2006
Proceedings

 Springer

Volume Editors

Bernd Mohr
Forschungszentrum Jülich GmbH
Zentralinstitut für Angewandte Mathematik
52425 Jülich, Germany
E-mail: b.mohr@fz-juelich.de

Jesper Larsson Träff
C&C Research Laboratories NEC Europe Ltd.
Rathausallee 10, 53757 Sankt Augustin, Germany
E-mail: traff@ccrl-nece.de

Joachim Worringen
Dolphin Interconnect Solutions ASA
R&D Germany
Siebengebirgsblick 26, 53343 Wachtberg, Germany
E-mail: joachim@dolphinics.com

Jack Dongarra
University of Tennessee
Computer Science Department
1122 Volunteer Blvd, Knoxville, TN 37996-3450, USA
E-mail: dongarra@cs.utk.edu

Library of Congress Control Number: 2006931769

CR Subject Classification (1998): D.1.3, D.3.2, F.1.2, G.1.0, B.2.1, C.1.2

LNCS Sublibrary: SL 2 – Programming and Software Engineering

ISSN   0302-9743
ISBN-10   3-540-39110-X Springer Berlin Heidelberg New York
ISBN-13   978-3-540-39110-4 Springer Berlin Heidelberg New York

Springer is a part of Springer Science+Business Media

springer.com

© Springer-Verlag Berlin Heidelberg 2006
Printed in Germany

Typesetting: Camera-ready by author, data conversion by Scientific Publishing Services, Chennai, India
Printed on acid-free paper   SPIN: 11846802   06/3142   5 4 3 2 1 0

Preface

Since its inception in 1994 as a European PVM user's group meeting, EuroPVM/MPI has evolved into the foremost international conference dedicated to the latest developments concerning MPI (Message Passing Interface) and PVM (Parallel Virtual Machine). These include fundamental aspects of these message passing standards, implementation, new algorithms and techniques, performance and benchmarking, support tools, and applications using message passing. Despite its focus, EuroPVM/MPI is accommodating to new message-passing and other parallel and distributed programming paradigms beyond MPI and PVM. Over the years the meeting has successfully brought together developers, researchers and users from both academia and industry. EuroPVM/MPI has contributed to furthering the understanding of message passing programming in these paradigms, and has positively influenced the quality of many implementations of both MPI and PVM through exchange of ideas and friendly competition.

EuroPVM/MPI takes place each year at a different European location, and the 2006 meeting was the 13th in the series. Previous meetings were held in Sorrento (2005), Budapest (2004), Venice (2003), Linz (2002), Santorini (2001), Balatonfüred (2000), Barcelona (1999), Liverpool (1998), Cracow (1997), Munich (1996), Lyon (1995), and Rome (1994). EuroPVM/MPI 2006 took place in Bonn, Germany, 17 – 20 September, 2006, and was organized jointly by the C&C Research Labs, NEC Europe Ltd., and the Research Center Jülich.

Contributions to EuroPVM/MPI 2006 were submitted in May as either full papers or posters, or (with a later deadline) as full papers to the special session ParSim on "Current Trends in Numerical Simulation for Parallel Engineering Environments" (see page 356). Out of the 75 submitted full papers, 38 were selected for presentation at the conference. Of the 9 submitted poster abstracts, 6 were chosen for the poster session. The ParSim session received 11 submissions, of which 5 were selected for this special session. The task of reviewing was carried out smoothly within very strict time limits by a large program committee and a number of external referees, counting members from most of the American and European groups involved in MPI and PVM development, as well as from significant user communities. Almost all papers received 4 reviews, some even 5, and none fewer than 3, which provided a solid basis for the program chairs to make the final selection for the conference program. The result was a well-balanced and focused program of high quality. All authors are thanked for their contribution to the conference. Out of the accepted 38 papers, 3 were selected as *outstanding contributions* to EuroPVM/MPI 2006, and were presented in a special, plenary session:

- "Issues in Developing a Thread-Safe MPI Implementation" by William Gropp and Rajeev Thakur (page 12)
- "Scalable Parallel Suffix Array Construction" by Fabian Kulla and Peter Sanders (page 22)

- "Formal Verification of Programs That Use MPI One-Sided Communication" by Salman Pervez, Ganesh Gopalakrishnan, Robert M. Kirby, Rajeev Thakur and William Gropp (page 30)

"Late and breaking results", which were submitted in August as brief abstracts and therefore not included in these proceedings, were presented in the eponymous session. Like the "Outstanding Papers" session, this was a premiere at EuroPVM/MPI 2006.

Complementing the emphasis in the call for papers on new message-passing paradigms and programming models, the invited talks by Richard Graham, William Gropp and Al Geist addressed possible shortcomings of MPI for emerging, large-scale systems, covering issues on fault-tolerance and heterogeneity, productivity and scalability, while the invited talk of Katherine Yelick dealt with advantages of higher-level, partitioned global address space languages. The invited talk of Vaidy Sunderam discussed challenges to message-passing programming in dynamic metacomputing environments. Finally, with the invited talk of Ryutaro Himeno, the audience gained insight into the role and design of the projected Japanese peta-scale supercomputer.

An important part of EuroPVM/MPI is the technically oriented vendor session. At EuroPVM/MPI 2006 eight significant vendors of hard- and software for high-performance computing (Etnus, IBM, Intel, NEC, Dolphin Interconnect Solutions, Hewlett-Packard, Microsoft, and Sun), presented their latest products and developments.

Prior to the conference proper, four tutorials on various aspects of message passing programming ("Using MPI-2: A Problem-Based Approach", "Performance Tools for Parallel Programming", "High-Performance Parallel I/O", and "Hybrid MPI and OpenMP Parallel Programming") were given by experts in the respective fields.

Information about the conference can be found at the conference Web-site http://www.pvmmpi06.org, which will be kept available.

The proceedings were edited by Bernd Mohr, Jesper Larsson Träff and Joachim Worringen. The EuroPVM/MPI 2006 logo was designed by Bernd Mohr and Joachim Worringen.

The program and general chairs would like to thank all who contributed to making EuroPVM/MPI 2006 a fruitful and stimulating meeting, be they technical paper or poster authors, program committee members, external referees, participants, or sponsors.

September 2006

Bernd Mohr
Jesper Larsson Träff
Joachim Worringen
Jack Dongarra

Organization

General Chair

Jack Dongarra   University of Tennessee, USA

Program Chairs

Bernd Mohr   Forschungszentrum Jülich, Germany
Jesper Larsson Träff   C&C Research Labs, NEC Europe, Germany
Joachim Worringen   C&C Research Labs, NEC Europe, Germany

Program Committee

George Almasi   IBM, USA
Ranieri Baraglia   CNUCE Institute, Italy
Richard Barrett   ORNL, USA
Gil Bloch   Mellanox, Israel
Arndt Bode   Technical University of Munich, Germany
Marian Bubak   AGH Cracow, Poland
Hakon Bugge   Scali, Norway
Franck Cappello   Université de Paris-Sud, France
Barbara Chapman   University of Houston, USA
Brian Coghlan   Trinity College Dublin, Ireland
Yiannis Cotronis   University of Athens, Greece
Jose Cunha   New University of Lisbon, Portugal
Marco Danelutto   University of Pisa, Italy
Frank Dehne   Carleton University, Canada
Luiz DeRose   Cray, USA
Frederic Desprez   INRIA, France
Erik D'Hollander   University of Ghent, Belgium
Beniamino Di Martino   Second University of Naples, Italy
Jack Dongarra   University of Tennessee, USA
Graham Fagg   University of Tennessee, USA
Edgar Gabriel   University of Houston, USA
Al Geist   OakRidge National Laboratory, USA
Patrick Geoffray   Myricom, USA
Michael Gerndt   Tu München, Germany
Andrzej Goscinski   Deakin University, Australia
Richard L. Graham   LANL, USA
William D. Gropp   Argonne National Laboratory, USA
Erez Haba   Microsoft, USA

Program Committee (cont'd)

Rolf Hempel	DLR - German Aerospace Center, Germany
Dieter Kranzlmüller	Johannes Kepler Universität Linz, Austria
Rainer Keller	HLRS, Germany
Stefan Lankes	RWTH Aachen, Germany
Erwin Laure	CERN, Switzerland
Laurent Lefevre	INRIA/LIP, France
Greg Lindahl	QLogic, USA
Thomas Ludwig	University of Heidelberg, Germany
Emilio Luque	Universitat Autònoma de Barcelona, Spain
Ewing Rusty Lusk	Argonne National Laboratory, USA
Tomas Margalef	Universitat Autònoma de Barcelona, Spain
Bart Miller	University of Wisconsin, USA
Bernd Mohr	Forschungszentrum Jülich, Germany
Matthias Müller	Dresden University of Technology, Germany
Salvatore Orlando	University of Venice, Italy
Fabrizio Petrini	PNNL, USA
Neil Pundit	Sandia National Laboratories, USA
Rolf Rabenseifner	HLRS, Germany
Thomas Rauber	Universität Bayreuth, Germany
Wolfgang Rehm	TU Chemnitz, Germany
Casiano Rodriguez-Leon	Universidad de La Laguna, Spain
Michiel Ronsse	University of Ghent, Belgium
Peter Sanders	Universität Karlsruhe, Germany
Martin Schulz	Lawrence Livermore National Laboratory, USA
Jeffrey Squyres	Open System Lab, Indiana
Vaidy Sunderam	Emory University, USA
Bernard Tourancheau	Université de Lyon/INRIA, France
Jesper Larsson Träff	C&C Research Labs, NEC Europe, Germany
Carsten Trinitis	TU München, Germany
Jerzy Wasniewski	Danish Technical University, Denmark
Roland Wismueller	University of Siegen, Germany
Felix Wolf	Forschungszentrum Jülich, Germany
Joachim Worringen	C&C Research Labs, NEC Europe, Germany
Laurence T. Yang	St. Francis Xavier University, Canada

External Referees

(excluding members of the Program Committee)

Dorian Arnold	Ron Brightwell	Rafael Corchuelo
Christian Bell	Michael Brim	Karen Devine
Boris Bierbaum	Carsten Clauss	Frank Dopatka

Gábor Dózsa	Frederic Loulergue	vera
Renato Ferrini	Ricardo Peña Marí	Nathan Rosenblum
Rainer Finocchiaro	Torsten Mehlan	John Ryan
Igor Grobman	Frank Mietke	Carsten Scholtes
Yuri Gurevich	Alexander Mirgorodskiy	Silke Schuch
Torsten Höfler	Francesco Moscato	Stephen F. Siegel
Andreas Hoffmann	Zsolt Nemeth	Nicola Tonellotto
Ralf Hoffmann	Raik Nagel	Gara Miranda Valladares
Sascha Hunold	Raffaele Perego	Salvatore Venticinque
Mauro Iacono	Laura Ricci	John Walsh
Adrian Kacso	Rolf Riesen	Zhaofang Wen
Matthew Legendre	Francisco Fernández Ri-	

For the ParSim session the following external referees provided reviews.

Georg Acher	Michael Ott	Max Walter
Tobias Klug	Daniel Stodden	Josef Weidendorfer

Conference Organization

Bernd Mohr
Jesper Larsson Träff
Joachim Worringen

Sponsors

The conference would have been substantially more expensive and much less pleasant to organize without the generous support of a good many industrial sponsors. Platinum and Gold level sponsors also gave talks at the vendor session on their latest products in parallel systems and message passing software. EuroPVM/MPI 2006 gratefully acknowledges the contributions of the sponsors to a successful conference.

Platinum Level Sponsors

Etnus, IBM, Intel, and NEC.

Gold Level Sponsors

Dolphin Interconnect Solutions, Hewlett-Packard, Microsoft, and Sun.

Standard Level Sponsor

QLogic.

Table of Contents

Invited Talks

Tutorials

Outstanding Papers

Collective Communication

Communication Protocols

Debugging and Verification

Fault Tolerance

Metacomputing and Grid

Parallel I/O

Implementation Issues

Object-Oriented Message Passing

Limitations and Extensions

Performance

ParSim

Poster Abstracts

Too Big for MPI?

Al Geist

Oak Ridge National Laboratory
Oak Ridge, Tennessee, USA
gst@ornl.gov

In 2008 the National Leadership Computing Facility at Oak Ridge National Laboratory will have a petaflop system in place. This system will have tens of thousands of processors and petabytes of memory. This capability system will focus on application problems that are so hard that they require weeks on the full system to achieve breakthrough science in nanotechnology, medicine, and energy. With long running jobs on such huge computing systems the question arises: Are the computers and applications getting too big for MPI? This talk will address several reasons why the answer to this question may be yes.

The first reason is the growing need for fault tolerance. This talk will review the recent efforts in adding fault tolerance to MPI and the broader need for holistic fault tolerance across petascale machines. The second reason is the potential need by these applications for new features or capabilities that don't exist in the MPI standard. A third reason is the emergence of new languages and programming paradigms on the horizon.

This talk will discuss the DARPA High Productivity Computing Systems project and the new languages, Fortress, Chapel, Fortress, and X10 being developed by Cray, Sun, and IBM respectively.

B. Mohr et al. (Eds.): PVM/MPI 2006, LNCS 4192, p. 1, 2006.

Approaches for Parallel Applications Fault Tolerance

Richard L. Graham

Advanced Computing Laboratory
Los Alamos National Laboratory
Los Alamos, NM 87544, USA
rlgraham@lanl.gov

System component failure - hardware and software, permanent and transient - are an integral part of the life cycle of any computer system. The degree to which a system suffers from these failures depends on factors such as system complexity, system design and implementation, and system size. These errors may lead to catastrophic application failure (termination of an application run with a CPU failure), silent application errors (such as network data corruption), or application hangs (such as when network interface card (NIC) malfunction), all wasting valuable computer time. For certain classes of computer systems, dealing with these failures is a requirement to provide a simulation environment reliable enough to meet end-user needs. Also, the more automated these solutions are, requiring minimal or no end-user intervention, the more likely they are to be used to achieve the required application stability. Dealing with failure, or fault tolerance, while minimizing application performance degradation, is an active research area, with no consensus as to what are optimal solution strategies, or even what failures need to be considered. Errors include items such as transient data transmission errors (dropped or corrupt packets), transient and permanent network failures (NIC), and process failure, to list a few. The current MPI standard addresses a limited number of failure scenarios, with application termination being the default response to failure. While the standard provide a mechanism for users to override this default response, it does not define error codes that provide information on system level failures - hardware or software. None-the-less, these need to be addressed to provide end-users with systems that meet their computing needs. Building on experience gained in the LA-MPI, FT-MPI, and LAM/MPI projects, the Open MPI collaboration has implemented, and is continuing to implement optional solutions that deal with a number of failure scenarios, to decrease the application mean-time-to-failure rate, to acceptable rates. The types of errors currently being dealt with include transient network data transmission errors, transient and permanent NIC failures, and process failure. The talk will discuss fault detection, fault recovery methods, and the degree to which applications need to be modified to benefit from these, if any. In addition, the performance impact of these solutions on several applications will be discussed.

B. Mohr et al. (Eds.): PVM/MPI 2006, LNCS 4192, p. 2, 2006.

Where Does MPI Need to Grow?

William D. Gropp

Mathematics and Computer Science Division
Argonne National Laboratory
Argonne, Illinois, USA
gropp@mcs.anl.gov

MPI has been a successful parallel programming model. The combination of performance, scalability, composability, and support for libraries has made it relatively easy to build complex parallel applications. However, MPI is by no means the perfect parallel programming model. This talk will review the strengths of MPI with respect to other parallel programming models and discuss some of the weaknesses and limitations of MPI in the areas of performance, productivity, scalability, and interoperability. The talk will conclude with a discussion of what extensions (or even changes) may be needed in MPI, and what issues should be addressed by combining MPI with other parallel programming models.

B. Mohr et al. (Eds.): PVM/MPI 2006, LNCS 4192, p. 3, 2006.

Peta-Scale Supercomputer Project in Japan and Challenges to Life and Human Simulation in Japan

Ryutaro Himeno

RIKEN Advanced Center for Computing and Communication
Hirosawa 2-1, Wako, Saitama 351-0198, Japan
himeno@riken.jp

After two-year-long discussion, we are about to start the Peta-Scale Supercomputer project. The target performance is currently 10 Peta FLOPS for a few selected codes and one Peta FLOPS for major applications. It will start operation in March, 2011, and then its capability will be enlarged in 2011. Finally, it will be completed in March, 2012. The project will end in March, 2013.

This project includes two important items in software development: grid middleware and application software in Nano Science and Life Science. The development in grid middleware is planed because the supercomputer center which will operate the Peta-scale supercomputer is planed to provide services not for a specific institute or application area like the Earth Simulator but for general uses as a national infrastructure. Nano and Life sciences are the major application areas we are going to put emphases on as well as industrial applications.

We are starting to select the target applications to make a benchmark suite in various scientific and industrial applications. We are also discussing concept design and will finalize it in Summer, 2006. I will introduce the project plan and application area, especially in Life science, in detail at the conference.

B. Mohr et al. (Eds.): PVM/MPI 2006, LNCS 4192, p. 4, 2006.

Resource and Application Adaptivity in Message Passing Systems

Vaidy Sunderam

Department of Mathematics and Computer Science
Emory University
Atlanta, Georgia, USA
vss@emory.edu

Clusters and MPP's are traditional platforms for message passing applications, but there is growing interest in more dynamic metacomputing environments. The latter are characterized by dynamicity in availability and available capacity – of both nodes and interconnects. This talk will discuss fundamental challenges in executing message passing programs in such environments, and analyze the issue of adaptivity from the resource and application points of view. Pragmatic solutions to some of these challenges will then be described, along with new approaches to dealing with the aggregation of multidomain computing platforms for distributed memory concurrent computing.

B. Mohr et al. (Eds.): PVM/MPI 2006, LNCS 4192, p. 5, 2006.

Performance Advantages of Partitioned Global Address Space Languages

Katherine Yelick

Computer Science Division
University of California at Berkeley
Berkeley, California, USA
yelick@cs.berkeley.edu

For nearly a decade, the Message Passing Interface (MPI) has been the dominant programming model for high performance parallel computing, in large part because it is universally available and scales to thousands of processors. In this talk I will describe some of the alternatives to MPI based on a Partitioned Global Address Space model of programming, such as UPC and Titanium. I will show that these models offer significant advantages in performance as well as programmer productivity, because they allow the programmer to build global data structures and perform one-sided communication in the form of remote reads and writes, while still giving programmers control over data layout. In particular, I will show that these languages make more effective use of cluster networks with RDMA support, allowing them to outperform two-sided communication on both microbenchmarks and bandwidth-limited computational problems, such as global FFTs. The key optimization is overlap of communication with computation and pipelining communication. Surprisingly, sending smaller messages more frequently can be faster than a few large messages if overlap with computation is possible. This creates an interesting open problem for global scheduling of communication, since the simple strategy of maximum aggregation is not always best. I will also show some of the productivity advantages of these languages through application case studies, including complete Titanium implementations of two different application frameworks: an immersed boundary method package and an elliptic solver using adaptive mesh refinement.

B. Mohr et al. (Eds.): PVM/MPI 2006, LNCS 4192, p. 6, 2006.
© Springer-Verlag Berlin Heidelberg 2006

Using MPI-2: A Problem-Based Approach

William D. Gropp and Ewing Lusk

Mathematics and Computer Science Division
Argonne National Laboratory
Argonne, Illinois, USA
{gropp, lusk}@mcs.anl.gov

MPI-2 introduced many new capabilities, including dynamic process management, one-sided communication, and parallel I/O. Implementations of these features are becoming widespread. This tutorial shows how to use these features by showing all of the steps involved in designing, coding, and tuning solutions to specific problems. The problems are chosen for their practical use in applications as well as for their ability to illustrate specific MPI-2 topics. Complete examples will be discussed and full source code will be made available to the attendees.

B. Mohr et al. (Eds.): PVM/MPI 2006, LNCS 4192, p. 7, 2006.

Performance Tools for Parallel Programming

Bernd Mohr and Felix Wolf

Research Centre Jülich
Jülich, Germany
{b.mohr, f.wolf}@fz-juelich.de

Extended Abstract. Application developers are facing new and more complicated performance tuning and optimization problems as architectures become more
complex. In order to achieve reasonable performance on these systems, HPC users
need help from performance analysis tools. In this tutorial we will introduce the
principles of experimental performance instrumentation, measurement, and analysis, with an overview of the major issues, techniques, and resources in performance
tools development, as well as an overview of the performance measurement tools
available from vendors and research groups.

The focus of this tutorial will be on experimental performance analysis, which
is currently the method of choice for tuning large-scale, parallel systems. The
goal of experimental performance analysis is to provide the data and insights required to optimize the execution behavior of applications or system components.
Using such data and insights, application and system developers can choose to
optimize software and execution environments along many axes, including execution time, memory requirements, and resource utilization. In this tutorial we will
present a broad range of techniques used for the development of software for performance measurement and analysis of scientific applications. These techniques
range from mechanisms for simple code timings to multi-level hardware/software
measurements. In addition, we will present state of the art tools from research
groups, as well as software and hardware vendors, including practical tips and
tricks on how to use them for performance tuning.

When designing, developing, or using a performance tool, one has to decide on
which instrumentation technique to use. We will cover the main instrumentation
techniques, which can be divided into either static, during code development, compilation, or linking, or dynamic, during execution. The most common instrumentation
approach augments source code with calls to specific instrumentation libraries. During execution, these library routines collect behavioral data. One example of static
instrumentation systems that will be covered in details is the MPI profiling interface, which is part of the MPI specification, and was defined to provide a mechanism
for quick development of performance analysis system for parallel programs. In addition, we will present similar work (POMP, OPARI) that has been proposed in the
context of OpenMP. In contrast to static instrumentation, dynamic instrumentation allows users to interactively change instrumentation points, focusing measurements on code regions where performance problems have been detected. An example
of such dynamic instrumentation systems is the DynInst project from the University of Maryland and University of Wisconsin, which provides an infrastructure to

B. Mohr et al. (Eds.): PVM/MPI 2006, LNCS 4192, pp. 8–9, 2006.
© Springer-Verlag Berlin Heidelberg 2006

help tools developers to build performance tools. We will compare and contrast these instrumentation approaches.

Regardless of the instrumentation mechanism, there are two dimensions that need to be considered for performance data collection: when the performance collection is triggered and how the performance data is recorded. The triggering mechanism can be activated by an external agent, such as a timer or a hardware counter overflow, or internally, by code inserted through instrumentation. The former is also known as sampling or asynchronous, while the latter is sometimes referred as synchronous. Performance data can be summarized during runtime and stored in the form of a profile, or can be stored in the form of traces. We will present these approaches and discuss how each one reflects a different balance among data volume, potential instrumentation perturbation, accuracy, and implementation complexity. Performance data should be stored in a format that allows the generality and extensibility necessary to represent a diverse set of performance metrics and measurement points, independent of language and architecture idiosyncrasies. We will describe common trace file formats (Vampir, CLOG, SLOG, EPILOG), as well as profile data formats based on the eXtensible Markup Language (XML), which is becoming a standard for describing performance data representation.

Hardware performance counters have become an essential asset for application performance tuning. We will discuss in detail how users can access hardware performance counters using application programming interfaces such as PAPI and PCL, in order to correlate the behavior of the application to one or more of the components of the hardware.

Visualization systems should provide natural and intuitive user interfaces, as well as, methods for users to manipulate large data collections, such that they could grasp essential features of large performance data sets. In addition, given the diversity of performance data, and the fact that performance problems can arise at several levels, visualization systems should also be able to provide multiple levels of details, such that users could focus on interesting yet complex behavior while avoiding irrelevant or unnecessary details. We will discuss the different visualization and presentation approaches currently used on state of the art research tools, as well as tools from software and hardware vendors.

The tutorial will be concluded with discussion on open research problems. Given the complexity of the state of the art of parallel applications, new performance tools must be deeply integrated, combining instrumentation, measurement, data analysis, and visualization. In addition, they should be able to guide or perform performance remediation. Ideally, these environments should scale to hundreds or thousands of processors, support analysis of distributed computations, and be portable across a wide range of parallel systems. Also, performing a whole series of experiments (studies) should be supported to allow a comparative or scalability analysis. We will discuss research efforts in automating the process of performance analysis such as the projects under the APART working group effort. We conclude the tutorial with a discussion on issues related to analysis of grid applications.

High-Performance Parallel I/O

Robert Ross[1] and Joachim Worringen[2]

[1] Mathematics and Computer Science Division
Argonne National Laboratory
Argonne, Illinois, USA
`rross@mcs.anl.gov`
[2] Dolphin Interconnect Solutions R&D Germany
Wachtberg, Germany
`joachim@dolphinics.com`

Effectively using I/O resources on HPC machines is a black art. The purpose of this tutorial is to shed light on the state-of-the-art in parallel I/O and to provide the knowledge necessary for attendees to best leverage the I/O resources available to them.

In the first half of the tutorial we discuss the software involved in parallel I/O. We cover the entire I/O software stack from parallel file systems at the lowest layer, to intermediate layers (such as MPI-IO), and finally high-level I/O libraries (such as HDF-5). The emphasis is not just on how to use these layers, but ways to use them that result in high performance. As part of this discussion we will present benchmark results from current systems.

The second half of the tutorial will be hands-on, with the participants solving typical problems of parallel I/O using different approaches. The performance of these approaches will be evaluate on different machines at remote sites, using various types of file systems. The results are then compared to get a full picture of the performance differences and characteristics of the chosen approaches on the different platforms.

Basic knowledge of parallel (MPI) programming in C and/or Fortran is assumed. For the second half, each participant should bring his own notebook computer, running either Windows XP or Linux (x86). A limited number of loan notebook computers are available on request.

B. Mohr et al. (Eds.): PVM/MPI 2006, LNCS 4192, p. 10, 2006.

Hybrid MPI and OpenMP Parallel Programming

Rolf Rabenseifner[1], Georg Hager[2], Gabriele Jost[3], and Rainer Keller[1]

[1] High Performance Computing Center Stuttgart (HLRS)
Department Parallel Computing
Stuttgart, Germany
{keller, rabenseifner}@hlrs.de
[2] University of Erlangen-Nuremberg
Erlangen, Germany
georg.hager@rrze.uni-erlangen.de
[3] Sun Microsystems, Germany
gabriele.jost@sun.com

Most HPC systems are clusters of shared memory nodes. Such systems can be PC clusters with dual or quad boards, but also "constelation" type systems with large SMP nodes. Parallel programming must combine the distributed memory parallelization on the node inter-connect with the shared memory parallelization inside of each node.

This tutorial analyzes the strength and weakness of several parallel programming models on clusters of SMP nodes. Various hybrid MPI+OpenMP programming models are compared with pure MPI. Benchmark results of several platforms are presented. A hybrid-masteronly programming model can be used more efficiently on some vector-type systems, but also on clusters of dual-CPUs. On other systems, one CPU is not able to saturate the inter-node network and the commonly used masteronly programming model suffers from insufficient inter-node bandwidth. The thread-safety quality of several existing MPI libraries is also discussed. Case studies from the fields of CFD (NAS Parallel Benchmarks and Multi-zone NAS Parallel Benchmarks, in detail), Climate Modelling (POP2, maybe) and Particle Simulation (GTC, maybe) will be provided to demonstrate various aspect of hybrid MPI/OpenMP programming.

Another option is the use of distributed virtual shared-memory technologies which enable the utilization of "near-standard" OpenMP on distributed memory architectures. The performance issues of this approach and its impact on existing applications are discussed. This tutorial analyzes strategies to overcome typical drawbacks of easily usable programming schemes on clusters of SMP nodes.

B. Mohr et al. (Eds.): PVM/MPI 2006, LNCS 4192, p. 11, 2006.
© Springer-Verlag Berlin Heidelberg 2006

Issues in Developing a Thread-Safe
MPI Implementation

William Gropp and Rajeev Thakur

Mathematics and Computer Science Division
Argonne National Laboratory
Argonne, IL 60439, USA
{gropp, thakur}@mcs.anl.gov

Abstract. The MPI-2 Standard has carefully specified the interaction
between MPI and user-created threads, with the goal of enabling users to
write multithreaded programs while also enabling MPI implementations
to deliver high performance. In this paper, we describe and analyze what
the MPI Standard says about thread safety and what it implies for an implementation. We classify the MPI functions based on their thread-safety
requirements and discuss several issues to consider when implementing
thread safety in MPI. We use the example of generating new context ids
(required for creating new communicators) to demonstrate how a simple solution for the single-threaded case cannot be used when there are
multiple threads and how a naïve thread-safe algorithm can be expensive. We then present an algorithm for generating context ids that works
efficiently in both single-threaded and multithreaded cases.

1   Introduction

With SMP machines being commonly available and multicore chips becoming the
norm, the mixing of the message-passing programming model with multithread-
ing on a single multicore chip or SMP node is becoming increasingly important.
The MPI-2 Standard has clearly defined the interaction between MPI and user-
created threads in an MPI program [5]. This specification was written with the
goal of enabling users to write multithreaded MPI programs easily, without unduly burdening MPI implementations to support more than what a user might
need. Nonetheless, implementing thread safety in MPI without sacrificing too
much performance requires careful thought and analysis.

In this paper, we discuss issues involved in developing an efficient thread-safe
MPI implementation. We had to deal with many of these issues when designing
and implementing thread safety in MPICH2 [6]. We first describe in brief the
thread-safety specification in MPI. We then classify the MPI functions based on
their thread-safety requirements. We discuss issues to consider when implementing thread safety in MPI. In addition, we discuss the example of generating context ids and present an efficient, thread-safe algorithm for both single-threaded
and multithreaded cases.

B. Mohr et al. (Eds.): PVM/MPI 2006, LNCS 4192, pp. 12–21, 2006.

Thread safety in MPI has been studied by a few researchers, but none of them have covered the topics discussed in this paper. Protopopov et al. discuss a number of issues related to threads and MPI, including a design for a thread-safe version of MPICH-1 [8,9]. Plachetka describes a mechanism for making a thread-unsafe PVM or MPI implementation quasi-thread-safe by adding an interrupt mechanism and two functions to the implementation [7]. García et al. present MiMPI, a thread-safe implementation of MPI [3]. TOMPI [2] and TMPI [10] are *thread-based* MPI implementations, where each MPI process is actually a thread. A good discussion of the difficulty of programming with threads in general is given in [4].

2   What MPI Says About Thread Safety

MPI defines four "levels" of thread safety: MPI_THREAD_SINGLE, where only one thread of execution exists; MPI_THREAD_FUNNELED, where a process may be multi-threaded but only the thread that initialized MPI makes MPI calls; MPI_THREAD_SERIALIZED, where multiple threads may make MPI calls but not si-multaneously; and MPI_THREAD_MULTIPLE, where multiple threads may call MPI at any time. An implementation is not required to support levels higher than MPI_THREAD_SINGLE; that is, an implementation is not required to be thread safe. A fully thread-compliant implementation, however, will support MPI_THREAD_MULTIPLE. MPI provides a function, MPI_Init_thread, by which the user can indicate the desired level of thread support, and the implementation can return the level supported. A portable program that does not call MPI_Init_thread should assume that only MPI_THREAD_SINGLE is supported. In this paper, we focus on the MPI_THREAD_MULTIPLE (fully multithreaded) case.

For MPI_THREAD_MULTIPLE, the MPI Standard specifies that when multiple threads make MPI calls concurrently, the outcome will be as if the calls executed sequentially in some (any) order. Also, blocking MPI calls will block only the calling thread and will not prevent other threads from running or executing MPI functions. MPI also says that it is the user's responsibility to prevent races when threads in the same application post conflicting MPI calls. For example, the user cannot call MPI_Info_set and MPI_Info_free on the same info object concurrently from two threads of the same process; the user must ensure that the MPI_Info_free is called only after MPI_Info_set returns on the other thread. Similarly, the user must ensure that collective operations on the same communicator, window, or file handle are correctly ordered among threads.

3   Thread-Safety Classification of MPI Functions

We analyzed each MPI function (about 305 functions in all) to determine its thread-safety requirements. We then classified each function into one of several categories based on its primary requirement. The categories and examples of functions in those categories are described below; the complete classification can be found in [1].

None Either the function has no thread-safety issues, or the function has no thread-safety issues in correct programs and the function must have low overhead, so an optimized (nondebug) version need not check for race conditions. Examples: `MPI_Address`, `MPI_Wtick`.

Access Only The function accesses fixed data for an MPI object, such as the size of a communicator. This case differs from the "None" case because an erroneous MPI program could free the object in a race with a function that accesses the read-only data. A production MPI implementation need not guard this function against changes in another thread. This category may also include replacing a value in an object, such as setting the name of a communicator. Examples: `MPI_Comm_rank`, `MPI_Get_count`.

Update Ref The function updates the reference count of an MPI object. Such a function is typically used to return a reference to an existing object, such as a datatype or error handler. Examples: `MPI_Comm_group`, `MPI_File_get_view`.

Comm/IO The function needs to access the communication or I/O system in a thread-safe way. This is a very coarse-grained category but is sufficient to provide thread safety. In other words, an implementation may (and probably should) use finer-grained controls within this category. Examples: `MPI_Send`, `MPI_File_read`.

Collective The function is collective. MPI requires that the user not call collective functions on the same communicator in different threads in a way that may make the order of invocation depend on thread timing (race). Therefore, a production MPI implementation need not separately lock around the collective functions, but a debug version may want to detect races. The communication part of the collective function is assumed to be handled separately through the communication thread locks. Examples: `MPI_Bcast`, `MPI_Comm_spawn`.

Read List The function returns an element from a list of items, such as an attribute or info value. A correct MPI program will not contain any race that might update or delete the entry that is being read. This guarantee enables an implementation to use a lock-free, thread-safe set of list update and access operations in the production version; a debug version can attempt to detect improper race conditions. Examples: `MPI_Info_get`, `MPI_Comm_get_attr`.

Update List The function updates a list of items that may also be read. Multiple threads are allowed to simultaneously update the list, so the update implementation must be thread safe. Examples: `MPI_Info_set`, `MPI_Type_delete_attr`.

Allocate The function allocates an MPI object (may also need memory allocation such as with `malloc`). Examples: `MPI_Send_init`, `MPI_Keyval_create`.

Own The function has its own thread-safety management. Examples are "global" state such as buffers for `MPI_Bsend`. Examples: `MPI_Buffer_attach`, `MPI_Cart_create`.

Other Special cases. Examples: `MPI_Abort` and `MPI_Finalize`.

This classification helps an implementation determine the scope of the thread-safety requirements of various MPI functions and accordingly decide how to

Process 0		Process 1	
Thread 0	Thread 1	Thread 0	Thread 1
MPI_Recv(src=1)	MPI_Send(dest=1)	MPI_Recv(src=0)	MPI_Send(dest=0)

Fig. 1. An implementation must ensure that this example never deadlocks for any ordering of thread execution

implement them. For example, functions that fall under the "None" or "Access Only" category need not have any thread lock in them. Appropriate thread locks can be added to other functions.

4 Issues in Implementing Thread Safety

A straightforward implication of the MPI thread-safety specification is that an implementation cannot implement thread safety by simply acquiring a lock at the beginning of each MPI function and releasing it at the end of the function: A blocked function that holds a lock may prevent MPI functions on other threads from executing, which in turn might prevent the occurrence of the event that is needed for the blocked function to return. An example is shown in Figure 1. If thread 0 happened to get scheduled first on both processes, and MPI_Recv simply acquired a lock and waited for the data to arrive, the MPI_Send on thread 1 would not be able to acquire its lock and send its data, which in turn would cause the MPI_Recv to block forever.

In addition to using a more detailed strategy than simply locking around every function, an implementation must consider other issues that are described below. In particular, it is not enough to just lock around nonblocking communication calls and release the locks before calling a blocking communication call.

4.1 Updates of MPI Objects

A number of MPI objects, such as datatypes and communicators, have *reference-count* semantics. That is, the user can free a datatype after it has been used in a nonblocking communication operation even before that communication completes. MPI guarantees that the object will not be deleted until all uses have completed. A common way to implement this semantic is to maintain with each object a reference count that is incremented each time the object is used and decremented when the use is complete. In the multithreaded case, the reference count must be changed atomically because multiple threads could attempt to modify it simultaneously.

4.2 Thread-Private Memory

In the multithreaded case, an MPI implementation may sometimes need to use global or static variables that have different values on different threads. This cannot be achieved with regular variables because the threads of a process share a single memory space. Instead, one has to use special functions provided by the threads package for accessing thread-private memory (for example, pthread_getspecific).

For example, thread-private memory is needed for keeping track of the "nesting level" of MPI functions. MPI functions may be nested because the implementation of an MPI function may call another MPI function. For example, the collective I/O functions may internally call MPI communication functions. If an error occurs in the nested MPI function, the implementation must not invoke the error handler. Instead, the error must be propagated back up to the top-level MPI function, and the error handler for that function must be invoked. This process requires keeping track of the nesting level of MPI functions and not invoking the error handler if the nesting level is more than one. (The implementation cannot simply reset the error handler before calling the nested function because the application may call the same function from another thread and expect the error handler to be invoked.) In the single-threaded case, an implementation could simply use a global variable to keep track of the nesting level, but in the multithreaded case, thread-private memory must be used.

Since accessing thread-private data requires a function call, implementations must ensure that such access is minimized in order to maintain good efficiency.

4.3   Memory Consistency

Updates to memory in one thread may not be seen in the same order by another thread. For example, some processors require an explicit *write barrier* to ensure that all memory-store operations have completed in memory. The lock and unlock operations for thread mutexes typically also perform the necessary synchronization operations needed for memory consistency. If an implementation avoids using mutex locks for higher performance, however, and instead uses other mechanisms such as lock-free atomic updates, it must be careful to ensure that the memory updates happen as desired. This is a deep issue, a full discussion of which must include concepts such as sequential consistency and release consistency and is beyond the scope of this paper. Nonetheless, it suffices to say that an implementation must ensure that, for any object that multiple threads may access, the updates are consistent across all threads, not just the thread performing the updates.

4.4   Thread Failure

A major problem with any lock-based thread-safety model is what happens when a thread that holds a lock fails or is deliberately canceled (for example, with pthread_cancel). In that case, no other thread can acquire the lock, and the application may hang. One solution is to avoid using locks and instead use lock-free algorithms wherever possible (such as for the Update List category of functions described in Section 3).

4.5   Performance and Code Complexity

A tradeoff in performance and code complexity exists between using a single, coarse-grained lock and multiple, finer-grained locks. The single lock is relatively easy to implement but effectively serializes the MPI functions among threads. A finer-grained approach, using either multiple locks or a combination of locks and lock-free methods, risks the occurrence of deadly embrace (when two threads each hold one of the two locks that the other thread needs) as well as considerable code complexity. In addition, if the finer-grained approach requires multiple locks, each operation may be more expensive than if a single lock is used. MPI functions that can avoid using locks altogether by using lock-free methods, such as the functions in the Update List or Allocate categories, can provide a middle ground, trading a small amount of code complexity for more concurrency in execution.

4.6   Thread Scheduling

Another issue is avoiding "busy waiting" or "spin locks." In multithreaded code, it is common practice to have a thread that is waiting for an event (such as an incoming message for a blocking MPI_Recv) to yield to other threads, so that those threads can perform useful work. Thread systems provide various mechanisms for implementing this, such as condition variables. One difficulty is that not all events have the ability to wake up a thread; for example, if a low-latency method is being used to communicate between different processes in the same shared-memory node, there may be no easy way to signal the target process or thread. This situation often leads to a tradeoff between latency and effective scheduling.

5   An Algorithm for Generating Context Ids

In this section, we use the example of generating context ids to show how a simple solution for the single-threaded case cannot be used when there are multiple threads. We then present an efficient algorithm for generating context ids in the multithreaded case.

5.1   Basic Concept and Single-Threaded Solution

A communicator in MPI has a notion of a "context" associated with it, which is invisible to the user. This notion is implicit in a communicator and provides a safe communication space so that a message sent on a communicator is matched only by a receive posted on the same communicator (and not any other communicator).

Typically, the context is implemented as an integer that has the same value on all processes that are part of the communicator and is unique among all communicators on a given process. For example, if the context id of a communicator

'X' on a process is 42, all other processes that are part of X must use 42 as the context id for X, and no other communicator on any of these processes may use 42 as its context id. Processes that are not part of X, however, may use 42 as the context id for some other communicator.

Whenever a new communicator is created (for example, with MPI_Comm_create or MPI_Comm_dup), the processes in that communicator must agree on a context id for the new communicator, following the constraints given above. In the single-threaded case, generating a new context id is easy. One approach could be for each process to maintain a global data structure containing the list of available context ids on that process. In order to save memory space, the list can be maintained as a bit vector, with the bits indicating whether the corresponding context ids are available. A new context id can be generated by performing an MPI_Allreduce with the appropriate bit operator (MPI_BAND). The position of the lowest set bit can be used as the new context id.

5.2   Naïve Multithreaded Algorithm

The multithreaded case is more difficult. A process cannot simply acquire a thread lock, call MPI_Allreduce, and release the lock, because the threads on various processes may acquire locks in different order, causing the allreduce operation to hang because of a deadly embrace.

One possible solution is to acquire a thread lock, read the bit vector, release the lock, then do the MPI_Allreduce, followed by another MPI_Allreduce to determine whether the bit vector has been changed by another thread between the lock release and the first allreduce. If not, then the value for the context id can be accepted; otherwise, the algorithm must be repeated. This method is expensive, however, as it requires multiple MPI_Allreduce calls. In addition, two competing threads could loop forever, with each thread invalidating the other's choice of context value.

5.3   Efficient Algorithm for the Multithreaded Case

We instead present a new algorithm that works efficiently in both single-threaded and multithreaded cases. We have implemented this algorithm in MPICH2 [6]. For simplicity, we present the algorithm only for the case of intracommunicators. The pseudocode is given in Figure 2.

The algorithm uses a bit mask of context ids; each bit set indicates a context id available. For example, 32 32-bit integers will cover 1024 context ids. This mask and two other variables, lowestContextId and mask_in_use, are stored in global memory (shared among the threads of a process). lowestContextId is used to store the smallest context id among the input communicators of the various threads on a process that need to find a new context id. mask_in_use indicates whether some thread has acquired the rights to the mask.

The algorithm works as follows. A thread wishing to get a new context id first acquires a thread lock. If mask_in_use is set or the context id of the thread's input communicator is greater than lowestContextId, the thread uses 0 as the

```
/* global variables (shared among threads of a process) */
mask          /* bit mask of context ids in use by a process */
mask_in_use        /* flag; initialized to 0 */
lowestContextId   /* initialized to MAXINT */

/* local variables (not shared among threads) */
local_mask     /* local copy of mask */
i_own_the_mask /* flag */
context_id     /* new context id; initialized to 0 */

while (context_id == 0) {
    Mutex_lock()
    if (mask_in_use || MyComm->contextid > lowestContextId) {
        local_mask = 0
    i_own_the_mask = 0
    if (MyComm->contextid < lowestContextId) {
            lowestContextId = MyComm->contextid
        }
    }
    else {
        local_mask = mask
        mask_in_use = 1
    i_own_the_mask = 1
        lowestContextId = MyComm->contextid
    }
    Mutex_unlock()

    MPI_Allreduce(local_mask, MPI_BAND)

    if (i_own_the_mask) {
        Mutex_lock()
    if (local_mask != 0) {
            context_id =
                location of first set bit in local_mask
            update mask
            if (lowestContextId == MyComm->contextid) {
                lowestContextId = MAXINT;
            }
        }
        mask_in_use = 0
        Mutex_unlock()
    }
}
return context_id
```

Fig. 2. Pseudocode for generating a new context id in the multithreaded case (for intracommunicators)

local mask (for allreduce) and sets the flag i_own_the_mask to 0. Otherwise, it uses the current context-id mask as the local mask (for allreduce) and sets the flags mask_in_use and i_own_the_mask to 1. Then it releases the lock and calls MPI_Allreduce.

After MPI_Allreduce returns, if i_own_the_mask is 1, the thread acquires the lock again. If the result of the allreduce (local mask) is not 0, it means all threads that participated in the allreduce owned the mask on their processes and therefore the location of the first set bit in local mask can be used as the new context id. If the result of the allreduce is 0, it means that some thread did not own the mask on its process and therefore the algorithm must be retried. mask_in_use is reset to 0 before releasing the lock.

The logic for lowestContextId exists to prevent a livelock situation where the allreduce operation always contains some threads that do not own the mask, resulting in a 0 output. Since threads in our algorithm yield ownership of the mask to the thread with the lowest context id, there will be a time when all the threads of the communicator with the lowest context id will own the mask on their respective processes, causing the allreduce to return a nonzero result, and a new context id to be found. Those threads will disappear from the contention, and the same algorithm will enable other threads to complete their operation.

In this algorithm, the case where different threads of a process may have the same input context id does not arise because it is not legal for multiple threads of a process to call collective functions with the same communicator at the same time, and all the MPI functions that need to create new context ids (namely, the functions that return new communicators) are collective functions.

We note that, in the single-threaded case, this algorithm is as efficient as the basic algorithm described in Section 5.1, because the mutex locks can be commented out and no extra communication is needed as the first allreduce itself will succeed. Even in the multithreaded case, in most common circumstances, the first allreduce will succeed, and no extra communication will be needed.

Further Improvements. A refinement to this algorithm could be to allow multiple threads to have disjoint masks; if the masks are cleverly picked, most threads would find an acceptable value even if multiple threads were concurrently executing the algorithm. Another refinement could be to use a queue of pending threads ordered by increasing context id of the input communicator. Threads that are high in this queue could wait on a condition variable or other thread-synchronization mechanism that is activated whenever there is a change in the thread with the lowest context id, either because a thread has found a new context id and is removed from the queue or because a new thread with a lower context id enters the function.

6   Conclusions and Future Work

Implementing thread safety in MPI is not simple or straightforward. Careful thought and analysis are required in order to implement thread safety correctly and without sacrificing too much performance. In this paper, we have discussed several issues that an implementation must consider when implementing thread safety in MPI. Some of the issues are subtle, but nonetheless important.

The default ch3:sock channel (TCP) in the current version of MPICH2 (1.0.3) is thread safe. It, however, needs to be configured and built separately for thread safety, with the configure option `--enable-threads`. In the next release, 1.0.4, the default build of the ch3:sock channel will support thread safety, but thread safety will be enabled only if the user calls `MPI_Init_thread` with `MPI_THREAD_MULTIPLE`. If not, no thread locks will be called, so there is no penalty. We are also working on performance improvements to the thread support in MPICH2 and extending thread safety to all the communication channels.

Although many MPI implementations claim to be thread safe, no comprehensive test suite exists to validate the claim. We plan to develop a test suite that can be used to verify the thread safety of MPI implementations.

Acknowledgments

This work was supported by the Mathematical, Information, and Computational Sciences Division subprogram of the Office of Advanced Scientific Computing Research, Office of Science, U.S. Department of Energy, under Contract W-31-109-ENG-38.

References

1. Analysis of thread safety needs of MPI routines.
   http://www.mcs.anl.gov/mpi/mpich2/developer/design/threadlist.htm.
2. Erik D. Demaine. A threads-only MPI implementation for the development of parallel programs. In *Proc. of the 11th International Symposium on High Performance Computing Systems*, pages 153–163, July 1997.
3. Francisco García, Alejandro Calderón, and Jesús Carretero. MiMPI: A multithread-safe implementation of MPI. In *Proc. of 6th European PVM/MPI Users' Group Meeting*, pages 207–214, September 1999.
4. Edward A. Lee. The problem with threads. *Computer*, 39(5):33–42, May 2006.
5. Message Passing Interface Forum. MPI-2: Extensions to the Message-Passing Interface, July 1997. http://www.mpi-forum.org/docs/docs.html.
6. MPICH2. http://www.mcs.anl.gov/mpi/mpich2.
7. Tomas Plachetka. (Quasi-) thread-safe PVM and (quasi-) thread-safe MPI without active polling. In *Proc. of 9th European PVM/MPI Users' Group Meeting*, pages 296–305, September 2002.
8. Boris V. Protopopov and Anthony Skjellum. A multithreaded message passing interface (MPI) architecture: Performance and program issues. *Journal of Parallel and Distributed Computing*, 61(4):449–466, April 2001.
9. Anthony Skjellum, Boris Protopopov, and Shane Hebert. A thread taxonomy for MPI. In *Proc. of the 2nd MPI Developers Conference*, pages 50–57, June 1996.
10. Hong Tang and Tao Yang. Optimizing threaded MPI execution on SMP clusters. In *Proc. of the 15th ACM International Conference on Supercomputing*, pages 381–392, June 2001.

Scalable Parallel Suffix Array Construction

Fabian Kulla and Peter Sanders

Universität Karlsruhe, 76128 Karlsruhe, Germany
sanders@ira.uka.de

Abstract. Suffix arrays are a simple and powerful data structure for text processing that can be used for full text indexes, data compression, and many other applications in particular in bioinformatics. We describe the first implementation and experimental evaluation of a scalable parallel algorithm for suffix array construction. The implementation works on distributed memory computers using MPI, Experiments with up to 128 processors show good constant factors and make it look likely that the algorithm would also scale to considerably larger systems. This makes it possible to build suffix arrays for huge inputs very quickly. Our algorithm is a parallelization of the linear time DC3 algorithm.

1 Introduction

The suffix array [1,2], a lexicographically sorted array of the suffixes of a string, has numerous applications, e.g., in string matching [1,2], genome analysis [3] and text compression [4]. For example, one can use it as full text index: To find all occurrences of a pattern P in a text T, do binary search in the suffix array of T, i.e., look for the interval of suffixes that have P as a prefix. A lot of effort has been devoted to efficient construction of suffix arrays, culminating recently in three direct linear time algorithms [5,6,7]. One of the linear time algorithms, *DC3* [8] is very simple and can also be adapted to different models of computation. An external memory version of the algorithm [9] already makes it possible to construct suffix array for huge inputs. However, this takes many hours and hence a scalable *parallel algorithm* might be more interesting. This is the subject of the present paper. We describe the algorithm, *pDC3*, in Section 2 and experimental results in Section 3. Section 4 concludes with an outline of possible additional questions.

Related Work

There are numerous theoretical results on parallel suffix *tree* construction (e.g., refer to the references given in [10,11]). Suffix trees can be easily converted to suffix arrays. However, these algorithms are fairly complicated. We are not aware of any implementations. Recently, a trend is to use simpler suffix array construction algorithms even as a means of constructing suffix trees. Parallel conversion algorithms are described in [10]. The basic ideas for parallel suffix array construction based on the DC3 algorithm are already given in [8,11] for several theoretical models of parallel computation. Here, we concentrate on the detailed description of a practical

B. Mohr et al. (Eds.): PVM/MPI 2006, LNCS 4192, pp. 22–29, 2006.

algorithm with particular emphasis on implementation and experimental evaluation. We are only aware of a single implemented parallel algorithm for suffix array construction [12]. This algorithm is practical but based on string sorting and thus needs quadratic work in the worst case. From experiments with sequential algorithms, it is also known that algorithms based on string sorting are not very fast even for some real world inputs with long common prefixes (e.g. [13]). Furthermore it seems that all processing elements (PEs) need access to the complete input. This is an impediment for scaling to large numbers of PEs and large inputs since there might not be enough space on distributed memory machines and since this implies an execution time of $\Omega(n)$, i.e., the maximal speedup is bounded by a constant independent of the number of PEs.

2   The pDC3 Algorithm

We use the shorthands $[i, j] = \{i, \ldots, j\}$ and $[i, j) = [i, j-1]$ for ranges of integers and extend to substrings as seen below. The **input** of a suffix array construction algorithm is a *string* $T = T[0, n) = t_0 t_1 \cdots t_{n-1}$ over the alphabet $[1, n]$, that is a sequence of n integers from the range $[1, n]$. For convenience, we assume that $t_j = 0$ for $j \geq n$. For $i \in [0, n]$, let S_i denote the *suffix* $T[i, n) = t_i t_{i+1} \cdots t_{n-1}$. We explain the algorithm using pseudocode manipulating sequences of tuples. For example, for $T = abcdef$, $\langle (T[i, i+2], i) : i \bmod 3 = 0 \rangle$ denotes $\langle (abc, 0), (def, 3) \rangle$. The goal is to sort the sequence $\langle S_0, \ldots, S_n \rangle$ of suffixes of T, where comparison of substrings or tuples assumes the lexicographic order throughout this paper. The **output** is the *suffix array* $SA[0, n)$ of T, a permutation of $[0, n)$ satisfying $S_{SA[0]} < S_{SA[1]} < \cdots < S_{SA[n-1]}$. Let p denote the number of processors (PEs). PEs are numbered from 0 to $p - 1$.

At the most abstract level, the DC3 Algorithm is very simple and completely independent of the model of computation: It first constructs the suffix array of the suffixes starting at positions $i \bmod 3 \neq 0$. This is done by reduction to the suffix array construction of a string of two thirds the length, which is solved recursively. Then this information is used to annotate the original input. With this annotation, two *arbitrary* suffixes S_i and S_j can be compared by looking at $T[i, i+2]$ and the annotations at positions $[i, i+2]$. For a more detailed explanation refer to [11].

Fig. 1 gives a more detailed pseudocode which exposes parallelism and which we will then refine to the full parallel algorithm. Line 1 extracts the information needed for building the recursive subproblem which consists of two concatenated substrings of length $n/3$ representing the mod1 suffixes and mod2 suffixes respectively. This length reduction is achieved by finding *lexicographic names* for triples of characters, i.e., integers that reflect the lexicographic order of these character triples. To find these names, the triples (annotated with their position in the input) are sorted in Line 2 and named in Line 3 using a subroutine to be discussed. If all triples are unique, no recursion is necessary (Line 4). Otherwise, Line 5 assembles the recursive subproblem, Line 6 solves it, and Line 7 brings it into a form compatible with the output of the naming routine. Line 8 permutes

Function $pDC3(T)$

$\quad S:= \langle((T[i, i+2]), i) : i \in [0, n), i \bmod 3 \neq 0\rangle$ $\hspace{2cm} 1$

$\quad$ *sort S by the first component* $\hspace{4.5cm} 2$

$\quad P:= name(S)$ $\hspace{6.5cm} 3$

$\quad$ **if** *the names in P are not unique* **then** $\hspace{2.7cm} 4$

$\quad\quad$ *permute the $(r, i) \in P$ such that they are sorted by $(i \bmod 3, i \operatorname{div} 3)$* $\hspace{0.3cm} 5$

$\quad\quad SA^{12}:= pDC3(\langle c : (c, i) \in P\rangle)$ $\hspace{3.5cm} 6$

$\quad\quad P:= \langle(j+1, SA^{12}[j]) : j \in [0, 2n/3)\rangle$ $\hspace{2.6cm} 7$

$\quad$ *permute P such that it is sorted by the second component* $\hspace{1.2cm} 8$

$\quad S_0:= \langle(T[i], T[i+1], c', c'', i) : i \bmod 3 = 0, (c', i+1), (c'', i+2) \in P\rangle$ $\hspace{0.2cm} 9$

$\quad S_1:= \langle(c, T[i], c', i) \hspace{1.6cm} : i \bmod 3 = 1, (c, i), (c', i+1) \in P\rangle$ $\hspace{0.2cm} 10$

$\quad S_2:= \langle(c, T[i], T[i+1], c'', i) : i \bmod 3 = 2, (c, i), (c'', i+2) \in P\rangle$ $\hspace{0.2cm} 11$

$\quad S:= $ *sort* $S_0 \cup S_1 \cup S_2$ *using comparison function:* $\hspace{2cm} 12$

$\quad\quad (c, \ldots) \in S_1 \cup S_2 \leq (d, \ldots) \in S_1 \cup S_2 \Leftrightarrow c \leq d$

$\quad\quad (t, t', c', c'', i) \in S_0 \leq (u, u', d', d'', j) \in S_0 \Leftrightarrow (t, c') \leq (u, d')$

$\quad\quad (t, t', c', c'', i) \in S_0 \leq (d, u, \quad d', j) \in S_1 \Leftrightarrow (t, c') \leq (u, d')$

$\quad\quad (t, t', c', c'', i) \in S_0 \leq (d, u, u', d'', j) \in S_2 \Leftrightarrow (t, t', c'') \leq (u, u', d'')$

$\quad$ **return** $\langle$*last component of $s : s \in S$*$\rangle$ $\hspace{5cm} 13$

Fig. 1. High level pseudo code for pDC3

the resulting tuples into the order of the input string. Now, Lines 9–11 use the input string and the result of the recursion to build 5-tuples and 4-tuples that contain all the information needed to compare the suffixes they represent. These are sorted in Line 12. Line 13 extracts the suffix array from the result.

The basic idea behind parallelization is that input, output, and intermediate tuple sequences are uniformly or almost uniformly distributed over all PEs. Lines 1,7, 9–11, and 13 are then straight forward to parallelize. The only necessary communication is between PE i and PE $i+1$ to retrieve values that are one or two places to the right in the sequence currently processed. Permutations (Lines 5 and 8) are mapped to personalized all-to-all communications with variable message lengths but balanced or almost balanced total communication volume at each PE. Sorting (Lines 2 and 12) can be implemented using any parallel sorting algorithm. The naming step in Line 3 is interesting since its sequential implementation scans S assigning a fresh name to any new triple found. On the first glance this looks inherently sequential. However consider replacing the naming step by the following two lines.

$$\Delta:= \langle[S[i] \neq S[i+1]] : 0 \leq i < 2n/3\rangle$$
$$P:= \left\langle(1 + \textstyle\sum_{j<i} \Delta[j], i) : 0 \leq i < 2n/3\right\rangle$$

The first line is a simple local computation. The second line computes a *prefix sum*, an operation easily done in time $\mathcal{O}(n/p + \log p)$. Finally, to implement Line 4 in Fig. 1, PE $p-1$ just needs to check whether the total sum over Δ was n and broadcast this information to all PEs.

This level of abstraction is the most appropriate for an analysis of the algorithm.

Theorem 1. *The suffix array of a string of size n can be computed in time $\mathcal{O}(T_{\mathrm{parsort}}(n,p) + T_{\mathrm{allall}}(n/p,p) + f(p)\log(n))$ where $T_{\mathrm{parsort}}(n,p)$ is a bound on the execution time of sorting n elements on p processors with the property $T_{\mathrm{parsort}}(2n/3,p) \leq \frac{2}{3}T_{\mathrm{parsort}}(n,p) + f(p)$ and $T_{\mathrm{allall}}(\ell,p)$ is a bound on the execution time of personalized all-to-all communication such that no PE sends or receives more than ℓ words of data with the property that $T_{\mathrm{allall}}(2\ell/3,p) \leq \frac{2}{3}T_{\mathrm{allall}}(\ell,p) + f(p)$.*

The term $f(p) \in \Omega(\log p)$ in Theorem 1 is a bottleneck term that does not decrease when the input size decreases.

Proof. (Outline) The algorithm goes through $\mathcal{O}(\log n)$ levels[1] of recursion. The involved data volumes are decreasing geometrically. Thus, up to constant factors, we can bound the total execution time of sorting, all-to-all, and local operations by the cost of the first level of recursion, plus $\mathcal{O}(\log n)$ times the bottleneck term $f(p)$. Further communication operations all take time $\mathcal{O}(\log p) = \mathcal{O}(f(p))$ in each level of recursion. ∎

The usual implementation of all-to-all directly delivers all messages to their destination. It has $T_{\mathrm{allall}}(\ell,p) = \mathcal{O}(\ell T_{\mathrm{byte}} + pT_{\mathrm{start}})$ on a machine with full interconnection network and time $kT_{\mathrm{byte}} + T_{\mathrm{start}}$ for point-to-point communication of a message of size k.

In our implementation we have

$$T_{\mathrm{parsort}}(n,p) = \mathcal{O}\big((n/p + p^2)\log p\big) + T_{\mathrm{allall}}(n/p,p)$$

using a simple variant of comparison based sample sort [14]: The input is first sorted locally. Each PE takes $\mathcal{O}(p)$ sample elements. The sample is gathered and sorted at a single PE. The sorted samples are used to obtain splitter elements $s_1,\ldots, s_{p-1}$ that are equally spaced in the sorted sample. These splitters are broadcast to all other PEs. Define $s_0 = -\infty$ and $s_p = +\infty$. Now each processor partitions the elements into buckets where the i-th buckets gets elements between s_i and s_{i+1}. All Elements from bucket i are then sent to PE i using an all-to-all personalized communication. Finally, each PE merges the received pieces of its bucket. In summary, sorting is reduced to local sorting, multiway merging, and further standard communication operations: gather of a small sample, splitter broadcast, and a single personalized all-to-all communication.

We get a bottleneck term of $f(p) = \mathcal{O}\big(p^2\log p + p^2 T_{\mathrm{byte}} + pT_{\mathrm{start}}\big)$ and a total execution time of

$$\mathcal{O}\big((n/p\log p + (p^2(\log p + T_{\mathrm{byte}}) + pT_{\mathrm{start}})\log n\big)$$

---

[1] One can get a slight improvement of the theoretical bound by switching to a sequential algorithm after $\mathcal{O}(\log p)$ levels of recursion. But this is irrelevant from a practical perspective.

Asymptotically better bounds are obtained in [11] using more sophisticated implementations of sorting and all-to-all. However, these algorithms are considerably more complicated and in Section 3 we will give evidence that on machines with a moderate number of processors no significant improvements can be expected from these theoretical algorithms.

All the required communication operations (point-to-point, prefix sum, broadcast, all-to-all, gather) are available in communication libraries such as MPI [15].

3   Experiments

We have implemented pDC3 with deterministic sample sort using C++ and MPI [15]. Most measurements were performed on a HP Integrity rx2620 running under Linux with 64 dual processor nodes using Itanium 2 processors with 1.5 GHz and 6 MByte Cache. The machine has 64×12 GByte of main memory. The nodes are connected by a Quadrics QSNet II network with 800 MByte/s communication bandwidth.

We have used the big real world inputs from [9]: The human genome, 3.125 GByte of books from the Gutenberg project, and 522 MByte of source code. In addition, we use the artificial inputs a^n and $(abc)^{n/3}$. Timing is started when all PEs have initialized MPI and hold n/p characters of the input each.

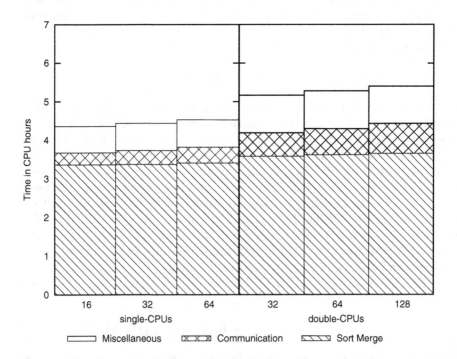

Fig. 2. The distribution of the execution time between sorting, communication and the remaining local operations for the Gutenberg instance

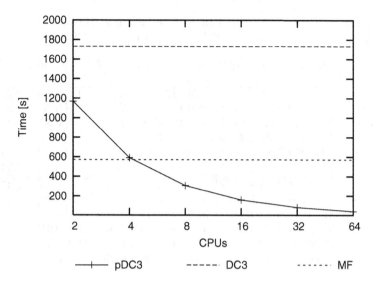

Fig. 3. Execution time of pDC3 compared to the sequential DC3 algorithm from [8] and to the sequential algorithm from [13]

Fig. 2 shows the work performed for the Gutenberg input using 16–128 PEs using one or two CPUs on each node. We see that sorting and merging takes most of the time. Communication time (mainly all-to-all) takes only a small fraction of the total execution time. However, it should be noted that low cost machines with switched Gigabit Ethernet have an order of magnitude smaller communication bandwidth than our machine. On such machines, communication would take about half of the time. (Which might still be acceptable considering that such machines are much cheaper). The overall work increases only slightly when increasing the number of processors. This indicates good scalability. As to be expected, using both CPUs increases internal work and total communication time since the CPUs have to share the main memory and the network interface.

We cannot give speedups for big inputs since no single node has enough memory to solve the problem. Therefore Fig. 3 compares pDC3 with two sequential algorithms for the source code instance. DC3 is the simple sequential linear time implementation from [5].[2] MF is one of the fastest practical algorithms [13]. With the minimal number of two processors our parallel algorithm already outperforms the simple sequential algorithm significantly although it has a factor $\Theta(\log n)$ disadvantage in its asymptotic execution time. The break even point to [13] is at four processors. The work per processor is about half as much as for the external algorithm from [9] on a 2GHz Intel Xeon processor. Unfortunately, a direct comparison with the parallel implementation from [12] is not possible since this paper does not specify the clock speed of the machine used.

---

[2] There are faster sequential implementations of DC3 by now [16] but they still do not beat implementations such as [13].

Table 1. Average (Ø) versus bottleneck (max) execution times of major parts of pDC3. Timings in second. Top part: 64 × 1 CPU. Bottom part: 64 × 2 CPUs.

Input	Size	Total	quicksort		mergesort		p-merge		All2all	Com	sample
			max	Ø	max	Ø	max	Ø	Ø	Ø	
Source	522	37.8	16.6	15.9	28.6	27.9	10.5	9.6	4.2	0.14	0.29
Genome	2928	282.0	160.3	115.0	182.6	178.7	62.8	58.0	22.2	0.36	1.24
Gutenberg	3125	254.6	124.0	119.5	197.4	195.6	68.1	66.5	22.2	0.36	1.30
a^n	3815	520.7	411.4	271.3	168.9	165.7	49.6	32.1	22.2	0.42	2.16
a^n	2000	259.7	202.2	130.6	85.2	83.4	25.8	16.6	11.5	0.37	1.78
$(abc)^{n/3}$	2000	263.7	198.2	98.5	85.2	83.2	33.3	16.4	13.8	0.38	1.54
Source	522	24.2	7.8	7.4	14.9	14.4	6.2	5.3	4.9	0.23	0.37
Genome	2928	180.8	94.3	53.8	99.6	95.7	39.0	37.2	21.9	0.67	1.25
Gutenberg	3125	151.8	58.7	55.8	107.5	105.7	44.2	40.9	21.1	0.53	1.31
a^n	3815	280.9	193.1	120.0	99.0	96.1	45.0	26.6	21.2	0.91	2.12
a^n	2000	140.7	93.4	56.9	49.4	47.8	23.2	13.7	11.2	0.53	1.76
$(abc)^{n/3}$	2000	146.1	92.3	42.7	49.5	47.8	30.9	13.5	13.4	0.56	1.49

Table 1 gives a more detailed breakdown of the execution time of pDC3 for different inputs. The STL quicksort used for local sorting shows considerable load imbalances, i.e., the slowest PE does much more work than the average PE. This is not due to significantly different amounts of data assigned to PEs but because quicksort has highly data dependent execution times in particular for the artificial inputs like a^n. In contrast, if we use mergesort, there is much less load imbalance. Here, the artificial inputs turn out to be *easier* to solve than the real world inputs. There is also some load imbalance for the p-way merging in sample sort for artificial inputs. However, this is not very critical since it only means that some PEs do less work than in the worst case.

4   Conclusions

We have demonstrated that pDC3 is a practicable and scalable way to build huge suffix arrays. Several practical improvements could be considered. pDC3 might scale even to machines with thousands of processors if we use parallel sorting for sorting the sample. The DC3 algorithm can be generalized to larger difference covers that imply a different recurrence relation. Using this scheme in the first level of recursion could save a constant factor of time for small alphabets. A $\log n$ term in the execution time could be removed by switching from comparison based sorting to integer sorting. However, we are not aware of an algorithm that would really remove the $\log n$ in the worst case and would bring improvements in practice. For example, the implementation from [8] gets slightly *faster* when its linear time sorting algorithm is replaced by quicksort. There are also further opportunities for tuning. For inputs that are so large that they do not even fit in the main memory of a parallel computer, a parallel external algorithm could be developed by combining the results of the present paper with [9].

References

1. Manber, U., Myers, G.: Suffix arrays: A new method for on-line string searches. SIAM Journal on Computing **22** (1993) 935–948
2. Gonnet, G., Baeza-Yates, R., Snider, T.: New indices for text: PAT trees and PAT arrays. In Frakes, W.B., Baeza-Yates, R., eds.: Information Retrieval: Data Structures & Algorithms. Prentice-Hall (1992)
3. Abouelhoda, M.I., Kurtz, S., Ohlebusch, E.: The enhanced suffix array and its applications to genome analysis. In: Proc. 2nd Workshop on Algorithms in Bioinformatics. Volume 2452 of LNCS., Springer (2002) 449–463
4. Burrows, M., Wheeler, D.J.: A block-sorting lossless data compression algorithm. Technical Report 124, SRC (digital, Palo Alto) (1994)
5. Kärkkäinen, J., Sanders, P.: Simple linear work suffix array construction. In: Proc. 30th International Conference on Automata, Languages and Programming. Volume 2719 of LNCS., Springer (2003) 943–955
6. Kim, D.K., Sim, J.S., Park, H., Park, K.: Linear-time construction of suffix arrays. In: Proc. 14th Annual Symposium on Combinatorial Pattern Matching. LNCS, Springer (2003) 186–199 To appear.
7. Ko, P., Aluru, S.: Space efficient linear time construction of suffix arrays. In: Proc. 14th Annual Symposium on Combinatorial Pattern Matching. Volume 2676 of LNCS., Springer (2003) 200–210
8. Kärkkäinen, J., Sanders, P.: Simple linear work suffix array construction. In: 30th International Colloquium on Automata, Languages and Programming. Number 2719 in LNCS (2003) 943–955
9. Dementiev, R., Kärkkäinen, J., Mehnert, J., Sanders, P.: Better external memory suffix array construction. In: Workshop on Algorithm Engineering & Experiments, Vancouver (2005) 86–97
10. Iliopoulos, C.S., Rytter, W.: On parallel transformations of suffix arrays into suffix trees. In: 15th Australasian Workshop on Combinatorial Algorithms (AWOCA). (2004)
11. Kärkkäinen, J., Sanders, P., Burkhardt, S.: Linear work suffix array construction. Journal of the ACM (2006) to appear.
12. Futamura, N., Aluru, S., Kurtz, S.: Parallel suffix sorting. In: Proc. 9th International Conference on Advanced Computing and Communications, Tata McGraw-Hill (2001) 76–81
13. Manzini, G., Ferragina, P.: Engineering a lightweight suffix array construction algorithm. In: Proc. 10th Annual European Symposium on Algorithms. Volume 2461 of LNCS., Springer (2002) 698–710
14. Shi, H., Schaeffer, J.: Parallel sorting by regular sampling. Journal of Parallel and Distributed Computing **14** (1992) 361–372
15. Snir, M., Otto, S.W., Huss-Lederman, S., Walker, D.W., Dongarra, J.: MPI – the Complete Reference. MIT Press (1996)
16. Smyth, B., Turpin, A.: The performance of linear time suffix sorting algorithms. In: IEEE Data Compression Conference. (2005)

Formal Verification of Programs That Use MPI One-Sided Communication

Salman Pervez[1], Ganesh Gopalakrishnan[1], Robert M. Kirby[1],
Rajeev Thakur[2], and William Gropp[2]

[1] School of Computing
University of Utah
Salt Lake City, UT 84112, USA
[2] Mathematics and Computer Science Division
Argonne National Laboratory
Argonne, IL 60439, USA

Abstract. We used formal-verification methods based on model check-ing to analyze the correctness properties of one existing and two new distributed-locking algorithms implemented by using MPI's one-sided communication. Model checking exposed an overlooked correctness issue with the first algorithm, which had been developed by relying only on manual reasoning. Model checking helped confirm the basic correctness properties of the two new algorithms, while also identifying the remain-ing problems in them. Our experience is that MPI-based programming, especially the tricky and relatively poorly understood one-sided commu-nication features, stand to gain immensely from model checking. Consid-ering that many other areas of concurrent hardware and software design now routinely employ model checking, our experience confirms that the MPI community can benefit greatly from the use of formal verification.

1 Introduction

Concurrent protocols are notoriously hard to design and verify. Experience has shown that virtually all nontrivial protocol implementations contain bugs such as deadlocks, livelocks, and memory leaks, despite extensive care taken during de-sign and testing. Most of these bugs are basic *design* errors due to "unexpected" (untested) concurrent behaviors. Therefore, it stands to reason that if finite-state models of these protocols are created and *exhaustively* analyzed for the desired formal properties, robust protocol implementations would result. The technol-ogy for such finite-state modeling, property description, and exhaustive analysis developed over the past three decades—known as *model checking* [2]—has been successfully applied to numerous software and hardware systems. Model check-ing is now an integral part of the Windows Device Driver Development Kit [1]. Virtually all cache-coherence protocols developed and deployed in modern mi-croprocessors have been verified by using model checking. However, although concurrency and concurrent-programming bugs in parallel scientific program-ming are similar to those in other areas, we find little evidence of model checking being applied to verify parallel scientific programs.

B. Mohr et al. (Eds.): PVM/MPI 2006, LNCS 4192, pp. 30–39, 2006.

In this paper, we conduct case studies that show the promise of the application of model checking in the area of parallel scientific programming using MPI. In particular, we focus on MPI one-sided communication [10]. Being (relatively) recently introduced and implemented, MPI one-sided communication is insufficiently understood and documented. One-sided communication involves shared-memory concurrency, which is known to be inherently harder to reason about than the message-passing concurrency of traditional MPI. One-sided communication exacerbates verification complexity because it guarantees only weak ordering semantics with respect to loads and stores, which can freely reorder within a given synchronization epoch. This paper demonstrates that, by using model checking, bugs in MPI programs that use one-sided communication can be caught easily, while expending only modest amounts of human and computer time.

After presenting background on MPI one-sided communication in Section 2, we provide an overview of model checking in Section 3. We then describe the design of an existing distributed byte-range locking algorithm [17] and its formal verification through model checking (Section 4). Model checking helped uncover the serious problem of a potential deadlock, which the authors of the algorithm were unaware of. Model checking also found a more benign problem of extra (zero-byte) sends in the algorithm, which might lend itself to an implementation-dependent correction using MPI_Iprobe and posted receives. However, this problem may well turn into a memory leak. We then present two other designs of the same algorithm, formally verify them using model checking, and provide empirical observations to interpret these model-checking results (Section 5). In Section 6, we conclude with a discussion of future work.

To our knowledge, nobody has applied model checking to analyze programs that use MPI one-sided communication. Siegel and Avrunin have used model checking to verify programs that use basic MPI point-to-point communication [13,14]. Kranzlmüller used a formal event-graph based method to help understand MPI program executions [6]. Matlin et al. used the SPIN model checker to verify parts of the MPD process manager used in MPICH2 [9].

2 MPI One-Sided Communication

For lack of space, we review only the features of MPI one-sided communication relevant to this paper. One feature in MPI one-sided communication allows processes to gain exclusive access to communication windows in a block of code bracketed by MPI_Win_lock and MPI_Win_unlock calls [10]. Read and write accesses can be performed by a process holding exclusive access to a window through MPI_Put and MPI_Get. The main semantic difficulty stems from these put and get calls being not required to obey their syntactic program order in terms of when they are performed. It is well known (see, e.g., [15]) that such ordering guarantees are crucial to the correctness of even simple concurrent protocols such as Peterson's mutual exclusion. The specification of one-sided communication in MPI further exacerbates the issue by introducing a complex set of informally stated rules that can easily lead to contradictory interpretations.[1] Common mistakes users make include nesting synchronization epochs on the same window object (such as a win_lock/unlock within a fence), doing read-modify-writes via a get-modify-put in the same synchronization epoch (even though gets and puts are defined to be nonblocking), and doing a put and a get to/from the same memory location in the same synchronization epoch. For example, the broadcast algorithms in Appendix B and C of [8] are incorrect because they rely on MPI_Get being a blocking function, which it is not. In implementations that take advantage of the nonblocking nature of MPI_Get, such as MPICH2 [16], the code will, indeed, go into an infinite loop. Since MPI one-sided communication can be implemented in a variety of ways [4], the result of making such mistakes is often implementation dependent: the program may work fine on some implementations and not on others.

3 Model Checking

Model checking is a term that has acquired an overloaded meaning. It essentially is the activity of exhaustively examining all possible behaviors of a *model* of a concurrent program (akin to wind-tunnel testing of scale models of airplanes). We consider *finite-state* model checking where the model of the concurrent system is expressed in a modeling language—Promela [5] in our case (all the pseudocodes expressed in this paper have an almost direct Promela encoding once the MPI constructs have been accurately modelled). By exhaustively executing the concurrent-system model, a model checker reveals its entire state-transition structure and is able to establish temporal properties, such as "always P" and "A implies eventually Q" with respect to this structure. The state graphs we generate are a result of the *interleaved* execution of various processes or threads. A fundamental problem with model checking is that reachable state graphs are exponential in the number of concurrent processes. The past three decades of research has, essentially, focused on getting a good handle on this exponential growth, so much so that astronomically large finite-state spaces—or often even many classes of infinite state spaces—can be handled by model checkers. Despite the very large state spaces of the MPI models discussed in this paper, our model checking runs finished within acceptable durations (often in minutes) on standard workstations.

4 Formal Verification of Byte-Range Locking

In [17], Thakur et al. present an algorithm implemented by using MPI one-sided communication (with passive-target lock-unlock synchronization) for coordinating a collection of parallel processes contending for byte-range locks. We first describe the algorithm briefly, followed by a description of how we model checked

---

[1] A collaborative project between the University of Utah and Argonne is addressing the issue of elucidating as well as formalizing this specification.

it. Because of space limits, we cannot present the full pseudocode of the original algorithm; the reader may refer to the original paper [17] for details.

4.1   The Byte-Range Locking Algorithm

Each process keeps in a single common memory window (`lockwin`) its state consisting of a `flag` (initialized to 0) and the `start` and `end` values for the byte range (initialized to -1). A flag of 0 indicates that the process does not have the lock, while 1 indicates that it either has acquired the lock or wants to acquire the lock. A process updates its state and reads others' states by acquiring exclusive access to `lockwin` and making `MPI_Put` and `MPI_Get` calls. Since the processes acquire exclusive access, the actions of any one process on `lockwin` are guaranteed to be atomic with respect to the actions of other processes.

In order to acquire the lock, a process sets its `flag` to 1, updates its `start` and `end` values, and gets the corresponding values of other processes. It then checks whether any other process has set a conflicting byte range and has a flag value of 1. If it does not find such a process, it assumes that it has acquired the lock. Otherwise, it assumes that it does not have the lock, resets its flag to 0 via another lock-put-unlock, and blocks on a zero-byte `MPI_Recv` call, waiting for a process that has the lock to wake it up with a zero-byte send. The process will retry the lock after receiving the message. To release a lock, a process again acquires exclusive access, resets its flag to 0 and its start and end offsets to -1, and gets the values of other processes. If it finds a process with a conflicting byte-range (ignoring the flag), it sends a zero-byte message (via `MPI_SEND`) to wake up that process.

4.2   Checking the Byte-Range Locking Algorithm

To model the algorithm, we first needed to model the MPI one-sided communication constructs used in the algorithm and capture their semantics precisely as specified in the MPI Standard [10]. For example, the MPI Standard specifies that if a communication epoch is started with `MPI_Win_lock`, it must end with `MPI_Win_unlock` and that the put/get/accumulate calls made within this epoch are not guaranteed to complete before `MPI_Win_unlock` returns. Furthermore, there are no ordering guarantees of the puts/gets/accumulates within an epoch. Therefore, in order to obtain adequate execution-space coverage, *all permutations of put/get/accumulate calls in the epoch must be examined*. However, the byte-range locking algorithm uses the `MPI_LOCK_EXCLUSIVE` lock type, which means that while a certain process has entered the synchronization epoch, no other process may enter until that process has left. This makes the synchronization epoch an atomic block and renders all permutations of the calls within it equivalent from the perspective of other processes. Modeling the byte-range locking algorithm itself was relatively straightforward. (This experience augurs well for the checking of other algorithms that use MPI one-sided communication, as one of the significant challenges in model checking lies in the ease of modeling constructs in the target domain using modeling primitives in the modeling language.) The complete Promela code used in our model checking can be found online [11].

When we model checked our model with three processes, our model checker, SPIN [5], discovered an error indicating an "invalid end state." Deeper probing revealed the following error scenario (explained through an example, which assumes that P1 tries to lock byte-range $\langle 1, 2 \rangle$, P2 tries to lock $\langle 3, 4 \rangle$, and P3 tries to lock $\langle 2, 3 \rangle$):

- P1 and P3 successfully acquire their byte-range locks.
- P2 then tries to acquire its lock, notices conflict with respect to both P1 and P3, and blocks on the `MPI_Recv`.
- P1 and P3 release their locks, both notice conflicts with P2, and both perform an MPI_Send, when only one send is needed.

Hence, while P2 ends up successfully waking up and acquiring the lock, the extra `MPI_Sends` may accumulate in the system. This is a subtle error whose severity depends on the MPI implementation being used. Recall that the MPI Standard allows implementors to decide whether to block on an MPI_Send call. In practice, a zero-byte send will rarely block. Nonetheless, an implementation of the byte-range locking algorithm can address this problem by periodically calling `MPI_Iprobe` and matching any unexpected messages with `MPI_Recvs`.

We then modeled the system as if these extra `MPI_Sends` do not exhaust the system resources and hence do not cause processes to block. In this case, model checking detected a far more serious deadlock situation, summarized in Figure 1. P1 expresses its intent to acquire a lock in the range $\langle 10, 20 \rangle$ (1), with P2 following suit (2). P1 acquires the lock (3), finishes using it and relinquishes it (4), and performs a send to unblock P2 (5). Before P2 gets a chance to change its global state, P1 tries to reacquire the lock (6). P1 reads P2's current flag value as 1, so it decides to block by carrying out events (10) and (12). At this point, P2 changes its global

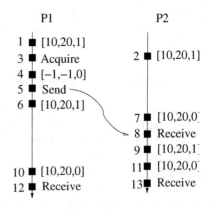

Fig. 1. A deadlock scenario found through model checking

state, receives the message sent by P1 (8), and proceeds to reacquire the lock (9). P2 reads P1's current flag value as 1, so it decides to block by carrying out events (11) and (13). Both processes now block on receive calls, and the result is deadlock. We note that the authors of the algorithm were unaware of this problem until the model checker found it!

5   Correcting the Byte-Range Locking Algorithm

We propose two approaches to fixing this deadlock problem and describe our experience with using model checking on these solutions.

Alternative 1. One way to eliminate deadlocks is to add a third state to the "flag" used in the algorithm. This is shown in the pseudocode in Figure 2. In the original algorithm, a flag value of '0' indicates that the process does not have the lock, while a flag value of '1' indicates that it either has acquired the lock or is in the process of determining whether it has acquired the lock. In other words, the '1' state is overloaded. In the proposed fix, we add a third state of '2,' with '0' denoting the same as before, '1' now denoting that the process has acquired the lock, and '2' denoting that it is in the process of determining whether it has acquired the lock. There is no change to the lock-release algorithm, but the lock-acquire algorithm changes as follows.

When a process wants to acquire a lock, it writes its flag value as '2' and its start and end values in the memory window. It also reads the state of the other processes from the memory window. If it finds a process with a conflicting byte range and a flag value of '1,' it knows that it does not have the lock. So it resets its flag value to '0' and blocks on an MPI_Recv. If no such process (with conflicting byte range and flag=1) is found, but there is another process with a conflicting byte range and a flag value of '2,' the process resets its flag to '0,' its start and end offsets to -1, and retries the lock from scratch. If neither of these cases is true, the process sets its flag value to '1' and considers the lock acquired. An assessment of this protocol using model checking is presented later in this section.

Alternative 2. This approach uses the same values for the flag as the original algorithm, but when a process tries to acquire a lock and determines that it does not have the lock, it picks a process (that currently has the lock) to wake it up and then blocks on the receive. For this purpose, we add a fourth field (the pick field) to the values for each process in the memory window (see Figure 3). The process about to block must now decide whether to block. This decision is based on two factors: (i) Has the process selected to wake it up already released the lock? and (ii) Is there a possibility of a deadlock caused by a cycle of processes that wait on each other to wake them up? The latter can be detected and avoided by using the algorithm in Figure 4. The former can be easily determined by reading the values returned by the MPI_Get on line 26. If the selected process has already released the lock, a new process must be picked in its place. We simply traverse the list of conflicting processes until we find one that has not yet released the lock. If no such process is found, the algorithm tries to reacquire the lock. Note the added complexity of going through the list of conflicting processes and doing put and get operations each time. However, if this loop is successful and the process blocks on MPI_Recv, we can save considerable processor time in the case of highly contentious lock requests as compared with Alternative 1.

```
 1   Lock_acquire (int start, int end)
 2   {
 3     val[0] = 2; /* flag */ val[1] = start; val[2] = end;
 4     while (1) {
 5       /* add self to locklist */
 6       MPI_Win_lock(MPI_LOCK_EXCLUSIVE, homerank, 0, lockwin);
 7       MPI_Put(&val, 3, MPI_INT, homerank, 3*myrank, 3, MPI_INT, lockwin);
 8       MPI_Get(locklistcopy, 3*(nprocs-1), MPI_INT, homerank, 0, 1,
 9               locktype1, lockwin);
10       MPI_Win_unlock(homerank, lockwin);
11       /* check to see if lock is already held */
12       conflict = flag1 = flag2 = 0;
13       for (i=0; i < (nprocs - 1); i ++) {
14         if ((flag == 1) && (byte ranges conflict with lock request))
15           { flag1 = 1; break; }
16         if ((flag == 2) && (byte ranges conflict with lock request))
17           { flag2 = 1; break; }
18       }
19       if (flag1 == 1) {
20         /* reset flag to 0, wait for notification, and then retry */
21         MPI_Win_lock(MPI_LOCK_EXCLUSIVE, homerank, 0, lockwin);
22         val[0] = 0;
23         MPI_Put(val, 1, MPI_INT, homerank, 3*myrank, 1, MPI_INT, lockwin);
24         MPI_Win_unlock(homerank, lockwin);
25         /* wait for notification from some other process */
26         MPI_Recv(NULL, 0, MPI_BYTE, MPI_ANY_SOURCE, WAKEUP, comm, &status);
27         /* retry the lock */
28         Lock_acquire(start, end); }
29       else if (flag2 == 1) {
30         /* reset flag to 0, start/end offsets to -1, and then retry */
31         MPI_Win_lock(MPI_LOCK_EXCLUSIVE, homerank, 0, lockwin);
32         val[0] = 0; /* flag */ val[1] = -1; val[2] = -1;
33         MPI_Put(val, 3, MPI_INT, homerank, 3*myrank, 3, MPI_INT, lockwin);
34         MPI_Win_unlock(homerank, lockwin);
35         /* retry the lock */
36         Lock_acquire(start, end); }
37       else {
38         MPI_Win_lock(MPI_LOCK_EXCLUSIVE, homerank, 0, lockwin);
39         val[0] = 1;
40         MPI_Put(val, 1, MPI_INT, homerank, 3*myrank, 1, MPI_INT, lockwin);
41         MPI_Win_unlock(homerank, lockwin);
42         /* lock is acquired */
43         break;
44       }
45     }
46   }
```

Fig. 2. Pseudocode for the deadlock-free byte-range locking algorithm (Alternative 1)

Assessment of the Alternative Algorithms. We model checked these algorithms using SPIN, which helped establish the following formal properties of these algorithms:

- Absence of deadlocks (both alternatives).
- Communal progress (that is, if a collection of processes request a lock, then someone will eventually obtain it). Alternative 2 satisfies this under all fair schedules (all processes are scheduled to run infinitely often), whereas Alternative 1 places a few additional restrictions to rule out a few rare schedules (details in [12]).

```
1    Lock_acquire (int start, int end)
2    {
3        int picklist[num_procs];
4        val[0] = 1; /* flag */ val[1] = start; val[2] = end;
5        val[3] = -1; /* pick */
6        while (1) {
7            /* add self to locklist */
8            MPI_Win_lock(MPI_LOCK_EXCLUSIVE, homerank, 0, lockwin);
9            MPI_Put(&val, 4, MPI_INT, homerank, 4*myrank, 4, MPI_INT, lockwin);
10           MPI_Get(locklistcopy, 4*(nprocs-1), MPI_INT, homerank, 0, 1,
11                  locktype1, lockwin);
12           MPI_Win_unlock(homerank, lockwin);
13           /* check to see if lock is already held */
14           cprocs_i = 0;
15           for (i=0; i < (nprocs - 1); i ++)
16               if ((flag == 1) && (byte range conflicts with Pi's request)) {
17                   conflict = 1; picklist[cprocs_i] = Pi; cprocs_i++; }
18           if (conflict == 1) {
19               for(j=0; j < cprocs_i; j++) {
20                   MPI_Win_lock(MPI_LOCK_EXCLUSIVE, homerank, 0, lockwin);
21                   val[0] = 0; val[3] = picklist[j];
22                   /* reset pick to 0, indicate pick and pick_counter */
23                   MPI_Put(&val, 4, MPI_INT, homerank, 4*myrank, 4, MPI_INT, lockwin);
24                   MPI_Get(locklistcopy, 4*(nprocs-1), MPI_INT, homerank, 0, 1,
25                          locktype1, lockwin);
26                   MPI_Win_unlock(homerank, lockwin);
27                   if (picklist[j] has released the lock || detect_deadlock())
28                       /* repeat for the next process in picklist */
29                       j++;
30                   else {
31                       /* wait for notification from picklist[j] */
32                       MPI_Recv (NULL, 0, MPI_BYTE, MPI_ANY_SOURCE, WAKEUP, comm,
33                              MPI_STATUS_IGNORE);
34                       break; /* retry the lock */ }
35               }
36               /* if the entire list has been traversed, retry the lock */
37           }
38           else
39               break;  /* lock is acquired */
40       }
41   }
```

Fig. 3. Pseudocode for the deadlock-free byte-range locking algorithm (Alternative 2)

```
1    detect_deadlock() {
2        cur_pick = locklistcopy[4 * myrank + 3];
3        while(i < num_procs) {
4            /* picking this process means a cycle is completed */
5            if(locklistcopy[4 * cur_pick + 3] == my_rank) return 1;
6            /* no cycle can be formed */
7            else if(locklistcopy[4 * cur_pick + 3] == -1) return 0;
8            else cur_pick = locklistcopy[4 * cur_pick + 3];
9        }
10   }
```

Fig. 4. Avoiding circular loops among processes picked to wake up other processes in Alternative 2

We note that neither of these alternatives eliminates the extra sends, but, as described in Section 4, an implementation can deal with them by using MPI_Iprobe. That said, Alternative 2 considerably reduces these extra sends, as it restricts the number of processes that can wake up a particular process compared with Alternative 1. The exact performance tradeoffs of these algorithms will be determined as part of our future work. We are still seeking algorithms that avoid deadlock, avoid extra sends, and are efficient.

6   Conclusions and Future Work

We have shown how formal verification based on model checking can be used to find actual deadlocks in published algorithms that use the MPI one-sided communication primitives. We have also discussed how this technology can help shed light on a number of related issues such as forward progress and the possibility of there being unconsumed messages. We presented and analyzed two deadlock-free algorithms for byte-range locking and verified their characteristics.

Nonetheless, our work in this field is still in its early stages. Capitalizing on the maxim that formal methods can have their biggest impact when applied to constructs that are relatively new or are under development, we plan to formalize the entire set of MPI one-sided communication primitives. This can help develop a comprehensive approach to verifying programs that use the MPI one-sided constructs. As future case studies, we will analyze other algorithms, such as the scalable fetch-and-increment algorithm described in [3]. We plan to explore the use of automated tools to extract models from MPI programs, instead of creating them by hand. We also plan to explore the advantages of using other modeling languages, such as +CAL [7], and investigate the possibility of directly model checking MPI programs, instead of their extracted formal models.

Acknowledgments

This work was supported by NSF award CNS-0509379 and by the Mathematical, Information, and Computational Sciences Division subprogram of the Office of Advanced Scientific Computing Research, Office of Science, U.S. Department of Energy, under Contract W-31-109-ENG-38.

References

1. Thomas Ball, Byron Cook, Vladimir Levin, and Sriram K. Rajamani. SLAM and static driver verifier: Technology transfer of formal methods inside Microsoft. In *Proceedings of IFM 04: Integrated Formal Methods*, pages 1–20. Springer, April 2004.
2. Edmund M. Clarke, Orna Grumberg, and Doron Peled. *Model Checking*. MIT Press, Cambridge, MA, 1999.
3. William Gropp, Ewing Lusk, and Rajeev Thakur. *Using MPI-2: Advanced Features of the Message-Passing Interface*. MIT Press, Cambridge, MA, 1999.
4. William Gropp and Rajeev Thakur. An evaluation of implementation options for MPI one-sided communication. In *Recent Advances in Parallel Virtual Machine and Message Passing Interface, 12th European PVM/MPI Users' Group Meeting*, pages 415–424. LNCS 3666, Springer, September 2005.

5. Gerard J. Holzmann. *The Spin Model Checker: Primer and Reference Manual.* Addison-Wesley, 2003.
6. Dieter Kranzlmüller. *Event Graph Analysis for Debugging Massively Parallel Programs.* PhD thesis, John Kepler University Linz, Austria, September 2000. http://www.gup.uni-linz.ac.at/~dk/thesis.
7. Leslie Lamport. http://research.microsoft.com/users/lamport/tla/pluscal.html.
8. Glenn R. Luecke, Silvia Spanoyannis, and Marina Kraeva. The performance and scalability of SHMEM and MPI-2 one-sided routines on a SGI Origin 2000 and a Cray T3E-600. *Concurrency and Computation: Practice and Experience,* 16(10):1037–1060, 2004.
9. Olga Shumsky Matlin, Ewing Lusk, and William McCune. SPINning parallel systems software. In *Model Checking of Software: 9th International SPIN Workshop,* pages 213–220. LNCS 2318, Springer, 2002.
10. Message Passing Interface Forum. MPI-2: Extensions to the Message-Passing Interface, July 1997. http://www.mpi-forum.org/docs/docs.html.
11. Salman Pervez. http://www.cs.utah.edu/~spervez/model.tar.gz.
12. Salman Pervez. Byte-range locks using MPI one-sided communication. Technical report, University of Utah, School of Computing, 2006. http://www.cs.utah.edu/formal_verification/OnesidedTR1/.
13. Stephen F. Siegel and George S. Avrunin. Verification of MPI-based software for scientific computation. In *Proceedings of the 11th International SPIN Workshop on Model Checking Software,* pages 286–303. LNCS 2989, Springer, April 2004.
14. Stephen F. Siegel, Anastasia Mironova, George S. Avrunin, and Lori A. Clarke. Using model checking with symbolic execution to verify parallel numerical programs. In *Proceedings of the ACM SIGSOFT 2006 International Symposium on Software Testing and Analysis,* July 2006.
15. Andrew S. Tanenbaum. *Modern Operating Systems.* Prentice-Hall, Inc., second edition, 2001.
16. Rajeev Thakur, William Gropp, and Brian Toonen. Optimizing the synchronization operations in MPI one-sided communication. *International Journal of High-Performance Computing Applications,* 19(2):119–128, Summer 2005.
17. Rajeev Thakur, Robert Ross, and Robert Latham. Implementing byte-range locks using MPI one-sided communication. In *Recent Advances in Parallel Virtual Machine and Message Passing Interface, 12th European PVM/MPI Users' Group Meeting,* pages 120–129. LNCS 3666, Springer, September 2005.

MPI Collective Algorithm Selection and Quadtree Encoding

Jelena Pješivac–Grbović, Graham E. Fagg,
Thara Angskun, George Bosilca, and Jack J. Dongarra

Innovative Computing Laboratory,
University of Tennessee Computer Science Department
1122 Volunteer Blvd., Knoxville, TN 37996-3450, USA
{pjesa, fagg, angskun, bosilca, dongarra}@cs.utk.edu

Abstract. Selecting the close-to-optimal collective algorithm based on
the parameters of the collective call at run time is an important step
in achieving good performance of MPI applications. In this paper, we
focus on MPI collective algorithm selection process and explore the applicability of the quadtree encoding method to this problem. We construct quadtrees with different properties from the measured algorithm
performance data and analyze the quality and performance of decision
functions generated from these trees. The experimental data shows that
in some cases, the decision function based on a quadtree structure with a
mean depth of 3 can incur as little as a 5% performance penalty on average. The exact, experimentally measured, decision function for all tested
collectives could be fully represented using quadtrees with a maximum of
6 levels. These results indicate that quadtrees may be a feasible choice for
both processing of the performance data and automatic decision function
generation.

1 Introduction

The performance of MPI collective operations is crucial for good performance
of MPI application which use them [1]. For this reason, significant efforts have
been put on design and implementation of efficient collective algorithms both
for homogeneous and heterogeneous cluster environments [2,3,4,5,6,7,8]. Performance of these algorithms varies with the total number of nodes involved in
communication, system and network characteristics, size of data being transferred, current load, and if applicable, the operation that is being performed as
well as the segment size which is used for operation pipelining. Thus, selecting
the best possible algorithm and segment size combination (*method*) for every
instance of collective operation is important.

To ensure good performance of MPI applications, collective operations can be
tuned for the particular system. The tuning process often involves detailed profiling of the system possibly combined with communication modeling, analyzing
the collected data, and generating a *decision function*. During run-time, the decision function selects close-to-optimal method for a particular collective instance.

B. Mohr et al. (Eds.): PVM/MPI 2006, LNCS 4192, pp. 40–48, 2006.

This approach relies on the ability of the decision function to accurately predict algorithm and segment size to be used for the particular collective instance. Alternatively, one could construct an in-memory decision system which could be queried/searched at the run-time to provide the optimal method information. In order for either of these approaches to be feasible, the memory footprint and the time it takes to make decisions need to be minimal.

This paper studies the applicability of the quadtree encoding method as a storage and optimization technique within the MPI collective method selection process. We assume that the system of interest has been benchmarked and that detailed performance information exists for each of available collective communication algorithm. With this information, we focus our efforts on investigating whether the quadtree encoding is a feasible way to generate static decision functions as well as, to represent the decision function in memory.

The paper proceeds as follows: Section 2 discusses existing approaches to the decision making/algorithm selection problem; Section 3 describes the quadtree construction and analysis of quadtree decision function in more detail; Section 4 presents experimental results; Section 5 concludes the paper with discussion of the results and future work.

2   Related work

The MPI collective algorithm selection problem has been addressed in many MPI implementations.

In the FT-MPI [9], the decision function is generated manually using visual inspection method augmented with Matlab scripts used for analysis of the experimentally collected performance data. This approach results in a precise albeit complex decision functions. In the MPICH-2 MPI implementation, the algorithm selection is based on bandwidth and latency requirements of an algorithm, and the switching points are predetermined by the implementers [5]. In the tuned collective module of the Open MPI [10], the algorithm selection can be done in either of the following three ways: via compiled decision function, via user-specified command line flags, or using rule-based run-length encoding scheme which can be tuned for particular system.

In this work, we treat the information about the optimal collective implementation on a system as a bit pattern which we encode using a similar technique to an image encoding process. We then use the encoded structure to generate decision function code. To the best of our knowledge, we are the only group which has approached the MPI collective tuning process in this way.

3   Quadtrees and MPI Collective Operations

We use the collective algorithm performance information on a particular system to extract the information about the optimal methods and construct a *decision map* for the collective on that system. An example of a decision map is displayed in Table 1. The decision map which will be used to initialize the quadtree must be a complete and square matrix with a dimension size that is a power of two, $2^k \times 2^k$. Complete decision map means that tests must cover all message and communicator sizes of interest. Neither of these requirements are real limitations, as the missing data can be interpolated and the size of the map can be adjusted by replicating some of the entries.

Table 1. Decision map example. The axis information relates to the decision maps in Figure 1.

Communicator size (y-axis)	Message size (x-axis)	Algorithm	Segment size	Method index
3	1	Linear	none	1
3	2	Linear	none	1
...	...	...	...	...
128	64KB	BinaryTree	8KB	13

Once a decision map is available, we initialize the quadtree from it using user specified constraints such as *accuracy threshold* and *maximum allowed depth* of the tree. The accuracy threshold is the minimum percentage of points in a block with the same "color", such that the whole block is "colored" in that "color". The quadtree with no maximum depth set and threshold of 100% is an *exact tree*. The exact tree truthfully represents the measured data. A quadtree with either threshold or maximum depth limit set allows us to reduce the size of the tree at the cost of prediction accuracy. Limiting the absolute tree depth limits the maximum number of tests we may need to execute to determine the method index for specified communicator and message size. Setting the accuracy threshold helps smooth the experimental data, thus possibly making the decision function more resistant to inaccuracies in measurements.

A property of any decision tree is that an internal node of the tree corresponds to an attribute test, and the links to children nodes correspond to the particular attribute values. In our encoding scheme, every non-leaf node in the quadtree corresponds to a test which matches both communicator and message size values. The leaf nodes contain information about the optimal method for the particular communicator and message size ranges. Thus, leaves represent the rules of the particular decision function. In effect, quadtrees allow us to perform a recursive binary search in a two-dimensional space.

3.1   Generating Decision Function Source Code

We provide functionality to generate decision function source code from the initialized quadtree. Recursively, for every internal node in the quadtree we generate the following code segment:

if (NW) {...} else if (NE) {...} else if (SW) {...} else if (SE) {...} else {error}.[1]

The current implementation is functional but lacks optimizations, i.e. ability to

---

[1] NW, NE, SW, and SE correspond to north-west, north-east, south-west, and south-east quadrants of the region.

merge conditions with same color[2]. The conditions for boundary points (minimum and maximum communicator and message sizes) are expanded to cover that region fully. For example, the rule for minimum communicator size will be used for all communicator sizes less than the minimum communicator size.

3.2 In-memory Quadtree Decision Structure

Alternative to generating the decision function source code is maintaining an in-memory quadtree decision structure which can be queried during the run time.

An optimized quadtree structure would contain 5 pointers and 1 method field, which could probably be a single byte or an integer value. Thus, the size of a node of the tree would be around 44B on 64-bit architectures[3]. Additionally, the system would need to maintain in memory the mapping of (algorithm, segment size) pairs to method indexes as well. The maximum depth decision quadtree we encountered in our tests had 6 levels. This means that in the worst case, the 6-level decision quadtree could take up to $\frac{4^7-1}{4-1} = 5461$ nodes, which would occupy close to 235KB of memory. However, our results indicate that the quadtrees with 3 levels can still produce reasonably good decisions. Three-level quadtree would occupy at most 3740B and as such could fit into 4 1KB pages of main memory. Even so, the smaller quadtree if cached would still occupy significant portion of the cache. Based on these memory requirements we decided not to implement the in-memory quadtree-based decision structure yet, and to focus our efforts on decision function source code generation.

4 Experimental Results and Analysis

Under the assumption that the collective operations parameters are uniformly distributed across communicator size and message size space, we expect that the average depth of the quadtree is the average number of conditions we need to evaluate before we can determine which method to use. In the worst case, we will follow the longest path in the tree to make the decision, and in the best case the shortest.

The performance data for broadcast and reduce collective algorithms was collected on Grig cluster located at the University of Tennessee at Knoxville and Nano cluster located at the Lawrence Berkeley National Laboratory.

4.1 Broadcast Decision Maps

Figure 1 shows six different quadtree decision maps for a broadcast collective on the Grig cluster. We considered five different broadcast algorithms (Linear,

---

[2] The code segment generated for each internal node contains at least 21 lines – 5 lines for conditional expressions, 10 lines for braces, a line for error handling, and at least a line per condition.

[3] In this analysis, we ignore data alignment issues which would lead to even larger size of the structure.

Binomial, Binary, Splitted-Binary, and Pipeline),[4] and four different segment sizes (no segmentation, 1KB, 8KB, and 16KB). The measurements covered all communicator sizes between 2 and 28 processes and message sizes in 1B to 384KB range.

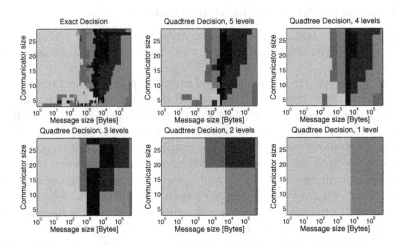

Fig. 1. Broadcast decision maps from Grig cluster. Different colors correspond to different method indexes. The trees were generated by limiting the maximum tree depth. The x-axis scale is logarithmic. The crossover line for 1-level quadtree is not in the middle due to the "fill-in" points used to adjust the original size of the decision map from 25×48 to 64×64 form.

The exact decision map in Figure 1 exhibits trends, but there is a considerable amount of information for intermediate size messages (between 1KB and 10KB) and small communicator sizes. Limiting the maximum tree depth smoothes the decision map and subsequently decreases the size of the quadtree. Table 2 shows the mean tree depth and related statistics for the decision maps presented in Figure 1.

4.2 Performance Penalty of Decision Quadtrees

One possible metrics of merit is the performance penalty one would incur by using a restricted quadtree instead of the exact one. To compute this, one can use the performance information for methods suggested by the restricted tree for particular set of communicator and message size values, and compare them to the performance results for methods suggested by the exact tree.

The reproducibility of measured results is out of scope of this paper, but we followed the guidelines from [11] to ensure good quality measurements. Even so, the "exact" decision function corresponds to a particular data set, and the

---

[4] For more details on these algorithms, refer to [8].

performance penalty of other decision functions was evaluated against the data that was used to generate them in the first place.

Figure 2 shows the performance penalty of decision quadtrees from Figure 1 and the Table 2 summarizes the properties and performance penalties for the same data. The analysis shows that even for noisy decision map in Figure 1, a 3-level quadtree would have less than 9% performance penalty on average, while the exact decision could be represented with a total of 6 levels.

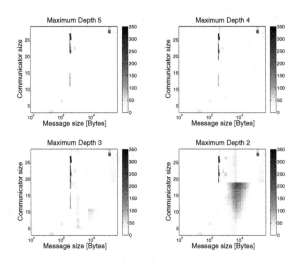

Fig. 2. Performance penalty of broadcast decision function from Grig cluster. Colorbar represents relative performance penalty in percents. White color means less than 5%, yellow/light gray is between 10% and 25%.

4.3   Quadtree Accuracy Threshold

In Section 3.1 we mentioned that an alternative way to limit the size of quadtree is to specify the tree accuracy threshold.

Figure 3 shows the effect of varying the accuracy threshold on the mean performance penalty of a reduce quadtree decision function on two different systems. On both systems, the mean performance penalty of the reduce decision was below 10% for an accuracy threshold of approximately 45%. This threshold corresponds to the quadtree structures of maximum depth 3. This means that the quadtree decision which would on average potentially cause a 10% performance penalty would be evaluated at most in 3 expressions.

4.4   Accuracy Threshold vs. Limiting Maximum Depth

Figure 4 shows the mean performance penalty of broadcast and reduce decisions on Grig cluster (See Figures 1, 2, and 3, and Table 2) as a function of the mean quadtree depth for quadtrees constructed by specifying accuracy threshold and

Table 2. Statistics for broadcast decision quadtrees in Figure 1. The number of leaves corresponds to the number of regions we divided the (communicator size, message size) space into. The number of lines in decision function includes lines containing only braces, error handling, etc.

Tree Depth			Performance Penalty	[%]			Number of	Function size
Max	Min	Mean	Min	Max	Mean	Median	Leaves	[# of lines]
1	1	1.0000	0.00	346.05	37.11	0.00	4	24
2	2	2.0000	0.00	436.02	18.63	0.00	16	82
3	2	2.9655	0.00	436.02	08.83	0.00	58	330
4	2	3.8554	0.00	391.53	06.29	0.00	166	932
5	2	4.7783	0.00	356.47	05.41	0.00	442	2,496
6	2	5.6269	0.00	000.00	00.00	0.00	973	5,505

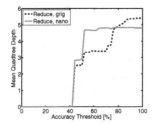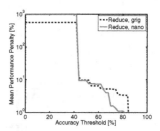

Fig. 3. Effect of the accuracy threshold on mean quadtree depth and performance penalty

maximum depth. The results indicate that in the cases we considered, constructing the decision quadtree by restricting the maximum depth of the tree directly incurs a smaller mean performance penalty than the tree of similar mean depth constructed by setting the accuracy threshold.

The results for the broadcast decision function show that when the quadtree is deep enough to cover almost the whole initial data set, the tree constructed using an accuracy thresholds achieves the smaller mean performance penalty. This is not the case for the quadtree-based reduce decision functions most likely due to the fact that this decision function was smoother to start with, so smoothing it with an accuracy threshold had no further positive effects. Still, we believe that

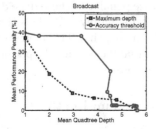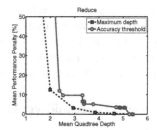

Fig. 4. Accuracy threshold vs. maximum depth quadtree construction

the example of the broadcast decisions indicates that the accuracy threshold setting could be used to avoid over-fitting the data when the tree depth is not a concern.

5   Discussion and Future Work

In this paper, we studied the applicability of a modified quadtree encoding method to the algorithm selection problem for the MPI collective function optimization. We analyzed the properties and performance of quadtree decision functions constructed by either limiting the maximum tree depth or specifying the accuracy threshold a the construction time.

Our experimental results for broadcast and reduce collectives, show that in some cases, the decision function based on a quadtree structure with a mean depth of 3, incurs less than a 5% performance penalty on the average. In other cases, deeper trees (5 or 6 levels) were necessary to achieve the same performance. However, in all cases we considered, a quadtree with 3 levels would incur less than a 10% performance penalty on average. Our results indicate that quadtrees may be a feasible choice for processing the performance data and decision function generation.

In this work we chose not to explore the performance of the in-memory quadtree decision systems due to relatively large memory requirements associated with storing the tree. The performance of an in-memory system will depend greatly on the implementation efficiency and the application access pattern. It is possible that in some cases and or in combination with other methods, it could achieve very good performance. We plan to explore this issue in more depth in the future.

One of the limitations of the quadtree encoding method is that since the decision is based on a 2D-region in communicator size - message size space, it will not be able to capture decisions which are optimal for single communicator values, e.g. communicator sizes which are power of 2. The same problem is exacerbated if the performance measurement data used to construct trees is too sparse.

The decision map reshaping process to convert measured data from $n \times m$ shape to $2^k \times 2^k$ affects encoding efficiency of the quadtree. In our current study, we did not address this issue, but in future work we plan to further improve the efficiency of the encoding regardless of initial data space.

The major focus of future research will be comparing the quadtree-based decision functions, to the ones generated using run-length encoding and standard decision tree algorithms such as C4.5.

Acknowledgement. This work was supported by Los Alamos Computer Science Institute (LACSI), funded by Rice University Subcontract #R7B127 under Regents of the University Subcontract #12783-001-05 49.

References

1. Rabenseifner, R.: Automatic MPI counter profiling of all users: First results on a CRAY T3E 900-512. In: Proceedings of the Message Passing Interface Developer's and User's Conference. (1999) 77–85
2. Worringen, J.: Pipelining and overlapping for MPI collective operations. In: 28th Annyal IEEE Conference on Local Computer Network, Boon/Königswinter, Germany, IEE Computer Society (2003) 548–557
3. Rabenseifner, R., Träff, J.L.: More efficient reduction algorithms for non-power-of-two number of processors in message-passing parallel systems. In: Proceedings of EuroPVM/MPI. Lecture Notes in Computer Science, Springer-Verlag (2004)
4. Chan, E.W., Heimlich, M.F., Purkayastha, A., van de Geijn, R.M.: On optimizing of collective communication. In: Proceedings of IEEE International Conference on Cluster Computing. (2004) 145–155
5. Thakur, R., Gropp, W.: Improving the performance of collective operations in MPICH. In Dongarra, J., Laforenza, D., Orlando, S., eds.: Recent Advances in Parallel Virtual Machine and Message Passing Interface. Number 2840 in LNCS, Springer Verlag (2003) 257–267 10th European PVM/MPI User's Group Meeting, Venice, Italy.
6. Kielmann, T., Hofman, R.F.H., Bal, H.E., Plaat, A., Bhoedjang, R.A.F.: MagPIe: MPI's collective communication operations for clustered wide area systems. In: Proceedings of the seventh ACM SIGPLAN symposium on Principles and Practice of Parallel Programming, ACM Press (1999) 131–140
7. Bernaschi, M., Iannello, G., Lauria, M.: Efficient implementation of reduce-scatter in MPI. Journal of Systems Architure **49**(3) (2003) 89–108
8. Pješivac-Grbović, J., Angskun, T., Bosilca, G., Fagg, G.E., Gabriel, E., Dongarra, J.J.: Performance analysis of mpi collective operations. In: Proceedings of the 19th IEEE International Parallel and Distributed Processing Symposium (IPDPS'05) - Workshop 15, Washington, DC, USA, IEEE Computer Society (2005) 272.1
9. Fagg, G.E., Gabriel, E., Bosilca, G., Angskun, T., Chen, Z., Pješivac-Grbović, J., London, K., Dongarra, J.: Extending the mpi specification for process fault tolerance on high performance computing systems. In: Proceedings of the International Supercomputer Conference (ICS) 2004, Primeur (2004)
10. Gabriel, E., Fagg, G.E., Bosilca, G., Angskun, T., Dongarra, J.J., Squyres, J.M., Sahay, V., Kambadur, P., Barrett, B., Lumsdaine, A., Castain, R.H., Daniel, D.J., Graham, R.L., Woodall, T.S.: Open MPI: Goals, concept, and design of a next generation MPI implementation. In: Proceedings, 11th European PVM/MPI Users' Group Meeting, Budapest, Hungary (2004) 97–104
11. Gropp, W., Lusk, E.L.: Reproducible measurements of MPI performance characteristics. In: Proceedings of the 6th European PVM/MPI Users' Group Meeting on Recent Advances in PVM and MPI, Springer-Verlag (1999) 11–18

Parallel Prefix (Scan) Algorithms for MPI

Peter Sanders[1] and Jesper Larsson Träff[2]

[1] Universität Karlsruhe
Am Fasanengarten 5, D-76131 Karlsruhe, Germany
sanders@ira.uka.de
[2] C&C Research Laboratories, NEC Europe Ltd.
Rathausallee 10, D-53757 Sankt Augustin, Germany
traff@ccrl-nece.de

Abstract. We describe and experimentally compare four theoretically well-known algorithms for the *parallel prefix operation* (*scan*, in MPI terms), and give a presumably novel, doubly-pipelined implementation of the in-order binary tree parallel prefix algorithm. Bidirectional interconnects can benefit from this implementation. We present results from a 32 node AMD Cluster with Myrinet 2000 and a 72-node SX-8 parallel vector system. The doubly-pipelined algorithm is more than a factor two faster than the straight-forward binomial-tree algorithm found in many MPI implementations. However, due to its small constant factors the simple, linear pipeline algorithm is preferable for systems with a moderate number of processors. We also discuss adapting the algorithms to clusters of SMP nodes.

Keywords: Cluster of SMPs, collective communication, MPI implementation, prefix sum, pipelining.

1 Introduction

The *parallel prefix* or *scan* operation is a surprisingly versatile primitive and a basic building block in massively parallel algorithms for a variety of different problems, as shown by research in the 80ties and 90ties [2,5]. Scan primitives are also included among the collective operations of the *Message Passing Interface* (MPI) [10], as an *inclusive* operation MPI_Scan, and with the MPI-2 standard [3], also as an *exclusive* operation MPI_Exscan.

The parallel prefix operation can be explained as follows. Let p be the number of Processing Element(PE)s numbered consecutively from 0 to $p - 1$, and let a sequence of p elements x_i with an associative, binary operation $\oplus$ be given.

The *inclusive parallel prefix operation* computes for each PE j, $0 \leq j < p$, the value $\bigoplus_{i=0}^{j} x_i = x_0 \oplus x_1 \oplus \cdots \oplus x_j$, with the convention that $\bigoplus_{i=j}^{j} x_i = x_j$ (a one element sum is just that one element).

The *exclusive parallel prefix operation* computes for each PE j, $0 < j < p$, *except* PE 0 the value $\bigoplus_{i=0}^{j-1} x_i$. With this definition, no neutral element for the operation $\oplus$ is required.

B. Mohr et al. (Eds.): PVM/MPI 2006, LNCS 4192, pp. 49–57, 2006.

For use with the scan (and other reduction) collectives, MPI provides a number of standard, binary operations like summation, maximum, Boolean and bitwise operations etc. on standard datatypes like integers, doubles, and so forth. In addition the user can define new associative operations on structured, possibly non-contiguous datatypes. Instead of a parallel prefix on a single element per process, the MPI scan operations work element-wise on vectors of elements. The number of elements in the vector is given by a `count` argument in the call of `MPI_Scan`/`MPI_Exscan`.

2   The Scan Algorithms

In this section we describe four standard algorithms for the parallel prefix operations, and discuss their implementation in MPI. We focus exclusively on the inclusive scan operation, but the discussion applies *mutatis mutandis* to `MPI_Exscan`. We assume single-ported communication in a fully connected network. Communication cost is $\alpha + \beta m$ for a communication involving m data elements. We use three variants of this model: a) *half-duplex* where each communicating PE can either send or receive a message, b) *telephone model*, where a matched pair of PEs can communicate bidirectionally, and c) *full-duplex* where a PE can simultaneously send data to one PE and receive data from a possibly different PE. An m element $\oplus$ computation takes γm time of local work.

All algorithms will work in (implicitly) synchronized rounds and exchange data packets of equal length. Hence, the number of communication rounds and the amount of data sent and received by each PE or the total communication volume normally suffice to characterize the algorithms.

We use the shorthand $\oplus[j..k]$ for $\bigoplus_{i=j}^{k} x_i$.

2.1   Binomial Tree

Let $n = \lfloor \log_2 p \rfloor$. The binomial tree algorithm consists of an *up-phase* and a *down-phase* each of n rounds. In round k, $k = 0, \ldots, n-1$ of the up-phase each PE j satisfying $j \wedge (2^{k+1} - 1) = 2^{k+1} - 1$ (where $\wedge$ denotes "bitwise and") receives a partial result from PE $j - 2^k$ (provided $0 \leq j - 2^k$). Afterwards, PE $j - 2^k$ is inactive for the remainder of the up-phase. The receiving PEs add the partial results, and after round k have a partial result of the form $\oplus[j - 2^{k+1} + 1..j]$. In the down-phase we count rounds downward from n to 1. A PE j with $j \wedge (2^k - 1) = 2^k - 1$ sends its partial result to PE $j + 2^{k-1}$ (provided $j + 2^{k-1} < p$) which can now compute its final result $\oplus[0..j + 2^{k-1}]$. The communication pattern is shown in Figure 1.

The number of communication rounds is $2\lfloor \log p \rfloor$, and the total communication volume is bounded by $2pm$ since in each round half the PEs become inactive. Since each PE is either sending or receiving data in each round, with no possibility for overlapping of sending and receiving due to the computation of partial results, the algorithm can be implemented in the half-duplex model.

2.2   Simultaneous Binomial Tree

Starting from round $k = 0$, in round k, PE j sends its partial result to PE $j + 2^k$ (provided $j + 2^k < p$) and receives a partial result from PE $j - 2^k$ (provided $0 \geq j - 2^k$). The partial results are added. It is easy to see that after round k, PE js partial result is $\oplus[\max(0, j - 2^{k+1} + 1)..j]$. PE j can terminate when both $j - 2^k < 0$ (nothing to receive) and $j + 2^k \geq p$ (nothing to send). This happens after $\lceil \log p \rceil$ rounds. This algorithm goes back (at least) to [4], and is illustrated in Figure 1.

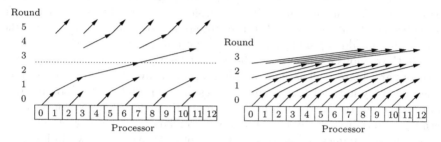

Fig. 1. The communication patterns for the simple binomial (left) and the simultaneous binomial (right) tree algorithm for $p = 13$

The total communication volume is bounded by $\lceil \log p \rceil pm$ since (almost) all PEs are active in all rounds. Since each PE is both sending and receiving data from two different PEs, the analysis assumes the full-duplex model. In [6] it is shown that the algorithm can be generalized to exploit k-ported communication.

A different algorithm with the same characteristics, but based on a butterfly communication pattern is used in the mpich2 MPI implementation. For this algorithm the telephone model of communication suffices but it has unbalanced computational load — in each round, half the PEs compute two partial results.

2.3   Linear Pipeline

The third algorithm arranges the PEs in a linear pipeline. PE j, $0 < j < p - 1$, receives the result $\oplus[0..j - 1]$ from PE $j - 1$, adds x_j, and sends the result $\oplus[0..j]$ to PE $j + 1$. The last PE finishes after $p - 1$ communication rounds at time $(p - 1)(\alpha + \beta m + \gamma m) + \gamma m$. Dividing the m elements into b blocks, which are sent along the linear pipeline one after the other, the time at which PE $p - 1$ finishes becomes $(p - 1)(\alpha + \beta' m/b) + (b - 1)(\alpha + \beta' m/b) + \gamma m/b$. After the first block has arrived at PE $p - 1$, a new block can be delivered in every round, assuming full-duplex communication, since PE j can send the current block and receive the next one simultaneously. Thus $b - 1$ rounds are required after the initial delay to finish the scan. Here β' is a constant between β and $\beta + \gamma$, which depends on the possible amount of overlap between communication and computation. Balancing α and β' terms depending on b, the optimal block size can be found as $b = \sqrt{\beta'/\alpha}\sqrt{m}$.

The linear pipeline was also discussed and implemented in [11].

2.4   Pipelined Binary Tree

The fourth algorithm arranges the PEs in a binary tree T with in-order numbering. This numbering has the property that the PEs in the subtree $T(j)$ rooted at j have consecutive numbers in the interval $[\ell, \ldots, j, \ldots, r]$ where ℓ and r denote the first and last PE in the subtree $T(j)$, respectively. The algorithm has two phases. In the *up-phase*, PE j first receives the partial result $\oplus[\ell..j - 1]$ from its left child and adds x_j to get $\oplus[\ell..j]$. This value is stored for the down-phase. PE j then receives the partial result $\oplus[j + 1..r]$ from its right child and computes the partial result $\oplus[\ell..r]$. PE j sends this value upward without keeping it. In the *down-phase*, PE j receives the partial result $\oplus[0..\ell - 1]$ from its parent. This is first sent down to the left child and then added to the stored partial result $\oplus[\ell..j]$ to form the final result $\oplus[0..j]$ for j. This final result is sent down to the right child.

With the obvious modifications, the general description covers also nodes that need not participate in all of these communications: Leaves have no children. Some nodes may only have a leftmost child. Nodes on the path between root and leftmost leaf do not receive data from their parent in the down-phase. Nodes on the path between rightmost child and root do not send data to their parent in the up-phase. The communication pattern and examples of trees are shown in Figure 2.

Let the *height* n of the tree denote the length of the longest root-to-leaf path. The number of rounds for both up- and down-phases are at most $2n - 1$ each. The total communication volume per phase is bounded by $(p - 1)m$. The algorithm assumes only half-duplex communication.

It is a standard observation (e.g. [7]) that each PE is (in each phase) only active in three consecutive rounds. Hence, successive up-phases (and down-phases) can be *pipelined*. More specifically, if the m element vectors can be divided into b blocks (and the operation $\oplus$ on the m element vectors can likewise be blocked), each phase can be done in $3(b - 1) + 2n - 1$ rounds: the $2n - 1$ rounds is the delay for the first block delivered at the root (or at the lowest leaf), with a new block delivered at every third round. Since the partial results computed by PE j in the up-phase is either needed by j or immediately sent upwards, there is no

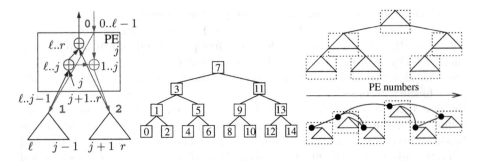

Fig. 2. From left to right: The basic schedule of the doubly pipelined algorithm. A balanced binary tree with in-order numbering. Two ways to build a binary tree from a cluster of six SMPs.

need for intermediate buffering between up- and down-phases. A single buffer of size m/b for receiving a single intermediary block therefore suffices also for the pipelined implementation. In our cost model, an m element prefix sum can be computed in time $O(n+m)$ using an optimal block size of $\Theta(\sqrt{m/n})$. Note that using a balanced binary tree we have $n = \lceil \log(p+1) \rceil - 1$.

For bidirectional communication networks in the telephone model the two pipelined phases can be combined. This can reduce the number of rounds by up to a factor of two. Depending on its position in the tree, a PE will first perform a certain number d of rounds working only on upward traffic while waiting for the first packet of downward data.[1] After this *fill phase*, it enters into a *steady state* in which it exchanges in each round one block of data either with its parent or with one of its children. After $3b - d$ rounds of the steady state, the up-phase blocks have been completed, and in d rounds of the *drain phase* the last blocks of the down-phase are processed. We call this algorithm the *doubly pipelined prefix algorithm*.

The largest *delay* is incurred for the rightmost leaf PE in the binary tree, which has to wait for $2(2n - 1)$ rounds for the first block to arrive. A new block arrives every third round, so the total number of rounds becomes $3(b-1)+4n-2$, or almost a factor two better than the $6(b - 1) + 4n - 2$ rounds required for the up- and subsequent down-phase of the two-phase algorithm.

Instead of pipelining, the same asymptotic running time is achieved by the algorithm in [1] which by repeated halving splits the m elements into p blocks, on which simultaneous scans are carried out by edge-disjoint binomial trees. This algorithm assumes that p is a power of two (with a trivial generalization), and also in terms of constant factors the algorithm is worse than the doubly pipelined prefix algorithm.

3 Performance Evaluation

The algorithms from Section 2 have currently been implemented for the case of one MPI process per node. The algorithms have been benchmarked on a 32-node AMD cluster with Myrinet 2000 (with the GM-library), and the 72 node SX-8 parallel vector supercomputer at HLRS (Hochleistungsrechenzentrum Stuttgart, Germany). We compare five algorithms, namely

- binomial tree
- simultaneous binomial trees
- linear pipeline
- pipelined binary tree
- doubly pipelined prefix algorithm

Results are shown in Figure 3 which shows the achieved throughput as a function of problem size m for fixed number of processes. For the pipelined

---

[1] In a complete binary tree, this waiting time is proportional to the number of parent connections one has to follow until reaching the leftmost root-leaf path in the tree.

algorithms, the block size has been chosen proportional to $\sqrt{m/n}$ with experimentally determined constants depending on α and β.

On both systems the simultaneous binomial tree algorithm is best for small message lengths (m up to about 50 KBytes for the Myrinet cluster, and up to about 1 MByte for the SX-8 system). For very small messages, it is up to a factor of two better than all other algorithms and it dominates the plain binomial tree algorithm for all input sizes. Beyond this threshold, on the Myrinet cluster the linear pipeline is superior, also to the pipelined binary tree algorithms. For 31 nodes it is about a factor 2 faster than the doubly pipelined prefix algorithm, which is again better than the simple pipelined binary tree algorithm, although by less than the factor of two predicted by the theoretical analysis. As the number of processors grow, the doubly pipelined algorithm can eventually be expected to outperform the linear pipeline, as shown by the scalability plot in Figure 4. The performance degradation for message sizes around 2MB is due to a protocol change at the size of the pipeline block m/b, and can be eliminated by more careful tuning.

On the NEC SX-8 the pipelined binary tree algorithms give significantly better throughput than binomial and simultaneous binomial tree algorithms, although the difference between the pipelined and the doubly pipelined algorithms (about a factor 1.4 faster for large problems) is smaller than expected from the theoretical analysis. Nevertheless, the capability for full-duplex communication can be exploited by the doubly pipelined prefix algorithm. But also on the SX-8 the highest throughput is achieved by the linear pipeline (about a factor 1.4 faster than the doubly pipelined algorithm).

4   Adaptation to the SMP Case

The parallel prefix algorithms were developed assuming a homogeneous communication network. For clusters of SMP nodes this assumption does not hold, and severe node contention can result if many PEs per SMP node must in the same round send and/or receive data from other nodes. In particular the simultaneous binomial tree algorithm will inevitably suffer from this kind of node contention.

Now we discuss several possible improvements for the case that there are P SMP (nodes) with p_i consecutively ranked PEs in SMP i.

For small inputs and/or very slow inter-SMP communication, a simple hierarchical decomposition works well: First compute a parallel prefix within each SMP. Then perform a parallel (exclusive) prefix over the SMPs using the result of the last PE on each SMP. Finally, within each SMP, add the global result to each local result. This algorithm has the advantage that at any time at most one PE per SMP is performing inter-SMP communication.

For large inputs it is better to arrange all the PEs into a single tree taking care that inter-SMP communication is small. This way, the time for intra-SMP prefix computation will not appear in the term of the execution time that depends on the input size m. We propose two basic ways to do this which are depicted in the right part of Figure 2: One is to build local trees of depth $O(\log p_i)$ on each

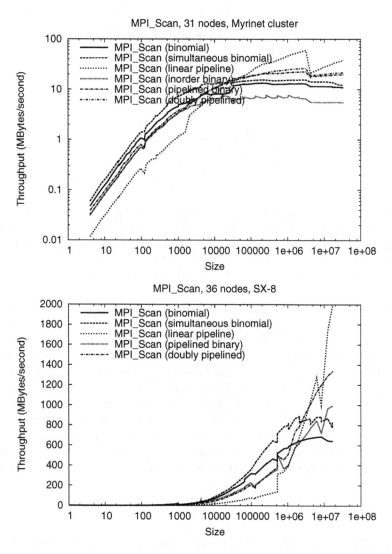

Fig. 3. The five scan algorithms: binomial tree, simultaneous binomial trees, linear pipeline, pipelined binary tree, doubly pipelined prefix algorithm. Top: 31-node Myrinet cluster (with add. measurements for non-pipelined binary tree, doubly logarithmic plot); Bottom: 36 nodes of the NEC SX-8 (logarithmic plot).

SMP and to build a binary tree of local trees as follows: The root PE of a left successor in the SMP tree has the leftmost PE of its parent SMP as its parent in the PE tree. Analogously, a right successor has a rightmost PE as its parent. Now suppose $p_i = p/P$ for all SMPs. We get a PE tree of height $(1 + o(1)) \log \frac{n}{p} \log P$. At most three PEs in each SMP perform inter-SMP communication. The total volume of inter-SMP communication is $\leq 2mP$.

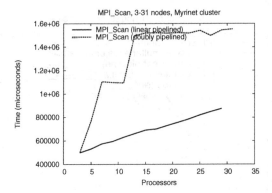

Fig. 4. Scalability of linear pipeline and doubly pipelined algorithm on the Myrinet cluster for a large problem of 36 MBytes. Completion time of the slowest PE is shown as a function of the number of processors for $p = 3, 5, \ldots, 31$. Running time of the linear pipeline increase linearly with the number of processors, whereas the doubly pipelined exhibits a jump each time a new level is added to the binary tree.

At the cost of increasing the inter-SMP communication to about $3mP$, we can decrease the height of the tree to $\lceil \log(P+1) \rceil + \max_{0 \le i < P} \lceil \log p_i \rceil - 1$ and reduce the number of PEs with inter-SMP communication to at most two per SMP: The leftmost PEs g_i of each SMP form a *global* balanced binary tree of height $h = \lceil \log(P+1) \rceil - 1$ with the following properties: An in-order traversal meets growing PE numbers. Only leaves have no left successor. All PEs without a right child are only on the rightmost path through the tree. The remaining PEs of each SMP form a *local* tree of height $\lceil \log p_i \rceil - 1$ rooted at some node r_i. If global tree PE g_i has no right successor in the global tree, its right successor is r_i. If g_i for $i > 0$ has no left successor in the global tree, its left successor will be r_{i-1}. It is easy to verify that the resulting tree has the claimed properties.

5   Summary

We described and implemented five algorithms for the MPI scan collectives. As shown by the performance evaluation, a production quality MPI should use a hybrid approach, using the simultaneous binomial tree algorithm for small problems, and for large problems switching to either the linear pipeline or the doubly pipelined algorithm depending on the number of processors. To the best of our knowledge our implementation is the first implementation of a pipelined binary tree scan algorithm. The doubly pipelined algorithm is new. Efficiently mapping communication trees to SMPs is already described for broadcasting in [8]. However our method to maintain the canonical numbering of the PEs as an in-order numbering of the tree is new.

For the design and determination of block sizes a linear cost function was assumed. This is a simplified assumption, and more accurate cost models could lead to better results. In [9] the prefix-sums problem is studied in the LogP model from a different perspective (what is the largest number of x_is that can be reduced in a given time?). The resulting algorithms are complex, so there is a trade-off between accuracy and implementation concerns.

References

1. S. Bae, D. Kim, and S. Ranka. Vector prefix and reduction computation on coarse-grained, distributed memory machines. In *International Parallel Processing Symposium/Symposium on Parallel and Distributed Processing (IPPS/SPDP 1998)*, pages 321–325, 1998.
2. G. E. Blelloch. Scans as primitive parallel operations. *IEEE Transactions on Computers*, 38(11):1526–1538, 1989.
3. W. Gropp, S. Huss-Lederman, A. Lumsdaine, E. Lusk, B. Nitzberg, W. Saphir, and M. Snir. *MPI – The Complete Reference*, volume 2, The MPI Extensions. MIT Press, 1998.
4. W. D. Hillis and J. Guy L. Steele. Data parallel algorithms. *Communications of the ACM*, 29(12):1170–1183, 1986.
5. J. JáJá. *An Introduction to Parallel Algorithms*. Addison-Wesley, 1992.
6. Y.-C. Lin and C.-S. Yeh. Efficient parallel prefix algorithms on multiport message-passing systems. *Information Processing Letters*, 71:91–95, 1999.
7. E. W. Mayr and C. G. Plaxton. Pipelined parallel prefix computations, and sorting on a pipelined hypercube. *Journal of Parallel and Distributed Computing*, 17:374–380, 1993.
8. P. Sanders and J. F. Sibeyn. A bandwidth latency tradeoff for broadcast and reduction. *Information Processing Letters*, 86(1):33–38, 2003.
9. E. E. Santos. Optimal and efficient algorithms for summing and prefix summing on parallel machines. *Journal of Parallel and Distributed Computing*, 62(4):517–543, 2002.
10. M. Snir, S. Otto, S. Huss-Lederman, D. Walker, and J. Dongarra. *MPI – The Complete Reference*, volume 1, The MPI Core. MIT Press, second edition, 1998.
11. J. Worringen. Pipelining and overlapping for MPI collective operations. In *28th Annual IEEE Conference on Local Computer Networks (LCN 2003)*, pages 548–557, 2003.

Efficient Allgather for Regular SMP-Clusters

Jesper Larsson Träff

C&C Research Laboratories, NEC Europe Ltd.
Rathausallee 10, D-53757 Sankt Augustin, Germany
`traff@ccrl-nece.de`

Abstract. We show how to adapt and extend a well-known *allgather* (all-to-all broadcast) algorithm to parallel systems with a hierarchical communication system such as clusters of SMP nodes. For small problem sizes, the new algorithm requires a logarithmic number of communication rounds in the number of SMP nodes, and *gracefully degrades* towards a linear algorithm as problem size increases. The algorithm has been used to implement the `MPI_Allgather` collective operation of MPI in the MPI/SX library. Performance measurements on a 72 node SX-8 system shows that graceful degradation provides a smooth transition from logarithmic to linear behavior, and significantly outperforms a standard, linear algorithm. The performance of the latter is furthermore highly sensitive to the distribution of MPI processes over the physical processors.

1 Introduction

An important and well-studied collective communication primitive for message-passing systems is the *allgather* or *all-to-all broadcast* operation [6], in which each processor has data which have to be distributed (i.e. broadcast) to all other processors. This primitive has been extensively studied in a variety of settings and, correspondingly, is known also as (for instance) *total exchange* [5,4], *catenation* [3], and *gossip* [9]. We will use the term *allgather* here.

The allgather primitive is incorporated as a *collective communication operation* in the *Message-Passing Interface* (MPI) standard [11] in two flavors. The `MPI_Allgather` collective is *regular* in the sense that the size of the data to be broadcast by each MPI process must be the same for all processes. The more general, *irregular* `MPI_Allgatherv` collective does not have this restriction, and each process may contribute data of different size. A peculiarity of both MPI primitives, however, is that the size of the data contributed by each process is known by all processes in advance.

There has recently been much interest in improving the collective operations in various MPI libraries, see for instance [1,10,12] (and the references therein). Various allgather algorithms for MPI were discussed and evaluated in [2]. However, the collective operations in many MPI libraries are not adapted to systems with hybrid, hierarchical communication systems such as clusters of SMP nodes (see [7,8,10] for exceptions).

B. Mohr et al. (Eds.): PVM/MPI 2006, LNCS 4192, pp. 58–65, 2006.

In this paper we give an improvement to a well-known allgather algorithm which makes it suitable to the SMP case. In this context an SMP cluster is simply a collection of multi-processor nodes interconnected by a communication network. Communication between processors on the same node (typically via shared memory) is assumed to be faster (lower latency, higher bandwidth) than between processors on different nodes. Most importantly, the number of processors per node that can simultaneously communicate with processors on different nodes is restricted, typically to only one processor, although some modern high-performance interconnects offer multiple communication ports. We assume that communication within nodes is homogeneous, and likewise that the interconnect over the nodes is homogeneous, that is the cost of communication between any two processors on two different nodes is independent of the location of the two processors.

MPI is a process based model. Sets of processes are represented by so-called *communicators*. The semantics of the MPI collectives is defined in terms of the numbering of the processes in the given communicator. Since new communicators can be defined arbitrarily from existing ones, no assumptions about the numbering of MPI processes residing on an SMP node can be made. In particular, it cannot be assumed that the processes on a node form a consecutively numbered subset. Since allgather is a symmetric operation, it is desirable that the performance of MPI_Allgather be independent of the numbering of the processes.

In this paper we are concerned with *regular* SMP clusters, where the SMP nodes have the same number of processors. Additionally, the performance bounds for the allgather algorithm requires each node to run the same number of MPI processes. Again, since MPI allows arbitrary creation of new communicators, also for regular clusters it is possible to create communicators that do not fulfill this assumption. The algorithm can be used for the general case also, but can incur load imbalance. Better performance could possibly be achieved by a dedicated, non-regular algorithm. The algorithm can also be used for the irregular MPI_Allgatherv collective, but for very irregular problems better performance could possibly be achieved by a dedicated, irregular allgather algorithm.

2   An Allgather Algorithm with Graceful Degradation

We first present the regular allgather algorithm independent of MPI for systems with a homogeneous communication system (non-SMP case). The new feature which makes the algorithm better suited to clusters of SMP nodes is a smooth transition from logarithmic to linear behavior as the problem size grows. We term this feature *graceful degradation*.

We let p denote the number of processors, which are numbered from 0 to $p - 1$. Each processor r has a block of data block[r] of size b. For the regular allgather problem, b is the same for all processors. The *total size* of the allgather problem at hand is $m = pb$. The task of the allgather operation is to collect all blocks block[0], block[1], ..., block[$p - 1$] on all processors (in that order). By convention, for $i \leq j$, we let block[i, j] denote the consecutive sequence of blocks block[i], block[$i + 1$], ..., block[j], and for $j < i$, we let block[i, j] denote the "wrapped" consecutive sequence of blocks block[i], block[$i + 1$], ..., block[$p - 1$], block[0], ..., block[j].

The algorithm consists of a *logarithmic phase*, a *linear phase*, and a *last round*, either of which can be empty. In the logarithmic phase, each processor in each round doubles the number of blocks that it has collected. The algorithm used in this phase is the catenation algorithm of [3] (whose communication pattern is a regular, so-called *circulant graph*). The number of rounds of the logarithmic phase is determined by K, which can be any integer less than or equal to $\lfloor \log p \rfloor$. In the linear phase, larger, consecutive chunks consisting of 2^K input blocks are pipelined through rings of processors, until in the last round a last chunk of size strictly less than 2^K blocks is sent and received by each processor. In each round each processor sends and receives the same number of blocks. Below follows a more precise description of the *combined algorithm*.

r $r+1$	$r+4$	$s = r+8$		
		linear round 0	linear round 1	last round

$r + 2$

Fig. 1. The three phases of the combined allgather algorithm illustrated from a single processors point of view. In the *logarithmic phase* processor r receives blocks of size $1, 2, 4$ from processors $r + 1, r + 2, r + 4$ respectively. In the *linear phase* processor r receives two blocks of size 8 from processor $s = r + 8$. In the *last round* the remaining smaller block of size 5 is finally received from processor s.

1. In round k of the *logarithmic phase*, for $k = 0, \ldots K - 1$, processor r receives blocks block[$s, (s + 2^k - 1)$ mod p] from processor $s = (r + 2^k)$ mod p and sends blocks block[$r, (r + 2^k - 1)$ mod p] to processor $(r - 2^k)$ mod p.
2. Let $s = (r + 2^K)$ mod p, and $t = (r - 2^K)$ mod p, where K is the number of rounds of the logarithmic phase. In round k of the *linear phase*, $k = 0, \ldots, \lfloor p/2^K \rfloor - 2$, processor r receives blocks block[$(s + k2^K)$ mod $p, (s + (k+1)2^K - 1)$ mod p] from processor s and sends blocks block[$(r + k2^K)$ mod $p, (r + (k + 1)2^K - 1)$ mod p] to processor t.
3. In the case that $\lceil p/2^K \rceil 2^K > p$, in the *last round* processor r receives blocks block[$(s + (\lfloor p/2^K \rfloor - 1)2^K)$ mod $p, (s + p - 2^K - 1)$ mod p] from processor s and sends blocks block[$(r + (\lfloor p/2^K \rfloor - 1)2^K)$ mod $p, (r + p - 2^K - 1)$ mod p] to processor t.

The three phases of the combined algorithm are illustrated in Figure 1. Correctness follows, since in each round, each processor receives a consecutive segment of new blocks, and sends a consecutive segment of blocks received in the previous round.

The number of rounds required is $K + \lceil p/2^K \rceil - 1$, and the total number of blocks sent and received per processor is $p - 1$ for a total communication volume per processor of $(p - 1)b = m - b$. Each round entails either two communication steps (a send and a receive) for uni-directional interconnects, or one communication step (a combined send-receive) for interconnects supporting full bi-directional communication.

For $K = \lfloor \log p \rfloor$ the algorithm coincides with the algorithm in [3], and for $K = 0$ with a trivial, linear time ring algorithm. By choosing $K = \lfloor \log(B/b) \rfloor$ for some fixed *intermediate buffer size* the algorithm switches from logarithmic to linear behavior before the size of the consecutive segments received and sent in a round exceeds B. Thus, with increasing block size b the algorithm gracefully changes from purely logarithmic to linear behavior.

3 Implementation on SMP Clusters

The allgather operation allows a simple *hierarchical decomposition* to exploit the faster communication between processors on the same SMP node. The hierarchical allgather algorithm looks as follows.

1. Choose a local root on each SMP node
2. On all nodes gather input blocks to local root
3. Perform allgather over local roots
4. On all nodes broadcast result from local root

A straightforward implementation of this scheme would be inefficient for medium and large problems, since non-root processes would sit idle throughout the allgather step. For the implementation of MPI_Allgather an additional complication is caused by the fact that MPI processes are not necessarily consecutively numbered within the SMP nodes. Thus the blocks gathered at the local roots in the second step will either be nonconsecutive at the local root, or will have to be stored consecutively in an intermediate buffer. Both solutions have undesirable drawbacks.

For the broadcast (and the gather) operation, an SMP implementation would presumably use shared memory. In many cases, shared memory used for communication between MPI processes has to be specially allocated outside of process memory, and cannot be arbitrarily large.

By using the allgather algorithm of Section 2 each of these problems can be effectively addressed for allgather problems up to a certain size. For now we consider *regular* SMP systems with the same number of MPI processes per node. We let N denote number of nodes, and n the number of processes per node such that $p = nN$.

A shared memory buffer is used for the gather and broadcast operations, and is chosen to be of a fixed, maximum size B. The number of logarithmic rounds of the allgather algorithm is chosen as $K = \min(\lfloor \log N \rfloor, \lfloor B/nb \rfloor)$, where nb is the total size of the input blocks on each SMP node.

The hierarchical allgather algorithm is implemented as follows. A local root process r is chosen for each SMP node, and allocates a shared memory communication buffer of size B. For each node the blocks block[i] for the processes on the node are packed consecutively into the shared memory buffer using a node local consecutive numbering of the processes. The local roots execute the allgather algorithm of Section 2 with the modification that after each round of the linear phase, the blocks sent to processor t are unpacked into the result buffer of *all* processes on the SMP node. After the last round, the broadcast is completed by unpacking the last segment of blocks. This implementation effectively pipelines the allgather and the broadcast step of the hierarchical algorithm. We note that for communication systems that support concurrent communication and computation, unpacking of the blocks being sent to t into the result buffer of the local root can be performed concurrently with sending these blocks and receiving the next blocks from process s.

This algorithm can be used for allgather problems for which $nb \leq B$, i.e. for problems where the blocks of the processes on each SMP node can fit into the shared memory buffer. For larger problems a *linear ring algorithm* can be used. This should be implemented as follows. The MPI processes are sorted according to their SMP node id. The index of each process in this sorted sequence is used as virtual rank. In $p - 1$ rounds, each process with virtual rank r receives a block from virtual process $(r + 1) \bmod p$ and sends a block to virtual process $(r - 1) \bmod p$.

For systems with large SMP nodes (say, more than 8 processors per node) the transition from SMP algorithm to linear ring (which occurs when $nb > B$) may be too coarse. This can be avoided by introducing a linear algorithm similar to the linear phase for "medium sized" problems. The number of input blocks that can be kept in the shared memory buffer is $\lfloor B/b \rfloor$, so the p processors are divided into $p/\lfloor B/b \rfloor$ *virtual nodes* each of size $\lfloor B/b \rfloor$. A local root is chosen for each virtual node, and the linear phase of the allgather algorithm is executed over the virtual roots.

4   Performance Evaluation

The SMP allgather algorithm has been incorporated into the MPI/SX implementation for NEC's SX series of parallel vector computers [10]. In this section we evaluate the implementation using the 72 node SX-8 system at HLRS (Hochleistungsrechenzentrum, Stuttgart).

The basic performance of the combined algorithm for $N = 36$ nodes and $n = 1, 4, 8$ processes per node is illustrated in Figure 2 by comparing it to a linear ring algorithm. Running time is given as a function of the block size per process b. For small blocks up to a few KBytes the improvement over the linear algorithm is more than a factor of 3 for $n = 1$ process per node, and more than 13 for $n = 8$ processes per node. As block size increases the performance of the combined algorithm converges towards that of the linear algorithm. For $n = 8$ processes per node the switch to linear ring occurs after 64 KBytes (per process; the maximum shared memory buffer size is set to $B = 1$ MByte), and incurs a performance decrease by a factor of two (thus, the additional linear algorithm over virtual nodes described above would be worth considering).

The effect of graceful degradation towards the linear performance is further illustrated in Figure 3. This compares the combined algorithm to an SMP implementation of the logarithmic algorithm of [3], which switches to a linear ring

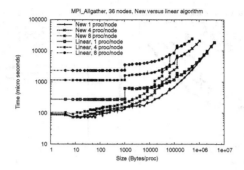

Fig. 2. The combined allgather algorithm with graceful degradation over ordered (`MPI_COMM_WORLD`) communicator compared to the linear ring algorithm for $N = 36$ nodes and $1, 4, 8$ processes per node

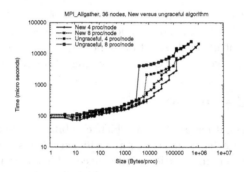

Fig. 3. The combined allgather algorithm over `MPI_COMM_WORLD` compared to an algorithm with without graceful degradation for $N = 36$ nodes and $n = 4, 8$ processes per node

as soon as the gathered result cannot fit into the shared memory buffer. This hybrid algorithm exhibits a very sharp jump in running time, which for $n = 8$ processes per node occurs at 8 KBytes, and is about a factor 7.

The potential sensitivity of a non-SMP algorithm to the numbering of of the MPI processes over the SMP nodes is illustrated in Figure 4. The combined algorithm (shown left) is compared to a linear ring algorithm over the MPI ranks (shown right) for the ordered `MPI_COMM_WORLD` communicator and a communicator in which the processes have been randomly permuted. In the latter case, the successor and predecessor of process r (namely $(r-1) \bmod p$, and $(r+1) \bmod p$) are almost always on a different SMP node, so in each communication round, almost all n processes on each node attempt to communicate with a process on another node, leading to serialization at the nodes. As expected, for $n = 4$ the random communicator performance is from a factor 2 for small block sizes up to almost a factor 4 for large block sizes worse than the ordered communicator. For $n = 8$ the performance degradation ranges from a factor of 3 up to a factor of 6. For somewhat larger block sizes, the combined algorithm is insensitive to the

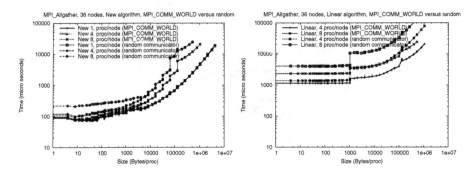

Fig. 4. The combined allgather algorithm (left) compared to non-SMP aware linear algorithm (right) for ordered `MPI_COMM_WORLD` and random communicator for $N = 36$ nodes and $n = 1, 4, 8$ processes per node

MPI process distribution. The difference for $n = 8$ for small block sizes is due to the fact that for the random communicator the gathered result does not form a consecutive segment of blocks, and must be unpacked as p individual blocks. On a vector machine like the NEC SX-8, copying of small blocks is penalized.

5   Concluding Remarks

We presented an allgather algorithm which combines well-known logarithmic and linear round allgather algorithms, and efficiently makes use of potentially limited intermediate communication buffer space. This makes the new algorithm suitable for use in *regular* SMP clusters, in which the number of processors (and MPI processes) per SMP node is the same for all nodes. The algorithm can be implemented also for non-regular SMP clusters, and has been used for the implementation of both `MPI_Allgather` and `MPI_Allgatherv` collectives. However, for very irregular problem instances, dedicated irregular algorithms might give better performance.

The combined algorithm was developed assuming single-port communication of the SMP nodes. It is worth pointing out that both the logarithmic and the linear phase can easily be generalized to the case where the SMP nodes have $k > 1$ communication ports.

References

1. G. Almási, P. Heidelberger, C. Archer, X. Martorell, C. C. Erway, J. E. Moreira, B. D. Steinmacher-Burow, and Y. Zheng. Optimization of MPI collective communication on BlueGene/L systems. In *19th ACM International Conference on Supercomputing (ICS 2005)*, pages 253–262, 2005.
2. G. D. Benson, C.-W. Chu, Q. Huang, and S. G. Caglar. A comparison of MPICH allgather algorithms on switched networks. In *Recent Advances in Parallel Virtual Machine and Message Passing Interface. 10th European PVM/MPI Users' Group Meeting*, volume 2840 of *Lecture Notes in Computer Science*, pages 335–343, 2003.

3. J. Bruck, C.-T. Ho, S. Kipnis, E. Upfal, and D. Weathersby. Efficient algorithms for all-to-all communications in multiport message-passing systems. *IEEE Transactions on Parallel and Distributed Systems*, 8(11):1143–1156, 1997.

4. P. Fraigniaud and E. Lazard. Methods and problems of communication in usual networks. *Discrete Applied Mathematics*, 53(1–3):79–133, 1994.

5. S. M. Hedetniemi, T. Hedetniemi, and A. L. Liestman. A survey of gossiping and broadcasting in communication networks. *Networks*, 18:319–349, 1988.

6. S. L. Johnsson and C.-T. Ho. Optimum broadcasting and personalized communication in hypercubes. *IEEE Transactions on Computers*, 38(9):1249–1268, 1989.

7. N. T. Karonis, B. R. Toonen, and I. T. Foster. MPICH-G2: A grid-enabled implementation of the message passing interface. *Journal of Parallel and Distributed Computing*, 63(5):551–563, 2003.

8. T. Kielmann, R. F. H. Hofman, H. E. Bal, A. Plaat, and R. A. F. Bhoedjang. MagPIe: MPI's collective communication operations for clustered wide area systems. In *Symposium on Principles and Practice of Parallel Programming (PPoPP'99)*, volume 34 of *ACM Sigplan Notices*, pages 131–140, 1999.

9. D. W. Krumme, G. Cybenko, and K. N. Venkataraman. Gossiping in minimal time. *SIAM Journal on Computing*, 21(1):111–139, 1992.

10. H. Ritzdorf and J. L. Träff. Collective operations in NEC's high-performance MPI libraries. In *International Parallel and Distributed Processing Symposium (IPDPS 2006)*, page 100, 2006.

11. M. Snir, S. Otto, S. Huss-Lederman, D. Walker, and J. Dongarra. *MPI – The Complete Reference*, volume 1, The MPI Core. MIT Press, second edition, 1998.

12. R. Thakur, W. D. Gropp, and R. Rabenseifner. Improving the performance of collective operations in MPICH. *International Journal on High Performance Computing Applications*, 19:49–66, 2004.

Efficient Shared Memory and RDMA Based Design for MPI_Allgather over InfiniBand*

Amith R. Mamidala, Abhinav Vishnu, and Dhabaleswar K. Panda

Department of Computer Science and Engineering
The Ohio State University
{mamidala, vishnu, panda}@cse.ohio-state.edu

Abstract. MPI_Allgather is an important collective operation which is used in applications such as matrix multiplication and in basic linear algebra operations. With the next generation systems going multi-core, the clusters deployed would enable a high process count per node. The traditional implementations of Allgather use two separate channels, namely network channel for communication across the nodes and shared memory channel for intra-node communication. An important drawback of this approach is the lack of sharing of communication buffers across these channels. This results in extra copying of data within a node yielding sub-optimal performance. This is true especially for a collective involving large number of processes with a high process density per node. In the approach proposed in the paper, we propose a solution which eliminates the extra copy costs by sharing the communication buffers for both intra and inter node communication. Further, we optimize the performance by allowing overlap of network operations with intra-node shared memory copies. On a 32, 2-way node cluster, we observe an improvement upto a factor of two for MPI_Allgather compared to the original implementation. Also, we observe overlap benefits upto 43% for 32x2 process configuration.

Keywords: MPI, MPI_Allgather, RDMA, Shared Memory.

1   Introduction

Clusters of commodity PCs are being increasingly deployed for high-end computing owing to their high performance-to-price ratios. Infact, many top 500 supercomputers are large scale clusters. These high-end systems are typically equipped with more than one processor per node such as a 2-way/4-way/8-way SMP or NUMA architecture. Also, the next generation systems feature multi-core support enabling more processes to run per processor. Already systems with dual-core and quad-core support have entered the high performance computing

* This research is supported in part by Department of Energy's Grant #DE-FC02-01ER25506; National Science Foundation's grants #CCR-0204429, #CCR-0311542 and #CNS-0403342; grants from Intel and Mellanox; and equipment donations from Intel, Mellanox, AMD, Apple, Advanced Clustering and Sun Microsystems.

B. Mohr et al. (Eds.): PVM/MPI 2006, LNCS 4192, pp. 66–75, 2006.

arena. This is expected to increase in future with even higher multi-cores being inducted to build ultra-scale clusters.

Message Passing Interface (MPI) [9] has evolved as the de-facto programming model for writing parallel applications. MPI provides many point-to-point and collective primitives which can be leveraged by these applications. Many parallel applications [7] employ these collective operations. MPI_Allgather is one such important operation which is used in applications involving matrix multiplication, solving differential equations and in basic linear algebra operations. Thus, optimizing the performance of this operation on the emerging next generation cluster architecture presents an important problem.

Recently, InfiniBand has emerged as one of the leaders in the high performance networking domain [4]. It provides RDMA which enables a process to directly write data to a remote process's address space. We have shown the benefits of using this feature for various collective operations such as MPI_Barrier, MPI_Allgather, MPI_AlltoAll [5] [8] [11] [10]. But, these approaches are optimal for one process running per node. With the next-generation systems going multi-core, it is essential to choose the fastest communication methods offered by the underlying system and network interconnect for efficient collective operations. In the existing approaches, collective communication is performed by utilizing two different channels, shared memory channel for intra-node communication and network channel for communication across the nodes. The major drawback of this approach is that as these two channels do not share the communication buffers, multiple copies are involved in the whole operation. This significantly degrades the performance on large clusters especially with multiple processes running on a single node. Also, since these channel do not have common buffers, overlapping communication across shared memory and network is difficult to accomplish.

In this paper, we propose a combined shared memory and RDMA based design which overcomes the problem outlined above. The copy costs are eliminated in our design by allowing the data buffers to be shared for both communication within the node and across the nodes. Our design extends the traditional recursive doubling algorithm for Allgather to accommodate more processes per node. Also, since the communication buffers in our design are shared, there is a benefit of overlapping of intra- and inter-node communication. We have implemented our designs and integrated them into MVAPICH [6] which is a popular MPI implementation for InfiniBand used by more than 365 organizations worldwide. We have evaluated our designs on two different cluster configurations. For a 32x2 configuration, our design improves the latency of the MPI_Allgather by a factor of two. Further, we observe that the overlapping network and shared memory communication improves the performance upto 43% in the latency of MPI_Allgather.

The rest of the paper is organized in the following way. In Section 2, we provide the background of our work. In Section 3, we explain the motivation for our scheme. In Section 4, we discuss detailed design issues. We evaluate our designs in Section 5 and talk about the related work in Section 6. Conclusions and Future work are presented in Section 7.

2   Background

2.1   Recursive Doubling Algorithm for Allgather

In this algorithm, a pair of processes exchange data for every step. The total number of steps in the algorithm is of the order of $log(p)$ where p is the number of processes in the operation. Also, the data involved for each step doubles as the operation progresses, hence the name recursive doubling. The total communication time of this algorithm is:

$$T_{rd} = t_s * log(p) + (p - 1) * m * t_w \qquad (1)$$

Where,
t_s = Message transmission startup time, t_w = Time to transfer one byte, m = Message size in bytes and p = Number of processes. MPICH [3] [12] uses Recursive doubling algorithm for power of two and up to medium size messages. For non-power of two processes, Bruck's Algorithm [2] is used for small messages. In this paper, we consider only the power of two case and hence we focus on the recursive doubling technique.

2.2   InfiniBand Overview

The InfiniBand Architecture [6] defines a switched network fabric for interconnecting processing and I/O nodes. In an InfiniBand network, hosts are connected to the fabric by Host Channel Adapters (HCAs). InfiniBand utilities and features are exposed to applications running on these hosts through a Verbs layer. InfiniBand Architecture supports both channel semantics and memory semantics. In channel semantics, send/receive operations are used for communication. In memory semantics, InfiniBand provides Remote Direct Memory Access (RDMA) operations, including RDMA Write and RDMA Read. RDMA operations are one-sided and do not incur software overhead at the remote side. Regardless of channel or memory semantics, InfiniBand requires that all communication buffers to be registered. This buffer registration is done in two stages. In the first stage, the buffer pages are pinned in memory (i.e. marked unswappable). In the second stage, the HCA memory access tables are updated with the physical addresses of the pages of the communication buffer.

2.3   Point-to-Point MPI Operations in MVAPICH

The two main protocols used for MPI point-to-point primitives are the eager and rendezvous protocols. In the eager protocol, the message is copied into communication buffers at the sender and destination process before it is copied into the user buffer. These copies are not present if rendezvous protocol is used. However, in this case an extra handshake is required to exchange user buffer information for zero-copy of the message. In this paper we deal with small to medium messages which are sent using the eager protocol and thus copy operation is involved at both the sender and the receiver. For intra-node communication, a separate shared memory channel is used for communication. In MVAPICH, the shared memory channel involves each MPI process on a local node attaching itself to a shared memory region at the initialization phase. This shared memory region can then be used amongst the local processes to exchange messages and other control information. Each pair of the local processes has its own send and receive queues. Small and medium messages are sent eagerly, where as a packetization approach is used for large messages.

3 Motivation

The traditional implementation of MPI_Allgather for multi-way SMP-based clusters uses MPI point-to-point operations. Depending on the pair of processes communicating, these operations use either the network channel or the shared memory channel for communication. Consider a scenario where eight processes are involved in Allgather with two processes per node as shown in Figure 1. The total number of steps involved in the operation is 3 which is $log(8)$.

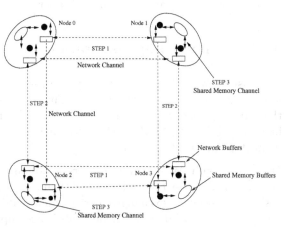

Fig. 1. Separate communication buffers

Depending on which pair is communicating at a step, the communication either proceeds over one of the two channels. Also, these two channels are designed separately and consequently do not allow sharing of buffers across the channels.

In the example considered, the first two steps involve the inter-node communication over the network channel. The third step involves the shared memory channel. Please note that we have taken this sequence of operations to illustrate the main idea. The network and shared memory operations can be scheduled in a different order depending on how the processes are launched on these nodes. As seen from the figure, separate sets of pinned buffers are associated with each channel for transmitting the data. As a result, though all the data for a given step has arrived at a node from other processes, it cannot be copied to every process local to the node. This is important in Allgather which involves an All-to-All broadcast of data. The reason why the data cannot be copied is because the network buffers are exclusive to a network channel and only the process communicating via this channel can access these buffers. Hence, a separate shared memory channel is needed resulting in extra copying of the data. Also the total amount of data exchanged increases linearly with the total number of processes participating in the operation. Thus, on a large cluster with more than one process per node, the copy costs play a dominant role degrading the performance of the collective operation. Another aspect to be taken into consideration is that since the buffers are not shared across the channels, overlapping shared memory and network communication becomes difficult to do. This further degrades the performance of this all-to-all operation.

This leads us to the following two questions:

1) *What mechanisms are needed to optimize MPI_Allgather for the emerging multi-core/multi-way InfiniBand Clusters?*

2) *How can we schedule the operations so as to easily allow overlap of network and shared memory operations?*

We address these questions in this paper.

4   Design and Implementation

The basic idea used in our approach is to use a common memory segment both for intra and network communication. This memory segment is shared across all the processes local to the node. Further, this segment is pinned so that it can be accessed directly by the NIC for the network operation. We now outline the main steps involved in our approach.

Our Approach: We extend the recursive doubling algorithm discussed earlier to be performed across the nodes rather than across the processes. In this fashion, a single message is exchanged per a pair of nodes irrespective of how many processes are scheduled on a node. This is accomplished by making all the local processes write their data into the shared memory segment in the initial step. This is the step 0 as shown in the Figure 2. Once all the processes have written the data into this buffer, the data exchange starts over the network. In the first step, node pairs 0, 1 and 2, 3 exchange the data. Note that the data exchanged in this step is one fourth the size of the total data. After this step, the second step as shown in the Figure 3 begins. The size of the data exchanged in this step is doubled as seen from the figure. The pairs which are involved in this exchange are now 0, 2 and 1, 3. Once this step is completed, each node has the data from all the processes. In the final step, which is the step four, the data is copied out of the shared memory segment.

As can be seen from the above example, in our approach the data is exchanged across the nodes in a recursive pair-wise fashion with a single data transfer operation between each pair of nodes. The number of steps would be equal to $log(n)$ where n is the number of nodes involved in the operation. In the example considered, the number of steps is $log(4)$ which is two. Note that by providing a common set of buffers for both network and intra-node data transfers, we eliminate the extra copying that would otherwise occur.

Overlap benefits: The main benefits of having a shared buffer is the potential of overlap between the network operations and the memory copy operations. By referring to the same Figures 2 and 3, it can be observed that the data arrived in step 1 of the operation can be copied to the processes' buffers concurrently with network operation in step 2. Thus, we need not wait till all the network operations are completed before the data is copied out of the shared memory segment. For a large scale cluster, this benefit is significant as both the size of the data involved is large and also there are more steps involved in the algorithm.

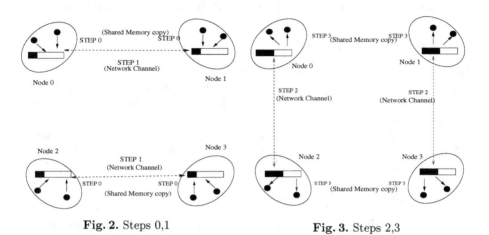

Fig. 2. Steps 0,1   **Fig. 3.** Steps 2,3

Implementation Details: The initial implementation step in our approach is creating a shared memory segment per node. This is done by making all the processes local to a node do a mmap of a shared file. After this step, this shared segment is pinned so that data can be accessed directly by the NIC for the network operation. In our design, the shared buffer is pinned by all the processes. This enables all the processes to issue network operations from this memory segment. RDMA is used for network data transfers as it is proven to be an efficient method for inter-node communication. In our implementation, we let one given process issue the network operations from a node. This can be easily accomplished as the processes have local ranks ranging from 0 to p-1 where p is the total number of processes per node. We choose the process with local rank 0 to issue network operations. Note that the addresses of this memory segment are exchanged before the Allgather is initiated. The data notification is done by doing a RDMA write of a one byte flag. These flags are also shared within a node and thus all the processes local to the node can poll for data arrival. This is useful for achieving overlap between network and shared memory copy operations. For synchronizing between the processes within a node another separate set of flags are used.

5   Performance Evaluation

In this section we compare the performance of the new scheme proposed in the paper with the already existing approach. The comparison is made by measuring the *Allgather latency* for the two schemes across different message sizes and for two different cluster configurations. The test was conducted for 1000 iterations for each message size. The abbreviations used for the comparison are as follows:

– new: The new shared-memory and RDMA based solution proposed in the paper.
– original: The original algorithm using MPI point-to-point operations

5.1 Experimental Testbed

We have carried tests on two different clusters:

1) Cluster A: Each node in this testbed has dual Opteron 2.4 GHz processors, 1024 KB L2 cache. They are equipped with MT25204 InfiniBand HCAs with PCI-Express interfaces.

2) Cluster B: Each node in this cluster is a Xeon 2.66 GHz processor with 512 KB L2 cache. Each node is connected with MT23108 InfiniBand HCA with PCI-X interface.

5.2 Latency of MPI_Allgather

As the results indicate our approach outperforms the original approach for the different cluster configurations considered. For Cluster A we observe benefits upto a factor of 1.47 and 1.39 for 32 and 64 processes as indicated by Figures 4 and 5 respectively. On cluster B, we observe an improvement by a factor of 1.97 and 1.82 for the considered configurations, 16x2, 32x2. These are shown in Figures 6 and 7 respectively.

We have also measured the impact of overlap of network operations and shared memory communication on these clusters. The non-overlap approach is implemented by making the processes copy the data from the shared buffers at the end after the network operations are completed. But, for the overlap case the processes copy the data as soon as it arrives and concurrently issue network operations. This is the approach taken in this paper. With the shared buffer RDMA design proposed the overlap improves the performance of the collective upto 30% for Cluster A and 43% for Cluster B as shown in the Figures 8 and 9.

6 Related Work

Utilizing shared memory for implementing collective communication has been a well studied problem in the past. In [13], the authors proposed to use remote memory operations across the cluster and shared memory within the cluster to develop efficient collective operations. They apply their solutions to Reduce, Bcast and Allreduce operations on IBM SP systems. In our approach we consider a different collective Allgather which has different communication pattern and present the results on commodity clusters. In [1], the authors implement collective operations over Sun systems. In [14], the authors improve the performance of send and recv operations over shared memory and also apply the techniques for group data movement. We have also designed and implemented

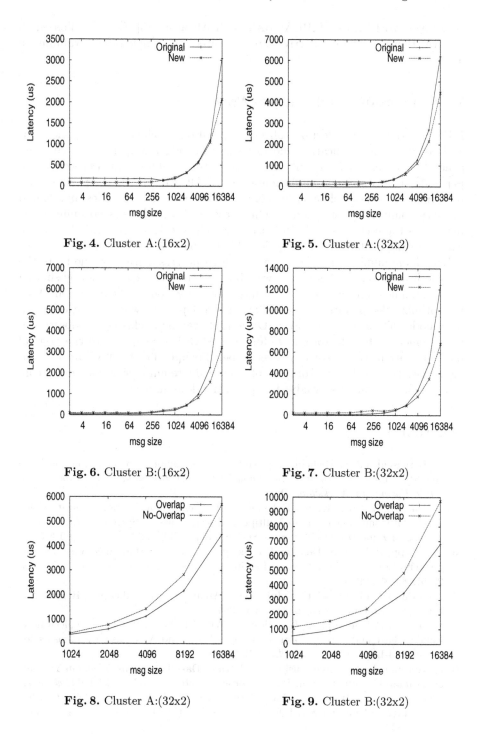

Fig. 4. Cluster A:(16x2)

Fig. 5. Cluster A:(32x2)

Fig. 6. Cluster B:(16x2)

Fig. 7. Cluster B:(32x2)

Fig. 8. Cluster A:(32x2)

Fig. 9. Cluster B:(32x2)

collectives, MPI_Barrier, MPI_AlltoAll, MPI_Allgather, [5] [8] [11] [10]based on RDMA. However, these collectives are optimized for a single process running on a node.

7   Conclusions and Future Work

MPI_Allgather is an important collective operation which is used in applications such as matrix multiplication and in basic linear algebra operations. The next generation systems feature multi-core architecture enabling a high process count per node. The traditional implementations of Allgather use two separate channels, namely network channel for communication across the nodes and shared memory channel for intra-node communication. Since there is no buffer sharing across these channels, the performance achieved is sub-optimal due to the extra copying of data within a node. This is true especially for a collective involving large number of processes with a high process density per node. In the approach proposed in this paper, we eliminate the extra copy costs by sharing the communication buffers for both intra and inter node communication. Also, we optimize the performance by allowing overlap of network operations with intra-node shared memory copies. On a 32, 2-way node cluster, we observe an improvement upto a factor of two for MPI_Allgather compared to the original implementation. We also observe overlap benefits upto 43% for 32x2 process configuration. For our future work, we plan to evaluate our design with multi-core enabled clusters and also study the application-level impact.

References

1. M Bernaschi and G Richelli. Mpi collective communication operations on large shared memory systems. In *Parallel and Distributed Processing, 2001. Proceedings. Ninth Euromicro Workshop*, 2001.
2. J. Bruck, C.-T. Ho, S. Kipnis, E. Upfal, and D. Weathersby. Efficient Algorithms for All-to-All Communications in Multiport Message-Passing Systems. *IEEE Transactions in Parallel and Distributed Systems*, 8(11):1143–1156, November 1997.
3. W. Gropp, E. Lusk, N. Doss, and A. Skjellum. A High-Performance, Portable Implementation of the MPI Message Passing Interface Standard. *Parallel Computing*, 22(6):789–828, 1996.
4. InfiniBand Trade Association. InfiniBand Architecture Specification, Release 1.1. http://www.infinibandta.org, October 2004.
5. Sushmitha P. Kini, Jiuxing Liu, Jiesheng Wu, Pete Wyckoff, and Dhabaleswar K. Panda. Fast and Scalable Barrier using RDMA and Multicast Mechanisms for InfiniBand-Based Clusters. In *EuroPVM/MPI*, Oct. 2003.
6. Jiuxing Liu, Jiesheng Wu, Sushmitha P. Kinis, Darius Buntinas, Weikuan Yu, Balasubraman Chandrasekaran, Ranjit Noronha, Pete Wyckoff, and Dhabaleswar K. Panda. MPI over InfiniBand: Early Experiences. Technical Report, OSU-CISRC-10/02-TR25, Computer and Information Science, the Ohio State University, January 2003.
7. NASA. NAS Parallel Benchmarks. http://www.nas.nasa.gov/Software/NPB/.

8. Amith R.Mamidala, Jiuxing Liu, and Dhabaleswar K. panda. Efficient Barrier and Allreduce InfiniBand Clusters using Hardware Multicast and Adaptive Algorithms . In *Proceedings of Cluster Computing*, 2004.

9. Marc Snir, Steve Otto, Steve Huss-Lederman, David Walker, and Jack Dongarra. *MPI–The Complete Reference. Volume 1 - The MPI-1 Core, 2nd edition.* The MIT Press, 1998.

10. S. Sur, U.K.R. Bondhugula, A.R. Mamidala, H.-W. Jin, and D. K. Panda. High performance RDMA based All-to-All Broadcast for InfiniBand Clusters. In *(HiPC)*, 2005.

11. S. Sur, H.-W. Jin, and D. K. Panda. Efficient and Scalable All-to-All Exchange for InfiniBand-based Clusters. In *International Conference on Parallel Processing (ICPP)*, 2004.

12. R. Thakur, R. Rabenseifner, and W. Gropp. Optimization of Collective communication operations in MPICH. *Int'l Journal of High Performance Computing Applications*, 19(1):49–66, Spring 2005.

13. V Tipparaju, J Nieplocha, and D K Panda. Fast collective operations using shared and remote memory access protocols on clusters. In *International Parallel and Distributed Processing Symposium, 2003*, 2003.

14. Meng-Shiou Wu, R A Kendall, and K Wright. Optimizing collective communications on smp clusters. In *ICPP 2005*, 2005.

High Performance RDMA Protocols in HPC

Tim S. Woodall[1], Galen M. Shipman[1], George Bosilca[2],
Richard L. Graham[1], and Arthur B. Maccabe[3]

Los Alamos National Laboratory, Advanced Computing Laboratory
{twoodall, gshipman, rlgraham}@lanl.gov
University of Tennessee, Dept. of Computer Science
bosilca@cs.utk.edu
University of New Mexico, Dept. of Computer Science
maccabe@cs.unm.edu

Abstract. Modern network communication libraries that leverage Remote Directory Memory Access (RDMA) and OS bypass protocols, such as Infiniband [2] and Myrinet [10] can offer significant performance advantages over conventional send/receive protocols. However, this performance often comes with hidden per buffer setup costs [4]. This paper describes a unique long-message MPI [9] library 'pipeline' protocol that addresses these constraints while avoiding some of the pitfalls of existing techniques. By using portable send/receive semantics to hide the cost of initializing the pipeline algorithm, and then effectively overlapping the cost of memory registration with RDMA operations, this protocol provides very good performance for any large-memory usage pattern. This approach avoids the use of non-portable memory hooks or keeping registered memory from being returned to the OS. Through this approach, bandwidth may be increased up to 67% when memory buffers are not effectively reused while providing superior performance in the effective bandwidth benchmark. Several user level protocols are explored using Open MPI's PML (Point to point messaging layer) and compared/contrasted to this 'pipeline' protocol.

1 Introduction

RDMA capable interconnects are widely used in high performance computing (HPC) systems. While these interconnects provide for high bandwidth and low latency messaging, they also pose unique challenges to HPC software designers. The Message Passing Interface (MPI) standard [9] [7], one of the most widely used HPC messaging paradigms, generally abstracts these issues from the parallel programmer. However, implementations of this standard and other messaging middleware must address these challenges to achieve balanced performance across a wide variety of application communication patterns.

One such challenge is the requirement by most RDMA interconnects that memory be explicitly registered with the interconnect and pinned by the OS. Memory registration is often an expensive operation requiring a trap to the OS and an additional linear cost that is a function of the number of pages in the

B. Mohr et al. (Eds.): PVM/MPI 2006, LNCS 4192, pp. 76–85, 2006.

memory region. Various techniques exist to minimize the impact of memory registration but each make specific assumptions about application behavior or system usage. Typical approaches involve caching registrations for later reuse, but many applications do not effectively reuse communication buffers. Additionally, the application may invalidate the cached buffer, which places constraints on the return of cached pages to the OS.

In this paper we describe a high performance user level RDMA protocol which minimizes the impact of memory registration while avoiding the pitfalls of other techniques. This protocol provides good performance while minimizing resource usage. In addition, this protocol makes no assumption about application behavior thereby providing improved performance for certain applications. This protocol and others were implemented and evaluated in the context of Open MPI [6].

The remainder of this paper is organized as follows. Section 2 provides an overview of Infiniband, the RDMA interconnect used in our research. Next, section 3 discusses different approaches to minimize the impact of memory registration, while section 4 discusses our user-level pipeline protocol. Open MPI's support for multiple techniques is discussed in Section 5. Results are discussed in Section 6, followed by related work in Section 7. Conclusions and future work are discussed in Section 8.

2   Infiniband

Infiniband, similar to Myrinet GM and iWARP [5], provides both RDMA and OS bypass facilities. RDMA enables data transfer from the address space of an application process to a peer process across the network fabric without requiring involvement of the host CPU. Infiniband RDMA operations support both two-sided send/receive and one-sided put/get semantics. Each of these operations may be queued from the user level directly to the host channel adapter (HCA) for execution, bypassing the OS to minimize latency and processing requirements on the host CPU.

Infiniband does place some constraints on these operations. As data is moved directly between the host channel adapter (HCA) and user level source/destination buffers, these buffers must be registered with the HCA in advance of their use. Registration is a relatively expensive operation which locks the memory pages associated with the request, thereby preserving the virtual to physical mappings. Additionally, when supporting send/receive semantics, pre-posted receive buffers are consumed in order as data arrives on the host channel adapter (HCA). Since no attempt is made to match available buffers to the incoming message size, the maximum size of a message is constrained to the minimum size of the posted receive buffers.

Infiniband shares many characteristics with other common RDMA interconnects including Myrinet GM and emerging standards such as iWARP. The common requirement for explicit memory registration, local knowledge of peer registrations prior to initiating an RDMA operation, and the associated issues with effectively managing these registrations motivated the work described in the following sections.

3   RDMA

To overcome the expense of registering memory with the interconnect, application and library developers have used several techniques. A simple solution is to restrict all RDMA operations to a static memory region. This allows the application to register the memory region once and amortize this cost over a potentially large number of RDMA operations. While this does help in hiding the costs of the memory registration, it restricts the application to a static memory region. For many applications this usage model is inappropriate and results in copy in/out of the registered memory. For larger messages copy costs quickly become a bottleneck.

Another approach is to register memory on demand. The target and source buffers are registered prior to the RDMA operation and then deregistered upon completion of the operation. This approach allows the MPI library to RDMA from any memory region providing a true zero copy transfer but at a high memory registering cost prior to each RDMA operation.

A third approach, first explored in MPICH-GM [1] and later in MVAPICH [8], avoids the high cost of copying in/out of a static memory region and in some use cases allows the cost of registering the memory region to be amortized over multiple RDMA operations. Prior to the RDMA operation the memory region is registered and the registration is then cached locally. Subsequent RDMA operations first query the cache for a matching registration and if found uses this registration to immediately initiate the RDMA operation. For applications which regularly reuse source and target buffers for RDMA operations the cost of the initial registration is effectively amortized over these subsequent RDMA operations. A drawback to this approach is that some applications may not effectively reuse source and target buffers incurring a high cost for each RDMA operation. Additionally, cached buffers may be frequently invalidated by the application. Message buffers allocated by the application, registered and cached by the MPI layer, and later returned to the OS by the application must be removed from the cache at the MPI layer. Reuse of these cached registrations would result in silent corrupted data transfers due to changes in the page table mappings on subsequent allocations. Current approaches to addressing this issue involve either non-portable memory hooks and/or linker tricks to intercept sbrk, munmap, and/or free to insure that returned pages are removed from the cache, or simply disabling the return of pages to the OS by the allocator.

Hardware based approaches as found in Quadrics [3] eliminate the need to register memory entirely. While avoiding the high cost of registration this approach increases the complexity of the interconnect and therefore the cost, and also involves kernel modifications to support the ghost page tables resident on the NIC.

This paper describes an alternate approach which allows RDMA operations from arbitrary memory regions while maintaining high performance, and makes no assumption about the applications reuse of source and target buffers. Send/Recv operations are employed to cover the cost of initializing the RDMA pipleline, and memory is dynamically registered in smaller pieces and RDMA operations are overlapped with memory registration/deregistration. This pipeline protocol is described in further detail in the following section.

4   RDMA Pipeline Protocol

The pipeline protocol begins by eagerly sending the first part of the message data, up to a configurable eager limit, along with a MATCH header using send/receive semantics, as illustrated in Figure 1,

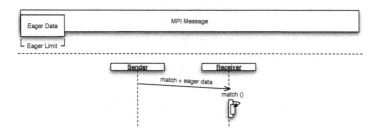

Fig. 1. Sending Eager Data with Match Header

Upon receiving and matching the header to a posted receive (figure 2, the receiver responds with an ACK to the source and begins registering, up to a configurable pipeline depth, blocks (RDMA fragments) of the target buffer across the available RDMA capable HCAs. The size of each block is constrained by the maximum configured RDMA size for each interconnect. As each registration in the pipeline completes, an RDMA target fragment READY control message is sent to the source to initiate a registration of the source RDMA fragment followed by an RDMA write on the block.

Next, send/receive semantics are used to send data from the eager limit up to the initial RDMA write offset, returned by the peer with the rendezvous ACK (figure 3), striping this data across all available interconnects. This unique pipeline feature hides the cost of initializing the pipeline.

From Figure 4 we see that as RDMA READY control messages are received at the source, the corresponding block of the source buffer is registered and an RDMA write operation is initiated on the current block. On local completion at the source, an RDMA FIN message is sent to the peer. Registered blocks are deregistered (released) upon local completion or receipt of the RDMA FIN message. If required, the receipt of an RDMA FIN messages may also further advance the RDMA pipeline.

This protocol effectively overlaps the cost of registration/deregistration with RDMA writes. Resources are released immediately and the high overhead of a single large memory registration is avoided. Additionally, this protocol results in improved performance for applications which may not reuse buffers for MPI operations effectively.

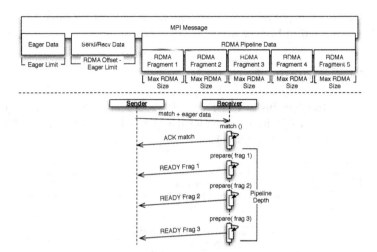

Fig. 2. Receiver Registers RDMA Target Fragments

5   Other RDMA Protocols

In addition to the RDMA pipeline protocol, which is enabled by default, Open MPI supports additional techniques for managing RDMA registrations. These approaches have been developed to contrast the relative merits of each, and allow the behavior to be tuned at run-time to best match the application characteristics.

5.1   RDMA Cache

Open MPI provides the capability to optionally register memory on first use and cache these registrations for later re-use. When this approach is used, the entire source/destination buffer is registered and a single RDMA operation is initiated on receipt of an ack from the peer. If multiple network interfaces are available, the message is divided across the available interfaces, and a single operation is initiated on each interface.

While other MPI implementations prevent physical pages from being released to the OS, Open MPI provides the capability to use memory hooks to intercept the deallocation of memory and it's return to the OS. When pages are returned to the OS via sbrk/munmap, the pages are checked and any matching entries de-registered. This prevents future use of an invalid memory registration while allowing memory to be returned to the host operating system. Intercepting memory deallocation introduces additional overhead and additional research into reducing this overhead is ongoing.

5.2   RDMA Caching Pipeline

A hybrid approach was developed to explore the benefit of caching individual registrations within the RDMA pipeline protocol. In this approach, the pipeline

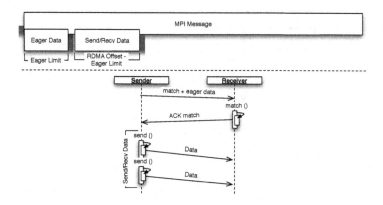

Fig. 3. Sender Sends Data up to the RDMA Offset to Cover Pipeline Initialization Costs

protocol described in section 4 is modified to cache each registration as it occurs in the pipeline. Subsequent pipeline RDMA operations from the same or overlapping buffer space then re-use the cached registrations. Registrations are aligned to the segment size to further promote reuse.

This hybrid approach leverages the pipeline protocol for good performance in the case of low buffer reuse, and achieves performance closer to the RDMA cache when existing registrations can be re-used. The drawback to the introduction of the cache is the added requirement for memory hooks to intercept the deallocation of memory as described above.

6   Results

This section presents a comparison of the different protocols in Open MPI. We first demonstrate the performance of the pipeline protocol as a function of buffer reuse in terms of bandwidth. Next, we examine effective bandwidth among multiple peers using different communication patters via the Effective Bandwidth Benchmark [11]. In general, our results provide comparisons of the pipeline protocol to:

copy in/out - Standard send/recv semantics with copy in/out of pre-registered buffers

leave pinned (memory hooks) - Registration cache described in section 5.1 with memory hooks to intercept memory deallocations.

leave pinned (disable sbrk) - Registration cache described in section 5.1 with return of pages to OS disabled.

pipeline - Pipeline protocol described in section 4

pipeline leave pinned (memory hooks) - Caching pipeline described in section 5.2 with memory hooks to intercept memory deallocations

pipeline leave pinned (disable sbrk) - Caching pipeline described in section 5.2 with return of pages to OS disabled.

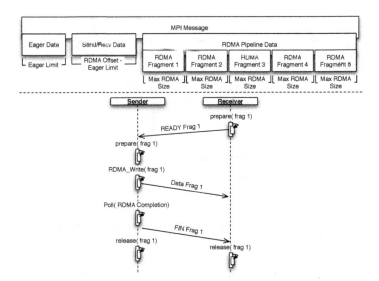

Fig. 4. Sender Prepares and RDMA Writes a Fragment

6.1   Experimental Setup

Our experiments were performed on a 256 node cluster consisting of dual In-
tel Xeon X86-64 3.4 GHz processors with a minimum 6GB of RAM, Mellanox
PCI-Express Lion Cub adapters connected via a Voltair switch. The Operating
System is Linux 2.6.9-11 and the MPI library is Open MPI 1.1 stable.

6.2   Bandwidth

The following graphs illustrate the performance of the pipeline protocol as a
function of buffer reuse. As Figure-5(a)[1] illustrates, with no buffer reuse, the
standard pipeline protocol achieves a speedup of up to 67 percent over the registration cache. This can be attributed to the pipeline protocol effectively overlapping the cost of registration with RDMA. In contrast, with no buffer reuse,
the caching protocol is limited by the high cost of registration. The copy in/out
protocol provides performance almost equal to that of the pipeline protocol but
it is important to note that this protocol limits the application ability to overlap
communication with computation by intensively using the CPU and memory
bus.

An interesting metric is the amount of reuse required for the caching protocol
to achieve performance comparable to the pipeline. Figure-5(b) illustrates the
bandwidth achieved for each protocol as a function of the number of times the
buffer is reused, for an arbitrary fixed message size (8 Mbytes). As the graph
illustrates, the buffer must be reused on the order of 40-50 times before the

---

[1] Lion Cub SDR HCA's in 4X PCI-Express slot.

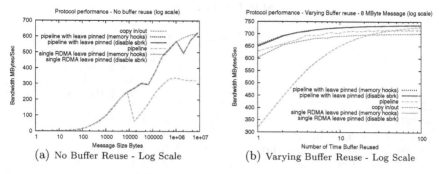

Fig. 5. Ping Pong Bandwidth

caching protocol achieves performance equivalent to the standard pipeline. Note that the hybrid caching pipeline ramps up much earlier, as the registrations are overlapped with RDMA and re-used on subsequent invocations of the pipeline. This approach improves bandwidth over the standard pipeline as we defer the cost of deregistration until memory is released by the application.

6.3   Effective Bandwidth Benchmark b_{eff}

The Effective Bandwidth Benchmark (b_{eff}) was used to examine protocol effect on bandwidth in more complex communication patterns. In this benchmark 8 nodes were used to communicate message sizes up to 8 MBytes in size (L_{max}) using different communication patterns. In this benchmark the benefits of the pipeline protocol are apparent. When the memory cache is used (leave pinned) in conjunction with the pipeline protocol the total bandwidth achieved outperforms all the other protocols by a significant margin, this is even the case for a single send/recv operation as the pipeline leave pinned protocol need not deregister any memory where the standard pipeline protocol does. The memory cache alone (using a single RDMA operation) does not perform as well because b_{eff} only reuses buffers 3 times by default. The high up front cost of registering the entire buffer is not effectively amortized over so few number of reuse. The pipeline protocol with memory cache is able to provide superior performance as buffer reuse is not required to achieve high bandwidth on the first operation, subsequent operations which reuse these buffers avoids memory registration thereby further enhancing performance.

7   Related Work

The pipeline protocol discussed in this paper shares some characteristics of the protocol developed in MPICH-GM although this did not influence the authors work and differs in several key ways. The PML protocol does not eagerly register memory upon transmission of the initial rendezvous as in MPICH-GM. Instead the receiver initializes the RDMA pipeline and the sender covers this cost via

Table 1. b_{eff} results

	b_{eff}	Lmax	b_{eff} at Lmax rings& random	b_{eff} at Lmax rings only	ping-pong bandwidth
	MByte/s		MByte/s	MByte/s	MByte/s
accumulated					
- pipeline	1536	8 MB	4794	4812	710
- pipeline leave pinned (memory hooks)	1815	8 MB	5299	5324	735
- pipeline leave pinned (disable sbrk)	1640	8 MB	5209	5216	714
- leave pinned (memory hooks)	1407	8 MB	4012	4029	599
- leave pinned (disable sbrk)	1418	8 MB	3992	3975	603
- copy in/out	1395	8 MB	3133	3130	683
per process					
- pipeline	192	599	601		
- pipeline leave pinned (memory hooks)	227	662	665		
- pipeline leave pinned (disable sbrk)	205	651	652		
- leave pinned (memory hooks)	176	502	504		
- leave pinned (disable sbrk)	177	499	497		
- copy in/out	174	392	391		

sending a portion of the buffer after receipt of the rendezvous acknowledgment. The PML protocol therefore conserves resources (registered memory) and delays the high cost of registration until the message is matched on the receiver. In addition the PML protocol will optionally cache memory registration which allows the MPI library to take advantage of buffer reuse while providing good performance in the absence of reuse.

ARMCI [12] also uses an RDMA pipeline although it differs significantly from the PML pipeline protocol. The ARMCI protocol overlaps RDMA operations with copy in/out of pre-registered buffers. This protocol does not provide the benefit of zero-copy and will result in a larger memory footprint due to the use of pre-registered buffers.

8   Conclusions and Future Work

RDMA capable interconnects pose unique challenges that require careful consideration to achieve balanced performance and scalability across a wide range of application communication patterns. The results of this work indicate the RDMA pipeline protocol effectively addresses these concerns, by hiding the cost of initializing the RDMA 'pipeline' protocol, and overlapping the dynamic registration and deregistration of memory buffers with data transfer. This approach avoids the issues associated with maintaining a registration cache, which requires non-portable memory hooks to either intercept deallocations, or disable the return of pages to the OS. Additionally, the pipeline protocol reduces the memory footprint and resource requirements of the application over the caching approach.

The hybrid pipeline protocol which cached registrations as they occurred in the pipeline provided promising results over the dynamic pipeline. However, the caching approach is still constrained by the above issues. Additional work to address these issues would involve effectively managing cache size, investigating the potential for efficient notification from the OS on changes to registered pages, and improving the performance of cache cleanup/deregistration.

This material is based upon work supported by Subcontract No. 12783-001-05 49 issued to Rice University from the Regents of the University of California (Los Alamos National Laboratory). Los Alamos National Laboratory is operated by the University of California for the National Nuclear Security Administration of the United States Department of Energy under contract W-7405-ENG-36.

Project support was provided through ASC/PSE and ASC/S&CS programs. LA-UR-06-1268.

References

[1] Performance of mpich-gm. http://www.myri.com/myrinet/performance/MPICH-GM/index.html.

[2] I. T. Association. Infiniband architecture specification vol 1. release 1.2. www.infinibandta.org, 2004.

[3] J. Beecroft, D. Addison, F. Petrini, and M. McLaren. QsNetII: An interconnect for supercomputing applications, 2003.

[4] R. Brightwell and A. Maccabe. Scalability limitations of VIA-based technologies in supporting MPI. In *Proceedings of the Fourth MPI Developer's and User's Conference*, March 2000.

[5] M. Chadalapaka, H. Shah, U. Elzur, P. Thaler, and M. Ko. A study of iscsi extensions for rdma (iser). In *NICELI '03: Proceedings of the ACM SIGCOMM workshop on Network-I/O convergence*, pages 209–219, New York, NY, USA, 2003. ACM Press.

[6] E. Garbriel, G. Fagg, G. Bosilica, T. Angskun, J. J. D. J. Squyres, V. Sahay, P. Kambadur, B. Barrett, A. Lumsdaine, R. Castain, D. Daniel, R. Graham, and T. Woodall. Open MPI: goals, concept, and design of a next generation MPI implementation. In *Proceedings, 11th European PVM/MPI Users' Group Meeting*, 2004.

[7] A. Geist, W. Gropp, S. Huss-Lederman, A. Lumsdaine, E. Lusk, W. Saphir, T. Skjellum, and M. Snir. MPI-2: Extending the Message-Passing Interface. In *Euro-Par '96 Parallel Processing*, pages 128–135. Springer Verlag, 1996.

[8] J. Liu, J. Wu, S. P. Kini, P. Wyckoff, and D. K. Panda. High performance RDMA-based MPI implementation over infiniband. In *ICS '03: Proceedings of the 17th annual international conference on Supercomputing*, pages 295–304, New York, NY, USA, 2003. ACM Press.

[9] Message Passing Interface Forum. MPI: A Message Passing Interface. In *Proc. of Supercomputing '93*, pages 878–883. IEEE Computer Society Press, November 1993.

[10] Myricom. Myrinet-on-VME protocol specification. http://www.myri.com/open-specs/.

[11] R. Rabenseifner and A. Koniges. The parallel communication and i/o bandwidth benchmarks: b_eff and b_eff_io. 2001.

[12] V. Tipparaju, G. Santhanaraman, J. Nieplocha, and D. K. Panda. Host-assisted zero-copy remote memory access communication on infiniband. *ipdps*, 01:31a, 2004.

Implementation and Shared-Memory Evaluation of MPICH2 over the Nemesis Communication Subsystem*

Darius Buntinas, Guillaume Mercier, and William Gropp

Mathematics and Computer Science Division
Argonne National Laboratory
{buntinas, mercierg, gropp}@mcs.anl.gov

Abstract. This paper presents the implementation of MPICH2 over the Nemesis communication subsystem and the evaluation of its shared-memory performance. We describe design issues as well as some of the optimization techniques we employed. We conducted a performance evaluation over shared memory using microbenchmarks as well as application benchmarks. The evaluation shows that MPICH2 Nemesis has very low communication overhead, making it suitable for smaller-grained applications.

1 Introduction

The Message Passing Interface (MPI) standard has been designed to enhance portability in parallel applications, as well as to bridge the gap between the performance offered by a parallel architecture and the actual performance delivered to the application. The level of achievable performance depends, however, on the implementation. Two critical areas determine the overall performance level of an MPI implementation. The first area is the low-level communication layer that the upper layers of an MPI implementation can use as foundations. The second area covers the communication progress and management. We designed an efficient communication subsystem, called Nemesis, that features very low overhead and is therefore suitable to serve as a basis for the MPICH2 software [1], an open source implementation of MPI.

The design and implementation of the Nemesis communication subsystem has been previously presented in [2]. In this paper, we describe how we ported MPICH2 over Nemesis and show the performance benefits of MPICH2 Nemesis. We also explain the improvements that have been made in the MPICH2 communication progress engine. The resulting MPICH2 software stack yields a very low latency and high bandwidth and compares favorably with competing software. The implementation also allows us to better assess both the overhead and the performance of MPI.

---

* This work was supported by the Mathematical, Information, and Computational Sciences Division subprogram of the Office of Advanced Scientific Computing Research, Office of Science, U.S. Department of Energy, under Contract W-31-109-ENG-38, and by a grant of computing time from the Ohio Supercomputer Center.

B. Mohr et al. (Eds.): PVM/MPI 2006, LNCS 4192, pp. 86–95, 2006.

Section 2 gives an overview of the Nemesis communication subsystem. Section 3 describes how this communication subsystem has been integrated in MPICH2 as a new CH3 channel. We detail how we implemented several important features of the MPI2 standard. The various optimizations that MPICH2 gained are also explained. Section 4 presents a performance evaluation using shared-memory communication; in particular, we compare our implementation with the MPICH2 shm channel and other MPI implementations. Section 5 concludes this paper and discusses future work.

2   Overview of the Nemesis Communication Subsystem

In this section, we briefly describe the Nemesis communication subsystem. See [2] for a complete description of the design and implementation.

The Nemesis communication subsystem was designed to be a scalable, high-performance, shared-memory, multinetwork communication subsystem for MPICH2. The goals for our design, in order of priority, were scalability, high-performance intranode communication, high-performance internode communication, and multinetwork internode communication. The implication of ranking the goals this way is that we strive to minimize the overhead for intranode communication, even if this comes at some penalty for internode communication.

To achieve the goals of high scalability and low intranode overhead, we designed Nemesis using lock-free queues in shared memory. Thus, each process needs only one receive queue, onto which other processes on the same node can enqueue messages without the overhead of acquiring a lock. Alternative designs would be to use a pair of receive queues per pair of processes or to use a single queue with a lock. On a large SMP, neither would be scalable, because of the $\mathcal{O}(N^2)$ number of queues needed or the contention on the lock, nor would they be efficient, because of the overhead of polling multiple queues or the overhead of acquiring and releasing a lock.

For internode communication, when a message is received from the network, a polling function for that *network module* enqueues the message onto the process's receive queue. A network module has a send queue onto which messages to be sent are enqueued. The send queue is analogous to a process's lock-free receive queue in that, when a process sends a message, it will enqueue the message onto the appropriate queue, whether it is a queue for another process on the same node, or a send queue for a network module. This simplifies the critical path when sending a message: No special action is taken when sending a message to a process on a remote node versus a process on the local node. Multiple networks can be supported by implementing additional network modules. Our current implementation supports internode communication over sockets and Myricom's GM message-passing system [3].

After analyzing our initial design, we applied several optimizations. To reduce latency, we optimized the placement of the receive queue *head* and *tail* pointers and added a *shadow head* pointer to reduce L2 cache misses. We also gathered variables that are often used together in the same cache line to reduce the number of L1 cache misses in the critical path. For small SMP nodes, we used a *fastbox* mechanism to bypass the queues. A pair of buffers is allocated between each pair of processes.

When sending a message, a process can bypass the queue by copying the message into the fastbox, if it is free, and setting a flag indicating a message is waiting. The receiving process then copies the message out of the fastbox and resets the flag. If the fastbox is full when a process is sending a message, the regular queue mechanism is used. This mechanism would not scale well for large SMPs and is used only for SMPs with a small number of processors. To improve bandwidth, we implemented architecture-specific memory copy functions. For ia32 and x86_64 architectures the memory copy function uses nontemporal store operations that bypass the cache. More details on these optimizations can be found in [2].

3   Integration into MPICH2

The communication portion of MPICH2 is implemented in several layers, as shown in Figure 1, and provides two ways to port MPICH2 to a communication subsystem. The ADI3 layer presents the MPI interface to the application layer above it, and the ADI3 interface to the device layer below it. MPICH2 can be ported to a new communication subsystem by implementing a device.

The figure shows the device for the Quadrics network. The figure also shows the CH3 device. The CH3 device presents the CH3 interface to the layer below it, and provides another way for MPICH2 to be ported to a new communication subsystem: by implementing a *channel*. This interface has fewer functions than the ADI3 interface, making it significantly simpler to implement. Because the interface is simpler, however, it may not be able to take advantage of certain features provided by the communication subsystem, such as RDMA or collective operations.

We chose to port MPICH2 over Nemesis by implementing a CH3 channel. While our intent is to eventually implement an ADI3 device for Nemesis, implementing a CH3 channel allowed us to rapidly create a prototype and evaluate the implementation of Nemesis. We did, however, modify the CH3 layer in order to allow certain optimizations of the Nemesis channel. In the rest of this section we describe the basic design of the Nemesis channel and key optimizations.

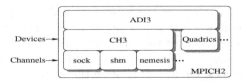

Fig. 1. Software layers of MPICH2

3.1   Basic Design of the Nemesis Channel

To send a message, the CH3 layer calls a send function implemented by the channel, passing in a pointer to the message header a description of the data to be sent and a pointer to an MPI *request* object. The description of the data consists of an array of pointers and lengths (i.e., an *IOV*) that can be used to describe noncontiguous data. The Nemesis channel copies the header and data into a Nemesis receive queue element, called a *cell*, and fills in a short Nemesis header, then enqueues it on the appropriate receive queue or fastbox, or sends it over the network to the appropriate remote node. If the CH3 message is larger than a cell, multiple cells can be used, since the cells are delivered in FIFO order.

If not enough free cells are available to send an entire message, the IOV describing the unsent data is saved in the request, which is then enqueued onto a pending-send queue. When free cells are available, the messages on the pending-send queue are sent out. When all the data described by the IOV has been sent, the channel makes an up-call to CH3 to see whether there is more data to be sent. If there is, the IOV is reloaded; otherwise the request is marked as complete.

To receive a message, the Nemesis channel polls the receive queue and fast-boxes. In order to reduce the overhead of unnecessarily polling too many fast-boxes, the Nemesis channel polls only *active* fastboxes, which are the fastboxes of processes for which this process has posted a receive. Because fastboxes intro-duce a second path for messages between two processes, sequence numbers are used to maintain the order of messages.

When a cell is found, either in the receive queue or the fastbox, and there are no pending receives for that source process, the channel makes an up-call to CH3 with a pointer to the message header. If there is data to receive, CH3 will return an IOV along with a pointer to a request. The channel then copies the data from the cell to the user buffer described by the IOV. If the IOV describes more data than is contained in the cell, the IOV for the unreceived data is saved in the request, and the request is saved as a pending-receive corresponding to the process that sent the message. When the next cell from that process is received, the channel gets the saved request, and the new data is copied from the cell to the user buffer described by the IOV in the request. When all of the data described by the IOV has been received, the channel makes an up-call to CH3 to either reload the IOV, if there is more data to receive, or to mark the request as complete.

Because cells are allocated in shared memory, they are a limited resource. Hence, it is important to process a cell and copy out its data as soon as possible, so that it can be freed. This means that an unexpected message should be copied out of its cells and into a temporary buffer, as opposed to leaving the data in the cells until the receive has been posted. Unexpected messages are handled by the CH3 layer in just this way. If an unexpected message is received, CH3 creates a new request and passes back an IOV pointing to a newly allocated temporary buffer. So, the channel takes the same action whether the received message is unexpected or not. The message is copied out of the temporary buffer into the user buffer once a receive matching the message has been posted.

3.2   Large Message Transfer Using Rendezvous

While the shared-memory queue is very efficient for transferring small- to medium-sized messages, transferring large amounts of data through the queue may not be the most efficient method. High-performance networks have RDMA capabilities where data can be transferred directly from the user's source buffer on one node to the user's destination buffer on another node, avoiding the data copies associated with using the queue. Some shared-memory machines, such as the SGI Altix, have similar mechanisms for processes on the same node. Even without special mechanisms, using a queue may not be the most efficient method of transferring large amounts of data between processes on the same node [4].

To support various mechanisms for transferring large messages, we defined the *Large Message Transfer* (LMT) interface and added it to CH3. Avoiding the queue can not only improve the bandwidth of the transfer but also reduce the impact on the application's data in the cache [4].

CH3 uses a rendezvous protocol when sending large messages, which ensures that a matching receive has been posted before the message data is sent. The rendezvous protocol is used primarily to avoid having to buffer the message if a matching receive has not been posted. We designed the LMT interface to be used together with the rendezvous protocol; the interface allows the channel to piggyback information on the CH3 rendezvous messages. The channel implements seven LMT functions, which are called by CH3.

For shared-memory communication, using the LMT interface, a shared-memory region is allocated and attached to by the sending and receiving processes. Then, using a double-buffering mechanism, the sending process copies the data into the shared-memory region while the receiving process copies it out. Because we used a memory copy function that uses nontemporal store operations, not only does this result in a high bandwidth transfer, but it has a very low impact on the application's data in the cache of the receiving process. The LMT optimization improves bandwidth for intranode communication by about 130 MiBps for large messages.

We have also used the LMT interface for the GM network module, which allows the use of RDMA operations. In the socket network module, we also used the LMT interface so that `read()` and `write()` operations can be issued to directly access the application's buffers, rather than copying the data through a cell.

3.3   Bypassing the Posted Receive Queue

We performed another optimization to improve the latency of small messages by bypassing the CH3 posted receive queue in certain cases. In the current implementation of CH3 when a receive is posted by the application, CH3 first searches the unexpected message queue to see whether it has already received a matching message. If a matching message is not found, the request is posted on the posted receive queue. CH3 then calls the progress engine to check for incoming messages. When a new message is received, CH3 looks for a matching receive request by searching the posted receive queue and enqueues the message in the unexpected queue if the message is not found.

Notice that if a receive is posted for which there is no matching message in the unexpected message queue, and the matching message is waiting to be received on the Nemesis receive queue or network, the receive request is queued on the posted receive queue, only to be matched and dequeued in the next step when the progress engine is called and the matching message received. Our optimization implements a new function to call the progress engine with a receive request. As messages are received from the Nemesis receive queue they are checked for a match with the receive request. Only when no matching messages are found on the receive queue, is the request enqueued onto the posted receive queue. Note that if there already is a request on the posted receive queue that can possibly match the same message as the new receive request, we cannot use the optimization and, instead, add the new request to the receive queue as in the original implementation. This optimization reduced latency by about 18%, or 62 ns.

4   Performance Evaluation of MPICH2 over Nemesis

In this section we evaluate the shared-memory performance of our implementation of MPICH2 over the Nemesis communication subsystem. First we present a microbenchmark evaluation on a 2 GHz dual-processor dual-core Opteron 280 machine with 2 GiB of memory. Then we present application benchmarks on an SGI Altix 350 machine with 16 1.4 GHz Itanium 2 processors and 32 GiB of memory. We configured MPICH2 with the `--enable-fast` option that disables error checking and configured OpenMPI to disable error checking and support for heterogeneous clusters, which should improve the performance for those implementations. All implementations were compiled using `-O3` optimization.

4.1   Latency and Bandwidth

We compare our implementation to LAM/MPI [5] version 7.1.2, OpenMPI [6] version 1.1, MPICH-GM [7] version 1.2.6..14b, and MPICH2 version 1.0.3 configured with the CH3 *shm* channel that communicates by using shared memory. All these MPI implementation use shared-memory intranode communication. Except where noted, the results for MPICH2 Nemesis have both the LMT and posted receive queue bypass optimizations applied. We measured latency and bandwidth using Netpipe [8]. Figure 2 shows these results.

The latency graph in Figure 2(a) shows two data series for MPICH2 Nemesis. The results shown by the data series labeled "MPICH2 Nemesis no BP" were taken without the posted receive queue bypass optimization. This optimization improves latency by about 62 ns, resulting in a zero-byte latency of 341 ns. With the optimizations applied, MPICH2 Nemesis has lower latency than the other MPI implementations. Even up to 128 bytes, the MPICH2 Nemesis latency is just over 500 ns.

Figure 2(b) shows the bandwidth comparison. Nemesis uses an optimized memory copy routine that uses nontemporal store operations. Using the nontemporal copy routine results in dramatically higher bandwidth for MPICH2 Nemesis compared to the other MPI implementations. The results shown by the data series labeled "MPICH2 Nemesis no-LMT" were taken without applying the LMT optimization to MPICH2 Nemesis. The LMT optimization improves bandwidth by about 130 MiBps for large messages, resulting in a peak bandwidth of over 1,500 MiBps. Notice that for MPICH2 Nemesis, at 16 KiB the bandwidth of the non-LMT implementation is a little higher than the implementation with LMT. The reason is that at 16 KiB, the communication protocol switches from eager to rendezvous and additional setup is performed for LMT. The figure shows that MPICH2 Nemesis has higher bandwidth than the other

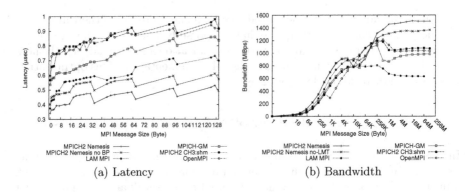

(a) Latency   (b) Bandwidth

Fig. 2. Shared-memory performance of MPI implementations

MPI implementations except for messages between about 16KiB and 256KiB. We intend to perform additional tuning to improve the medium-sized message bandwidth and find the optimal message size for the crossover from eager to rendezvous protocol.

4.2   Instruction Count

One of the goals of our implementation is to streamline the critical path. One way of measuring our success is by counting the number of instructions required to send or receive a message. Using the PAPI[9] performance counter interface, we measured the instruction count for send and receive eight-byte messages. When measuring the instruction count for the receive operations, we wanted to avoid counting instructions performed polling while waiting for the message to arrive because the waiting time can vary quite a bit. To do this we added a delay equal to the round trip time before starting to count instructions and performing the receive. This ensured that the incoming message had arrived and was waiting at the receive queue when MPI_Recv was called. The table in Figure 3 shows these results. All MPI implementations were compiled with the -O3 optimization level, except for MPICH-GM, where the unoptimized code had fewer instructions.

The row labeled "MPICH2 Nemesis no BP" shows the instruction counts when the posted receive queue bypass optimization was not applied. The results show that this optimization reduces the combined send and receive instruction count by almost half. With the optimization, the combined instruction count for MPICH2 Nemesis is less than 22% that of OpenMPI, less than 50% that of MPICH2 CH3:shm and MPICH-GM, and 55% that of LAM MPI. The instruction counts show that the critical path in our implementation is already quite efficient, however, we believe that we still can further streamline the critical path and improve cache utilization which will reduce overall latency for small messages.

4.3   The Halo Benchmark

One of the benchmarks we used to predict the application performance of MPICH2 Nemesis was the Halo benchmark [10]. This benchmark simulates a nearest neighbor exchange of a 1 to 2 row and column "halo" from a 2D array. The authors of

MPI Implementation	MPI_Send	MPI_Recv	Total
OpenMPI	550	1,745	2,295
MPICH-GM	455	617	1,072
LAM MPI	436	472	908
MPICH2 CH3:shm	311	748	1,059
MPICH2 Nemesis no BP	241	712	952
MPICH2 Nemesis	**241**	**259**	**500**

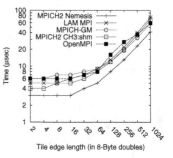

Fig. 3. Instruction count for sending and receiving a eight-byte message

Fig. 4. Results of the Halo benchmark using four processes

the Halo benchmark state that performance of the Halo benchmark correlates well with the performance of their layered ocean model application. We ran the benchmark on the Opteron machine using four processes.

The Halo benchmark performs the halo exchanges by using several different algorithms. The results in Figure 4 show the results for the algorithm that performed best for each MPI implementation. The algorithm which used `MPI_SendRecv()` performed best in MPICH2 Nemesis, MPICH-GM and OpenMPI. In MPICH2 CH3:shm, the algorithm using `MPI_ISend()` and `MPI_IRecv()` performed best. In LAM MPI, the best performance was seen when using the algorithm that used persistent sends and receives, where the receives are posted before the send operations are called. In the figure, we see that MPICH2 Nemesis performs considerably better than the other implementations for all tile sizes. Of the others, MPICH2 CH3:shm performs better than LAM MPI, MPICH-GM, and OpenMPI for small tile sizes. For larger tile sizes MPICH-GM performs better than MPICH2 CH3:shm, LAM MPI, and OpenMPI. The performance of this benchmark is dominated by latency for small tile sizes and by bandwidth for large tile sizes. The factor of improvement for MPICH2 Nemesis over the other implementations ranges from 1.5 to 2.6. This suggests that MPICH2 Nemesis should perform well on applications that are sensitive to latency or need high bandwidth.

4.4   The NAS Benchmarks

We evaluated the application-level performance of MPICH2 Nemesis using the NAS benchmarks [11]. We wanted to evaluate how the low latency and high bandwidth of MPICH2 Nemesis can benefit the parallel speedup of applications. To emphasize the communication cost over the computation time, we used smaller problem sizes, specifically, the class A problem size with the CG, MG, FT, SP, BT, and LU benchmarks and the class B problem size for the IS benchmark. For the IS benchmark the class A problem size was too small for 8 and 16 processes and resulted in too much variation in the results. We decided not to use the EP benchmark because there was very little communication.

To get results for a larger number of processes, we ran the benchmarks on a 16-processor SGI Altix at the Ohio Supercomputer Center (OSC). On that

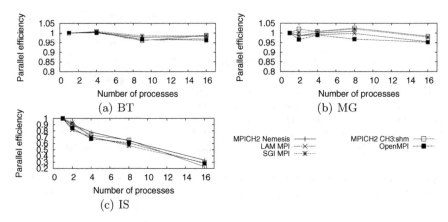

Fig. 5. Parallel efficiency for the NAS benchmarks

machine MPICH-GM was not available; instead, we evaluated the SGI MPI implementation. The Altix machine has features to allow one process to directly access another process's address space, which can allow for very efficient large message transfers. However, these features were not enabled on the OSC machine. It is not clear how much of an impact the lack of these features has on the performance of SGI MPI.

In our evaluation, all of the MPI implementations performed similarly. Figure 5 shows the parallel efficiency for the class A BT and MG and class B is benchmarks. We omit the graphs for the other results because of space limitations. We see that for the BT and MG benchmarks the parallel efficiency for all implementations is better than 0.95. For the IS benchmark, which has a higher communication to computation ratio than the other benchmarks [12], we see that the parallel efficiency decreases considerably with the number of processes. Here too, we see that all of the MPI implementations perform similarly. The parallel efficiency for any individual implementation differs less than 10% from the average for up to 8 processes, and less than 20% from the average for 16 processes.

5   Discussion and Future Work

In this paper we have presented our new implementation of MPICH2 over the Nemesis communication subsystem. We evaluated the shared-memory communication of our implementation on a 4-core Opteron machine using microbenchmarks. Our implementation achieved a zero-byte latency of 341 ns and a 128-byte latency of just over 500 ns. The peak bandwidth of our implementation was over 1,500 MiBps. We also measured the number of instructions required to send and receive MPI messages. MPICH2 Nemesis uses only 500 instructions to send and receive an eight-byte messages. To evaluate application-level performance, we used the Halo benchmark, which favors low-latency and high-bandwidth MPI implementations, and saw a factor of improvement from 1.5 to 2.6 compared to the other implementations we evaluated. Our evaluation using the NAS benchmarks on a 16-processor Altix machine did not show large differences in parallel efficiency between the different MPI implementations. These results show that MPICH2 Nemesis has an efficient implementation of shared-memory communication, which achieves low latency and high bandwidth. Moreover, the results indicate that MPICH2 Nemesis would be especially suitable for smaller-grained applications which are sensitive to latency and bandwidth.

Future work on MPICH2 Nemesis is to implement Nemesis as a full ADI3 device, which should further improve performance. We also intend to implement optimized collective communication operations that take advantage of shared memory, as well as collective operation primitives provided by network interfaces.

References

1. Argonne National Laboratory: MPICH2. (http://www.mcs.anl.gov/mpi/mpich2)
2. Buntinas, D., Mercier, G., Gropp, W.: Design and evaluation of Nemesis, a scalable low-latency message-passing communication subsystem. In: Proceedings of International Symposium on Cluster Computing and the Grid 2006 (CCGRID '06). (2006)
3. Brown, G.: The GM message-passing system. http://www.myri.com/news/02512/slides/Brown_gm.pdf (2002) Presented at the Myrinet User's Group Conference (MUG-2002).
4. Buntinas, D., Mercier, G., Gropp, W.: Data transfers between processes in an SMP system: Performance study and application to MPI. In: Proceedings of the 35th International Conference on Parallel Processing (ICPP 2006). (2006) To appear. Available at http://www-unix.mcs.anl.gov/ buntinas/papers/icpp06-smp-xfer.pdf.
5. Burns, G., Daoud, R., Vaigl, J.: LAM: An open cluster environment for MPI. In: Proceedings of Supercomputing Symposium. (1994) 379–386
6. Gabriel, E., Fagg, G.E., Bosilca, G., Angskun, T., Dongarra, J.J., Squyres, J.M., Sahay, V., Kambadur, P., Barrett, B., Lumsdaine, A., Castain, R.H., Daniel, D.J., Graham, R.L., Woodall, T.S.: Open MPI: Goals, concept, and design of a next generation MPI implementation. In: Proceedings of the 11th European PVM/MPI Users' Group Meeting, Budapest, Hungary (2004) 97–104
7. Myricom: (MPICH-GM) http://www.myri.com/scs/.
8. Snell, Q.O., Mikler, A.R., Gustafson, J.L.: Netpipe: A network protocol independent performace evaluator. In: Proceedings of Internation Conference on Intelligent InformationManagement and Systems. (1996)
9. Browne, S., Deane, C., Ho, G., Mucci, P.: PAPI: A portable interface to hardware performance counters. In: Proceedings of Department of Defense HPCMP Users Group Conference, Monterey, California. (1999)
10. Wallcraft, A.J.: The Halo benchmark. http://www.sdsc.edu/SciComp/PAA/Benchmarks/Portal/Halo/halo.html (1998)
11. Bailey, D.H., Barszcz, E., Barton, J., Browning, D., Carter, R., Dagum, L., Fatoohi, R., Fineberg, S., Frederickson, P., Lasinski, T., Schreiber, R., Simon, H., Venkatakrishnan, V., Weeratunga, S.: The NAS parallel benchmarks. Technical Report RNR-94-007, NASA Ames Research Center (1994)
12. Wong, F.C., Martin, R.P., Arpaci-Dusseau, R.H., Culler, D.E.: Architectural requirements and scalability of the NAS parallel benchmarks. In: Supercomputing '99: Proceedings of the 1999 ACM/IEEE conference on Supercomputing (CDROM), New York, ACM Press (1999) 41

MPI/CTP: A Reconfigurable MPI for HPC Applications

Manjunath Gorentla Venkata and Patrick G. Bridges*

University of New Mexico, Department of Computer Science, Albuquerque, USA
{manjugv, bridges}@cs.unm.edu

Abstract. Modern MPI applications have diverse communication requirements, with trends showing that they are moving from static communication requirements to more dynamic and evolving communication requirements. However, MPI libraries, which integrate MPI applications with the hardware, are not flexible enough to accommodate these diverse needs. This lack of flexibility leads to degraded performance of the applications. In this paper, we present the design of a protocol development framework and an MPI library implemented using our proposed framework that support compile-time and boot-time protocol configuration, as well as runtime protocol reconfiguration based on dynamic application requirements. Experimental results on the initial prototype of this design show that this prototype is able to dynamically reconfigure at runtime to optimize bandwidth under changing MPI requirements.

1 Introduction

Different MPI applications have diverse communication characteristics and thus varying protocol demands. The HPC machines on which these applications run similarly have networks with varying hardware interface requirements and again varying protocol demands. Further, MPI communication characteristics vary at runtime. For example, recent work has shown that different protocol implementations are appropriate based on the percentage of application preposted MPI receives [1]. Specifically, applications that prepost most of their large receives can gain substantial benefits from an eager rendezvous protocol; however eager rendezvous protocols waste substantial network bandwidth when most receives are not preposted. Application message preposting behavior can vary widely; this implies that no one protocol optimization decision is appropriate for all applications.

To optimize for these varying application and hardware demands, system software needs to provide application- and architecture-specific optimizations. In this paper, we describe a protocol architecture called MPI/CTP for application- and hardware-specific protocol reconfiguration in MPI, and present preliminary numbers from a prototype MPI implementation. These results illustrate the

---

* This work was supported under contract Sandia University Research Program contract number 190576 and DOE Office of Science grant DE-FG02-05ER25662.

ablity of this architecture to adapt to changing MPI protocol requirements at runtime. In the remainder of this paper, section 2 presents background on the framework we use to implement our prototype, and section 3 presents the design of our reconfigurable MPI protocol. Section 4 then presents performance results obtained using a prototype implementation of this framework, and sections 5 and 6 present related work, conclusions, and directions for future work.

2   Cactus and CTP

Our work on protocol reconfiguration in MPI is implemented as an enhancement to the Configurable Transport Protocol (CTP), which in turn is implemented in the Cactus composite protocol framework. In this section, we briefly describe Cactus and CTP. A detailed description of Cactus and its execution structure can be found in [2], of the message abstraction used by Cactus in [3], and of CTP itself in [4] and [5].

2.1   Cactus

Cactus is a system for constructing highly-configurable protocols for networked and distributed systems. Individual protocols in Cactus, generally termed *composite protocols*, are constructed from fine-grained software modules called *microprotocols* that interact using an event-driven execution paradigm. Each microprotocol implements a different function of the protocol. Instances of these microprotocols binds event handlers to protocol-specific events to effect protocol processing.

Processing of structured messages by microprotocol-defined event handlers comprises the basic programming model of Cactus. Events are used to signify state changes of interest, such as "message arrival from the network". When such an event occurs, all event handlers bound to that event are executed. Events can be raised explicitly by microprotocol instances or implicitly by the composite protocol runtime system.

The Cactus runtime system provides a variety of operations for managing events and event handlers. In addition to traditional blocking events, Cactus events can also be raised with a specified delay to implement time-driven execution, and can be raised asynchronously. Other operations are available for unbinding handlers, creating and deleting events, halting event execution, and canceling a delayed event. Finally, synchronization and coordination of execution activities in Cactus is accomplished through *event-based barriers* that may be associated with data items. These barriers are used to coordinate activities across multiple microprotocols.

The main features provided by the Cactus message abstraction are named *message attributes*, and the event-based barrier mechanism described above. These dynamically created message attributes are a generalization of traditional message headers. Messages are sent up or down the protocol stack and deallocated using event-based barriers associated with each message, in which context the barriers are generally referred to as send bits and deallocate bits, respectively.

2.2   CTP

CTP is a message-oriented configurable transport protocol written in the Cactus framework, primarily for use on local-area and wide-area Internet (e.g. Ethernet) connections. CTP includes a wide range of microprotocols for operating in this environment, including microprotocols implementing for acknowledgements (`PositiveAck`), retransmissions (`Retransmit`), forward error correction (`ForwardErrorCorrection`), and a range of congestion control mechanisms and policies (`WindowedFlowControl, TCPCongestionControl`, etc.). Using these and other microprotocols, researchers have implemented CTP configurations that support TCP-like, UDP-like, and SCTP-like semantics.

Microprotocols in CTP handle a set of predefined events, particularly those that indicate message availability from the network or an application. Two primary events are used for processing outgoing messages - `MessageFromUser` indicates that a new arbitrary-sized message is available for transmission, while `SegmentToNet` events are generated by fragmentation/reassembly microprotocols that fragment messages into segments for transmission over the network. Similarly, receive-side processing includes `SegmentFromNet` and `MessageFromNet` events, which again correspond to fragmented packets and reassembled messages. Each microprotocol can binds these handlers, set message attributes as appropriate, and sets send and/or deallocate bits to indicate that it is willing to permit the message to be transmitted (to the user or over the network) and/or deleted.

3   MPI/CTP Design

We have designed new microprotocols for CTP that enable it to be used for sending and receiving messages with MPI matching and ordering semantics. In addition, we implemented these changes to CTP in such a way that microprotocols can be selectively enabled and disabled in response to changing application needs. Initially, our work on reconfiguration has focused on changing between different rendezvous protocols based on various local and remote performance information, as recent work has shown that this can have a substantial impact on available MPI message-passing bandwidth [1].

3.1   Basic Functionality

Our implementation of MPI/CTP includes a variety of additions, particularly new microprotocols, new message attributes, and careful interaction with existing CTP microprotocols. The new microprotocols implement different MPI-specific protocol algorithms and the new message attributes are used to carry MPI-specific information for these microprotocols. The following sections describe these extensions. A diagram illustrating these changes is shown in figure 1.

MPI Support Microproto-col. Because CTP origi-nally used TCP-like mes-sage demultiplexing based on port numbers instead of MPI matching seman-tics, we first had to in-troduce protocols that cus-tomized CTP to support MPI matching semantics. The MPISupport micropro-tocol is responsible for implementing basic MPI matching semantics in CTP by receiving post requests from the applications through the CTP control interface and making posted and un-expected queues available to other microprotocols for their use.

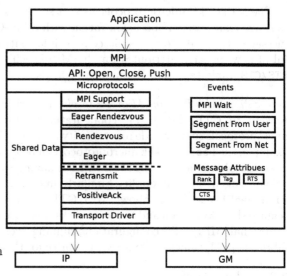

Fig. 1. MPI/CTP Architecture

An API to these lists is provided for other MPI/CTP microproto-cols to use as necessary. In addition, MPISupport handles miscellaneous local requests that do not require message generation and processing, for example calls to MPI_Wait(). Note that MPISupport introduces a new MPIWait event to CTP to signal threads blocked on synchronous MPI calls.

In the current MPI/CTP prototype implementation, MPISupport does not provide true zero-copy semantics because of limitations in the CTP protocol framework. In particular, CTP driver protocols always copy data into buffers prior to any demultiplexing decisions or MPI/CTP-level protocol processing be-ing done. We are currently extending the CTP framework with early demulti-plexing capabilities to address this limitation.

Message-Handling Microprotocols. MPI/CTP includes message-handling micro-protocols for sending MPI messages over the network. MPI/CTP currently in-cludes 3 microprotocols sending and receiving MPI-oriented messages: Eager, Rendezvous, and EagerRendezvous; these correspond to common techniques for sending short (Eager) and long (Rendezvous/ EagerRendezvous MPI messages.

Like most CTP microprotocols, each microprotocol implements handlers for the SegmentFromUser and SegmentFromNet events to enable them to process messages. In response to messages from the application to send, these micro-protocols may send the message immediately or send a request-to-send (RTS) to facilitate later transmission. Likewise, in response to SegmentFromNet events, they may doing nothing and rely on preexisting CTP microprotocols to handle acknowledgements, or they may send or schedule transmission of data to the requester if the received packet is an RTS or CTS.

The MPI/CTP microprotocols also set send and deallocate bits to coordinate message transmission and deallocation with other CTP microprotocols, and set message attributes to transmit control information. We have added a handful of new message attributes for the MPI-specific microprotocols, particularly RTS/CTS, rank, tag, and communicator fields.

Interactions with Existing Microprotocols. By writing MPI functionality as an extension to CTP, MPI/CTP configurations retain fill access to other CTP microprotocols that provide functionality that may be desirable in some cases. For example, the PositiveAck microprotocol can be used to acknowledge message receipt in a short-message protocol without having to reimplement and reoptimize acknowledgement functionality. Similarly, microprotocols such as Retransmit and WindowedFlowControl allow MPI/CTP protocol configurations to work seamlessly in long-haul and lossy networks. Because all such functionality in CTP is optional, MPI/CTP configurations running over standard high-speed reliable fabrics (e.g. Myrinet) need not pay the price for this functionality.

3.2 Adaptation

Protocol adaptation (sometimes referred to as the "protocol switch") in MPI/CTP is implemented by a combination of microprotocol reconfiguration and filtering code in message-passing microprotocols. Only those message-passing microprotocols that are configured into a given MPI/CTP configuration (and hence have bound appropriate event handlers) can process a message, allowing different message-transmission algorithms to be configured and reconfigured at a coarse scale. MPI/CTP uses this level of protocol switch, namely reconfiguration, between microprotocols that process similar messages, for example between the Rendezvous and EagerRendezvous message-processing microprotocols.

More fine-grained protocol adaptation on a message-by-message basis, specifically the message size-based protocol switch, is done by parameterization. In particular, each message-passing protocol is designated as either a long-message or short-message protocol, a global shared variable that designates the switch-point between long and short messages is exported by MPISupport, and each configured microprotocol only handles an outgoing message if it is of the appropriate size. Note that this requires that only one short and one long message protocol be configured into MPI/CTP at a given time.

4 Experimental Evaluation

4.1 Experimental Setup

To evaluate our MPI/CTP design, we implemented a simple prototype of the MPI point-to-point calls using the design described in section 3. This implementation runs on Myricom GM, supports all of the various MPI point-to-point calls, but does not yet support MPI collective communications. As mentioned in section 3.1, this prototype does not support zero-copy because of CTP framework limitations. Addressing both of these limitations are planned for future work.

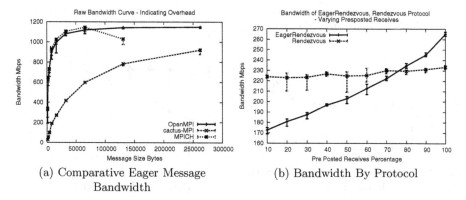

(a) Comparative Eager Message Bandwidth

(b) Bandwidth By Protocol

Fig. 2. Untuned MPI/CTP Message-passing Bandwidth; note that MPI/CTP currently includes a extra copy because of implementation framework limitations and that MPICH/GM fails in `gm_alloc` for messages larger than 128KB

We tested two different elements of our MPI/CTP prototype, namely basic message-passing bandwidths with different protocols and compared to those of current production-quality implementations, and message-passing bandwidth for fixed-size messages with protocol reconfiguration based on the percentage of messages preposted at the receiver. We tested these scenarios between two dual-processor 2.2 GHz Pentium III Xeon machines with Myrinet Lanai7 adapters. Each machine ran Linux kernel version 2.4.2 and GM 2.1.1. We compared bandwidths of our implementation versus those of OpenMPI 1.0.2 and MPICH/GM 1.2.6.

4.2 MPI/CTP Overhead

Figure 2 shows the basic bandwidth performance of our prototype implementation. As can be seen in figure 2(a), our prototype achieves approximately 81% of the point-to-point bandwidth of the OpenMPI or MPICH/GM implementations. The performance difference is due to the costs of extra copies that the existing CTP framework currently imposes on our prototype. Eliminating these copies should make MPI/CTP bandwidth-competitive with OpenMPI.

Figure 2(b) shows how MPI/CTP bandwidth varies by percentage of preposted receives with 32KB messages. As can be seen, the standard rendezvous protocol outperforms an eager rendezvous protocol when the 80% or less of receives are preposted. For carefully written applications where most receives are preposted, it is well known [1] that an eager large-message protocol can achieve better performance. This effect can be easily seen in MPI/CTP.

4.3 MPI/CTP Protocol Reconfiguration

To test the ability of MPI/CTP to optimize MPI behaviour through dynamic protocol reconfiguration, we enabled `MPISupport` to reconfigure which long message protocol MPI/CTP used based on feedback from the receiver on the average

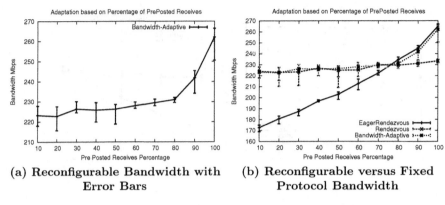

(a) **Reconfigurable Bandwidth with**
Error Bars

(b) **Reconfigurable versus Fixed**
Protocol Bandwidth

Fig. 3. Untuned MPI/CTP Message-passing Bandwidth

percentage of receives preposted there. `MPISupport` then dynamically changed between the `EagerRendezvous` and `Rendezvous` long message protocols by binding and unbinding handlers in each microprotocol at runtime; cutoffs for the protocol switch were determined ahead of time based on the information shown in figure 2(b).

Figure 3 shows that reconfiguration in MPI/CTP allows it to dynamically adjust its behavior based on remote application behavior, thereby optimizing available MPI protocol bandwidth. As MPI/CTP becomes more carefully tuned, we expect this to allow applications to achieve better MPI performance by dynamically reconfiguring protocol behavior based on application needs.

5   Related Work

There has been a variety of related work done on component-based MPI implementations and reconfiguration in protocol frameworks. Most recently, Open-MPI has implemented a component-based MPI framework that allows for easy customization of the MPI library at program startup based on, for example, hardware characteristics [6]. In OpenMPI's component architecture, however, components selected during initialization cannot easily be switched to other implementations, and configuration and componentization is relatively coursegrained. MPICH similarly supports an abstract device interface to enable MPI support for a range of different hardware devices. Neither of these systems, however, support the kind of fine-grained, online MPI protocol reconfiguration supported by our MPI/CTP design and prototype.

H-CTP [7] is, like MPI/CTP, a Cactus-based transport protocol aimed at high-end systems, in this case, Grid systems. In particular, H-CTP shows advantages of configurability in Grid computing environments by customizing transport protocol based on application requirements at link time. H-CTP cannot, however, change functional or QoS properties of the protocol at runtime based on dynamic communication characteristics as done in MPI-CTP. In addition, H-CTP does not directly support MPI semantics in, for example, matching.

6   Conclusion and Future Work

In this paper,we presented the design of a protocol architecture, called MPI/CTP, for application- and hardware-specific protocol reconfiguration in MPI, Using a prototype, we have shown how such protocol reconfiguration allows an MPI implementation to deal with dynamic application behavior, for example changing percentages of preposted receives, and to reconfigure at runtime based on this changing behavior. We are not aware of any other MPI implementation that can make such dramatic changes to protocol behavior at runtime in response to changing application behavior.

In the future, we plan to enhance and reimplement portions of MPI/CTP to optimize performance, as well as create a full-fledged MPI implementation that can be used for further research on dynamic protocol optimization in MPI. Such optimizations include changing collective implementations at runtime or offloading different portions of MPI/CTP based on different application and hardware demands. We also plan to use our prototype as a basis for implementing MPI in the configurable operating system we are currently developing in collaboration with Sandia National Labs and Louisiana State University [8].

Acknowledgements

The authors would like to thank Ron Brightwell and Rolf Riesen at Sandia National Labs for their support in providing benchmarks and testcases for our prototype MPI implementation.

References

1. Brightwell, R., Underwood, K.: Evaluation of an eager protocol optimization for MPI. In: Recent Advances in Parallel Virtual Machine and Message Passing Interface: Tenth European PVM/MPI Users' Group Meeting. (2003)
2. Hiltunen, M.A., Schlichting, R.D., Han, X., Cardozo, M., Das, R.: Real-time dependable channels: Customizing QoS attributes for distributed systems. IEEE Transactions on Parallel and Distributed Systems 10(6) (1999) 600–612
3. Hiltunen, M.A., Wong, G.T., Schlichting, R.D.: Dynamic messages: An abstraction for complex communication protocols. Software: Practice and Experience (2001) Submitted for publication.
4. Wong, G.T., Hiltunen, M.A., Schlichting, R.D.: A configurable and extensible transport protocol. In: Proceedings of IEEE INFOCOM '01, Anchorage, Alaska (2001) 319–328
5. Bridges, P.G., Hiltunen, M.A., Schlichting, R.D., Wong, G.T., Barrick, M.: A configurable and extensible transport protocol (2005) In revision for *ACM/IEEE Transactions on Networking*.

6. Graham, R.L., Woodall, T.S., Squyres, J.M.: Open MPI: A flexible high performance MPI. In: Proceedings, 6th Annual International Conference on Parallel Processing and Applied Mathematics, Poznan, Poland (2005)
7. Wu, R., Chien, A., Hiltunen, M., Schlichting, R., Sen, S.: A high performance configurable transport protocol for Grid computing. In: Proceedings of the Fifth IEEE International Symposium on Cluster Computing and the Grid (CCGrid 2005), Cardiff, Wales (2005)
8. Maccabe, A.B., Bridges, P.G., Brightwell, R., Riesen, R., Hudson, T.: Highly configurable operating systems for ultrascale systems. In: Proceedings of the First International Workshop on Operating Systems, Programming Environments, and Management Tools for High-Performance Computing on Clusters. (2004)

Correctness Checking of MPI One-Sided Communication Using Marmot

Bettina Krammer and Michael M. Resch

High Performance Computing Center Stuttgart
Nobelstrasse 19, D-70550 Stuttgart, Germany
{krammer, resch}@hlrs.de

Abstract. The MPI-2 standard defines functions for Remote Memory Access (RMA) by allowing one process to specify all communication parameters both for the sending and the receiving side, which is also referred to as *one-sided communication*. Having experienced parallel programming as a complex and error-prone task, we have developed the MPI correctness checking tool MARMOT covering the MPI-1.2 standard and are now aiming at extending it to support application developers also for the more frequently used parts of MPI-2 such as one-sided communication. In this paper we describe our tool, which is designed to check the correct usage of the MPI API automatically at run-time, and we also analyse to what extent it is possible to do so for RMA.

Keywords: MPI, Parallel Programming Tools, Analysis, One-sided communication, RMA.

1 Introduction

The Message Passing Interface (MPI) is a widely used standard for writing parallel programs in a convenient and efficient manner. Version 2 of the MPI standard [19] extends the functionality of the MPI-1.2 standard [18] significantly, adding about 200 functions to the already previously defined 129 functions. Several vendors offer implementations of MPI-2 and there are already open source implementations such as mpich2 or Open MPI [7,9,16], which cover at least some of the new features or even the full MPI-2 standard.

In order to facilitate the development of applications with dynamically changing data access patterns where the data distribution is fixed or slowly changing, the MPI-2 standard introduces the concept of the so-called *one-sided communication* for Remote Memory Access (RMA). Allowing one process to specify all the communication parameters for both the sender and the receiver avoids the need for global computations or explicit polling for potential communication requests. Thus, one process can access data in another process's own memory without the latter one knowing which data needs to be accessed or updated by which remote process.

Due to the complexity of parallel programming, in general, and the difficult semantics of one-sided communication, in particular, there is a demand for analysis and debugging tools to help users develop correct and portable MPI applications.

B. Mohr et al. (Eds.): PVM/MPI 2006, LNCS 4192, pp. 105–114, 2006.

2   Related Work

Fortunately, there are powerful tools to help application developers, be it (parallel) debuggers, memory checking or correctness tools, special MPI libraries or other tools that may perform a runtime or post-mortem analysis:

1. The freely available debugger gdb [3], which is also used with its graphical front-end ddd [1], has currently no support for MPI, but it can be attached to one or several, possibly already running MPI processes. The same can be done with special memory-checking debuggers like valgrind [5]. More convenient are parallel debuggers, such as the well-known commercial debuggers Totalview [4] or DDT [2].
2. The second approach is to provide a special debug version of the MPI library (e.g. mpich or NEC-MPI). This version is not only used to catch internal errors in the MPI library, but also to detect some incorrect usage of MPI by the user, e.g. a type mismatch of sending and receiving messages or mismatched collective operations [6,20].
3. Another possibility is to develop tools dedicated to finding problems within MPI applications at runtime, examples of which are the introduction of irreproducibility, deadlocks, incorrect management of resources such as communicators, groups, datatypes etc. or the use of non-portable constructs. At present, three different message-checking tools are under more or less active development: MPI-CHECK [17], Umpire [21] and MARMOT [12,11,14,13]. MPI-CHECK is currently restricted to Fortran code and performs argument type checking or finds problems such as deadlocks [17]. Like MARMOT, Umpire [21] uses the PMPI profiling interface.
4. The fourth approach is to perform a post-mortem analysis by collecting all information on MPI calls in a trace file. After program execution, this trace file is analysed by a separate tool or compared with the results from previous runs [15]. An example of this is the Intel Message Checker (IMC) [10].

3   Description of MARMOT

MARMOT is a library that uses the so-called PMPI profiling interface to intercept MPI calls and analyse them during runtime. It just has to be linked to the application in addition to the underlying MPI implementation, without any modification of the application's source code nor of the MPI library.

MARMOT supports the complete MPI-1.2 standard. Not all possible tests (such as consistency checks) are implemented yet as the development of our tool is still ongoing. MARMOT's output is a human-readable log file indicating errors and warnings, a graphical viewer is in progress. The tool can be used with any standard-conforming MPI implementation. MARMOT is tested on different Linux platforms, using different compilers and different MPI implementations (mpich, LAM/MPI, vendor MPIs, etc.). Functionality and performance tests are performed with test suites, microbenchmarks and real applications [11,14].

The following errors occur most frequently in MPI programming. Some of these errors may be tolerated by specific MPI implementations or by specific platforms. MARMOT tries to catch as many of them as possible:

- **Deadlocks:** In general, deadlocks are caused by the non-occurrence of something else, for example mismatched send/receive operations or mismatched collective calls. MARMOT contains a mechanism to automatically detect deadlocks and notify the user where and why they have occurred.
- **Data races:** Potential race conditions can be caused by various reasons, e.g. by the use of a receive call with wildcards, by the use of random numbers, or by the fact that nodes do not behave exactly the same. At present, MARMOT indicates the use of wildcards, but it does not construct dependency graphs to view the different possible executions nor does it use methods like record and replay to identify and track down bugs in parallel programs [15] or to compare different runs.
- **Mismatches:** Mismatches in arguments of one call can be detected locally, e.g. wrong type or number of arguments. Mismatches are also seen in arguments involving more than one call, e.g. in send/receive pairs or in collective calls, or in pairs of synchronising calls for one-sided communication.
- **Resource handling:** For the support of the MPI-1.2 standard, MARMOT has implemented its own book-keeping of the MPI resources (communicators, groups, datatypes, etc.). This is necessary for verifying the proper construction, usage and destruction of these MPI resources as they are "opaque" objects and therefore implementation-dependent. The MPI-2 standard introduces new opaque objects such as info objects MPI_Info or window handles MPI_Win to be used in the one-sided communication calls. New objects can be implemented in the same way as the MPI-1.2 objects.
- **Memory and other resource exhaustion:** Non-blocking calls such as MPI_Isend etc. can complete without issuing a matching test or wait call. However, the number of available request handles is limited (and implementation defined). Therefore requests should always be freed, as should allocated communicators, datatypes, etc. MARMOT gives a warning when a request is reused, and also when there are active or non-freed requests left at the MPI_Finalize.

  Another issue is reusing memory that is still in use, for example by reading/writing from/into a buffer by an unfinished send/receive operation. This type of error can also occur when using one-sided communication. MARMOT does currently not perform any checks whether a buffer can be reused safely because the transmission of data has completed. This kind of check is a subtle task that requires some insight into an MPI implementation.
- **Portability:** The MPI standard leaves many decisions to the implementors, for example how to implement opaque objects and handles to these objects, whether to implement collective calls as synchronising calls, whether to make the implementation thread-safe or not, whether RMA functions are blocking or not, etc. Relying on such non-portable constructs may resolve in deadlocks or other errors when using a different MPI implementation.

4   Description of MPI One-Sided Communication

Version 2 of MPI [19] provides a high-level interface to Remote Memory Access (RMA) that achieves two effects: *communication* of data from sender to receiver and *synchronisation* of sender with receiver. The first one is provided by the MPI_Put, MPI_Get and MPI_Accumulate functions for remote write, read and update, resp. (see 5.2). The second one is achieved through a number of synchronisation calls distinguishing between *active* and *passive target communication* (see 5.3)

For the RMA mechanism, the MPI-2 standard introduces a new kind of opaque object: MPI_Win, a handle to a *window* in a process's existing memory that is made accessible to remote processes. Therefore, a third category of RMA calls is needed for the construction and destruction of these objects (see 5.1).

In total, Chapter 6 on One-Sided Communication in the MPI-2 standard document lists 14 calls. However, there are also some calls hidden in other chapters of the standard that are relevant to RMA and have to be implemented in our tool to fully cover the functionality for one-sided communication, mainly error handlers for windows, attribute caching functions (see 5.4) and memory allocating calls, which finally results in about 30 functions to be implemented (see Table 1 and Table 2).

5   Possible Checks

In the following section, we consider the possible checks for the RMA functions in more detail (for a concise overview see also Table 1 and Table 2).

5.1   Initialisation

A process may specify a *window* of existing memory that is exposed to remote memory accesses from the other processes within the intracommunicator group. Windows consist of a number of bytes, starting at a base address, and are constructed using the MPI_Win_create and MPI_Win_free functions. Both these calls are collective and must therefore be called on all processes in our communicator to avoid deadlocks. Every process may specify a completely different target window concerning its location, size, displacement unit and info arguments. The same area in memory may also be associated with different windows. The attributes cached with a window can be retrieved with the MPI_Win_get_attr and MPI_Win_get_group functions (see 5.4). It is the user's responsability to ensure that the target window fits the specifications of the remote accesses and that there are no concurrent communications to distinct, overlapping windows.

We check the parameters of MPI_Win_create and MPI_Win_free for correctness, e.g. that, in the former call, the window size is a nonnegative integer, that the displacement unit size is a positive integer, that the info argument or the communicator are valid and, in the latter call, that the window argument is valid. The validity of communicators and windows can be checked similarly to

Table 1. Classification of RMA functions (initialisation, communication and synchronisation)

category	noncollective	collective	checks
initialisation		MPI_Win_create	collective error / deadlock; *parameters*: size nonneg. int, disp_unit pos. int, comm valid, info valid;
		MPI_Win_free	collective error / deadlock; *parameters*: window valid; pending RMA;
communication	MPI_Put		*parameters*: origin count / datatype, target rank / disp / count / datatype, window; access epoch for window;
	MPI_Get		*parameters*: origin count / datatype; target rank / disp / count / datatype, window; access epoch for window;
	MPI_Accumulate		*parameters*: origin count / datatype; target rank / disp / count / datatype, operator, window; access epoch for window;
synchronisation:		MPI_Win_fence	collective error /deadlock; *parameters*: assert, window;
active target	MPI_Win_start		matching pairs (origin: complete, target: post); *parameters*: group, assert, window;
	& MPI_Win_complete		matching pair (start); *parameters*: window;
	MPI_Win_post		matching pairs (target: wait or test, origin: start); *parameters*: group, assert, window;
	& MPI_Win_wait		matching pair (post); *parameters*: window;
	or MPI_Win_test		matching pair (post); called again after success; *parameters*: window;
passive target	MPI_Win_lock		matching pair (unlock); window exposed; window created with no_lock; *parameters*: lock type, rank, assert;
	& MPI_Win_unlock		matching pair (lock); *parameters*: rank, window;

Table 2. Classification of RMA functions (error handlers, etc)

category	noncollective	checks
error handlers	MPI_Win_create_errhandler	
	MPI_Win_get_errhandler	*parameters*: window;
	MPI_Win_set_errhandler	*parameters*: window, errhandler;
	MPI_Win_call_errhandler	*parameters*: window;
attribute caching	MPI_Win_get_group	*parameters*: window;
	MPI_Win_create_keyval	
	MPI_Win_free_keyval	*parameters*: keyval;
	MPI_Win_get_attr	*parameters*: window, keyval;
	MPI_Win_set_attr	*parameters*: window, keyval, attribute;
	MPI_Win_delete_attr	*parameters*: window, keyval;
	MPI_Win_get_name	*parameters*: window;
	MPI_Win_set_name	*parameters*: window, name;
transfer of handles	MPI_Win_f2c	*parameters*: window;
	MPI_Win_c2f	*parameters*: window;
memory allocation	MPI_Alloc_mem	*parameters*: size, info;
	MPI_Free_mem	

MPI-1 calls, e.g. whether they have been constructed properly or have already been freed.

MPI_Win_free can only be called after the RMA is completed, i.e. after the synchronisation calls MPI_Win_fence or MPI_Win_wait, MPI_Win_complete or MPI_Win_unlock have been called to match previous calls to MPI_Win_post, MPI_Win_start or MPI_Win_lock. MARMOT can verify whether there are any pending RMA function calls left when the window is to be freed.

Users may improve the performance of windows by using MPI_Alloc_mem and MPI_Free_mem for allocating and freeing memory, esp. on shared-memory systems [8]. For the alloc call, we can verify that the size of memory is a nonnegative integer and that the info argument is valid.

5.2 Communication

Three different RMA calls are supported: MPI_Get, MPI_Put and MPI_Accumulate take a reference to a window and a rank to address the target process for remote read, write and update. By *origin* we denote the process that performs the call, and by *target* we denote the process in which the memory is accessed. Target and origin may be identical. The get, put and accumulate calls are similar to the execution of a send by the target and receive by the origin process and vice versa, combining the data from sender and receiver in the case of an accumulate call. These three calls are non-blocking and complete both at the origin and at the target when a synchronisation call is issued on the involved window (see 5.3).

For all these calls we can check whether the window, the origin count or datatype and the target rank, displacement, count or datatype are valid. Additional requirements have to be fulfilled by the datatype arguments: For the put function, the target datatype may not specify overlapping entries in the target buffer, and the message must fit in the target buffer, which must fit in the target window. For the get function, the origin datatype may not specify overlapping entries in the origin buffer, and the message must fit in the origin buffer, the target buffer must be contained in the target window. For the accumulate call, each datatype argument must be a predefined datatype or a derived datatype, where all basic components are of the same predefined datatype. Both origin and target datatype must be derived from the same predefined datatype, and the target datatype must not specify overlapping entries. The target buffer must fit in the target window. For the get, put and accumulate calls, the target datatype must not contain absolute addresses, only relative displacements.

The `MPI_Accumulate` call takes an additional operator handle argument to specify the kind of update that is performed on the data: we verify that it is one of the predefined operations for `MPI_Reduce` or the newly defined `MPI_REPLACE` operation. On the other hand, the MPI-2 standard is unclear on whether `MPI_-REPLACE` is a valid reduction operator for MPI-1 functions such as `MPI_REDUCE` (and friends). Therefore we also implement a warning whether this operator is used in an MPI call other than `MPI_Accumulate`.

We also verify that the communication calls only occur within an *access epoch* for the window involved, i.e. within an epoch that is started and ended by synchronisation calls on the window. Distinct access epochs for a window at the same process must be disjoint whereas epochs pertaining to different windows may overlap.

It is erroneous to have conflicting accesses to the same memory location in a window, e.g. by concurrent RMA communication or local operations, with only one exception: Several concurrent accumulate operations may update the same location in memory, the outcome being as if the accumulate calls had appeared in some serial order.

5.3   Synchronisation

RMA communication can be synchronised using two modes:

- **active target** communication, where both the origin and the target process are explicitly involved in the communication, i.e. the target process participates in the synchronisation (thus not having truly one-sided communication anymore). In active target communication, a target window can only be accessed within an *exposure epoch*, i.e. an epoch that is started and completed by the target process. Access epochs on the origin side and exposure epochs on the target side match one-to-one. Distinct exposure epochs for the same window at a process must be disjoint but such an exposure epoch may overlap with exposure epochs on other windows or with access epochs for the same or other windows.
- **passive target** communication, where only the origin process is explicitly involved in the communication, i.e. the target process does not execute a synchronisation call and there is no concept of an exposure epoch.

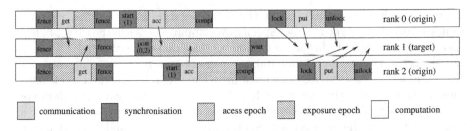

| | communication | | synchronisation | | acess epoch | | exposure epoch | | computation |

Fig. 1. Active (fence, start/complete/post/wait) and Passive (lock/unlock) Target Communication

Figure 1 illustrates the different synchronisation modes. The collective call MPI_Win_fence should both precede and follow communication calls (e.g. get) to delimit the access and exposure epochs on the origin and target processes for active target communication. For performance optimisation, it may be preferable to apply active target communication only to pairs of communicating processes. MPI_Win_start and MPI_Win_complete pairs start and terminate the access epochs while the MPI_Win_post and MPI_Win_wait or MPI_Win_test pairs mark the exposure epochs. It is erroneous to call MPI_Win_test again once the call has returned a true flag. The group arguments in the start and post calls specify which processes have remote memory access. The post and start calls must match as well as the start/complete and post/wait/test pairs.

For passive target communication, pairs of MPI_Win_lock and MPI_Win_unlock provide shared or exclusive access to the target window. It is erroneous to have a window locked and exposed at the same time. We can also verify whether a window is attempted to being locked although the no_locks info argument was provided at its creation time.

The assert argument in the post, start, fence and lock calls may be used for performance optimisation. It is erroneous to provide incorrect assert values (see Table 3). Implementations are, however, free to ignore the assert argument.

Table 3. Legal assert values for synchronisation calls

call	legal assert values (bit vector of zero or more of:)
MPI_Win_start	MPI_MODE_NOCHECK must be specified in start if and only if specified in each matching post
MPI_Win_post	MPI_MODE_NOCHECK, MPI_MODE_NOSTORE, MPI_MODE_NOPUT NOCHECK must be specified in post if and only if specified in each matching start
MPI_Win_fence	MPI_MODE_NOSTORE, MPI_MODE_NOPUT, MPI_MODE_NOPRECEDE, MPI_MODE_NOSUCCEED if NOPRECEDE or NOSUCCEED are specified on one process it must be specified on all processes in the group
MPI_Win_lock	MPI_MODE_NOCHECK

We also check, where applicable, whether the other arguments passed to the synchronisation calls, such as window, group, rank, lock type, are valid.

5.4 Error Handling and Attribute Caching

Table 2 shows an overview of the error handling and attribute caching calls. There is not much potential for possible errors in these calls. We can verify the correctness of arguments such as the window, keyval, errorhandler, etc.

6 Conclusions and Future Work

In this paper, we have presented the MARMOT tool, which analyses the behaviour of an MPI application during runtime and checks for errors frequently made in the use of the MPI API. We have unravelled some of the key features of the MPI RMA interface and have analysed it with regard to potential errors that can be made by application developers. In most cases these errors can be detected by tools such as MARMOT following the approach taken for MPI-1. Since there is currently no real application using RMA available to us our experience with the tool is limited to simple test cases. The lack of applications is, on one hand, probably due to the fact that the semantics of the RMA API are not easy to understand, and, on the other hand, that the performance of this new functionality may not be satisfying yet.

Future work on MARMOT includes an extension of its functionality to cover the complete MPI-2 standard and to support hybrid applications written in OpenMP and MPI. Another goal is to improve the performance and scalability of the tool, especially for communication-intensive applications.

Acknowledgments

The development of MARMOT has partially been supported by the European Commission through the IST-2001-32243 project "CrossGrid" and will be supported by its follow-up project "int.eu.grid". We thank Dr Matthias Mueller from University of Dresden for joint collaboration on MARMOT.

References

1. DDD. The Data Display Debugger. http://www.gnu.org/software/ddd.
2. DDT. The Distributed Debugging Tool.http://www.streamline-computing.com/.
3. gdb. The GNU Project Debugger. http://www.gnu.org/manual/gdb.
4. Totalview. http://www.etnus.com/Products/TotalView.
5. B. Carson and I. A. Mason. Clustergrind: Valgrinding LAM/MPI applications. In *12th European PVM/MPI*, volume LNCS 3666, pages 325–332. Springer, 2005.
6. E. Lusk, C. Falzone, A. Chan and W. Gropp. Collective Error Detection for MPI Collective Operations. In *12th European PVM/MPI*, volume LNCS 3666, pages 138–147. Springer, 2005.

7. E. Gabriel, G. E. Fagg, G. Bosilca, T. Angskun, J. J. Dongarra, J. M. Squyres, V. Sahay, P. Kambadur, B. Barrett, A. Lumsdaine, R. H. Castain, D. J. Daniel, R. L. Graham, and T. S. Woodall. Open MPI: Goals, Concept, and Design of a Next Generation MPI Implementation. In *11th European PVM/MPI*, volume LNCS 3241, pages 97–104. Springer, 2004.

8. E. Gabriel, G. E. Fagg, and J. J. Dongarra. Evaluating dynamic communicators and one-sided operations for current MPI libraries. *Inte'l Journal of High-Performance Computing Applications*, 19(1):67–80, 2005.

9. W. Gropp and R. Thakur. An Evaluation of Implementation Options for MPI One-Sided Communication. In *12th European PVM/MPI*, volume LNCS 3666, pages 415–424, September 2005.

10. B. Kuhn, J. DeSouza and B. R. de Supinski. Automated, scalable debugging of MPI programs with Intel Message Checker. In *SE-HPCS '05*, St. Louis, Missouri, USA, 2005. http://csdl.ics.hawaii.edu/se-hpcs/papers/11.pdf.

11. B. Krammer, K. Bidmon, M. S. Müller, and M. M. Resch. MARMOT: An MPI Analysis and Checking Tool. In *ParCO 2003*, Dresden, Germany, September 2003.

12. B. Krammer, M. S. Mueller, and M. M. Resch. MPI I/O analysis and error detection with MARMOT. In *11th European PVM/MPI*, volume LNCS 3241, pages 242–250. Springer, 2004.

13. B. Krammer, M. S. Mueller, and M. M. Resch. Runtime checking of MPI applications with MARMOT. In *ParCo 2005*, Malaga, Spain, September 2005.

14. B. Krammer, M. S. Müller, and M. M. Resch. MPI Application Development Using the Analysis Tool MARMOT. In *ICCS 2004*, volume LNCS 3038, pages 464–471. Springer, 2004.

15. D. Kranzlmüller. *Event Graph Analysis For Debugging Massively Parallel Programs*. PhD thesis, Joh. Kepler University Linz, Austria, 2000.

16. J. Liu, W. Jiang, P. Wyckoff, D. K. Panda, D. Ashton, D. Buntinas, W. D. Gropp, and B. R. Toonen. Design and Implementation of MPICH2 over InfiniBand with RDMA Support. In *IPDPS*. IEEE Computer Society, 2004.

17. G. Luecke, Y. Zou, J. Coyle, J. Hoekstra, and M. Kraeva. Deadlock Detection in MPI Programs. *Concurrency and Computation: Practice and Experience*, 14:911–932, 2002.

18. Message Passing Interface Forum. *MPI: A Message Passing Interface Standard*, June 1995. http://www.mpi-forum.org.

19. Message Passing Interface Forum. *MPI-2: Extensions to the Message Passing Interface*, July 1997. http://www.mpi-forum.org.

20. J.L. Träff and J. Worringen. Verifying Collective MPI Calls. In *11th European PVM/MPI*, volume LNCS 3241, pages 18–27. Springer, 2004.

21. J.S. Vetter and B.R. de Supinski. Dynamic Software Testing of MPI Applications with Umpire. In *SC 2000*, Dallas, Texas, 2000. ACM/IEEE. CD-ROM.

An Interface to Support the Identification of Dynamic MPI 2 Processes for Scalable Parallel Debugging

Christopher Gottbrath[1], Brian Barrett[2], Bill Gropp[3],
Ewing "Rusty" Lusk[3], and Jeff Squyres[4]

[1] Etnus, LLC, 24 Prime Park Way, Natick, MA 01760
Chris.Gottbrath@etnus.com
http://www.etnus.com
[2] 415 Lindley Hall, Computer Science Department, Bloomington, IN 47405
brbarret@osl.iu.edu
[3] Argonne National Laboratory, 9700 S. Cass Avenue, Argonne, IL 60439
{gropp, lusk}@mcs.anl.gov
[4] Cisco Systems, Inc., 225 East Tasman Dr., San Jose, CA 95134
jsquyres@cisco.com

Abstract. This paper proposes an interface that will allow MPI 2 dynamic programs – those using MPI SPAWN, CONNECT/ACCEPT, or JOIN – to provide information to parallel debuggers such as TotalView about the set of processes that constitute an individual application. The TotalView parallel debugger currently obtains information about the identity of processes directly from the MPI library using a widely accepted proctable interface. The existing interface does not support MPI 2 dynamic operations. The proposed interface supports MPI 2 dynamic operations, subset debugging, and helps the parallel debugger assign meaningful names to processes.

1 Introduction

MPI style parallel applications can comprise anywhere from one to many thousands of processes running on anything from a single user's workstation to the largest supercomputing clusters. Regardless of the scale of the application, when it fails to behave as expected and a developer sits down to debug it the first thing that they need to do is get their parallel debugger attached to their parallel program. This means that the debugger has to connect to not one but many processes running on both local and remote nodes. To the user this is a simple command or a simple 'click' in the interface of a parallel debugger like TotalView. The parallel debugger is able to fufill this request becuase the MPI library provides information about what processes running on both local and remote nodes constitute the parallel application.

This paper proposes a new interface between MPI processes and the TotalView parallel debugger that will enable users to debug applications taking advantage of the MPI 2 dynamic process capabilities to spawn new processes or combine

B. Mohr et al. (Eds.): PVM/MPI 2006, LNCS 4192, pp. 115–122, 2006.

two separately started parallel applications into one application. The dynamic nature of these applications provides a complex challenge to the debugger. Not only does the set of processes change over time but the performance focus of MPI leads vendors to favor highly asynchronous MPI library designs. The information that the debugger needs to get the debugging session started is something the MPI library is designed to keep distributed and balanced.

This new interface builds upon the foundation of and lessons learned from the current MPI -1 TotalView process acquisition interface[1] and the current MPI-1 TotalView message queue display interface[2] both of which have been almost universally adopted by MPI vendors over the past 9 years. To understand the new interface it helps to review the current process aquisition interface at a general level.

To attach to a parallel program TotalView first needs to attach to the starter process. This gives it the ability to halt and resume the process, set breakpoints and both read from and write to the program state. The debugger establishes – on the basis of these capabilities – an interface with the MPI library used by the application. If the debugger attaches to the starter before the starter program has created the parallel job the debugger runs the starter to the point that the parallel job exists but has not yet run user code. At this point the debugger reads a specified data structure in the starter process that holds a list of the MPI processes and information such as the network address of the node on which each process is running.

The user can then be prompted with a list of all the processes and can make an initial decision on which processes need to be actively debugged. In order to attach to the remote processes that the user selected TotalView needs a remote debugging agent, called a tvdsvr, on each node that hosts one or more selected processes. These are started by the debugger itself, often using ssh. The tvdsvr processes attach to each of the selected MPI processes and communicate back over the network to the debugger. At this point the user is attached to their parallel job. Any processes that were not chosen for debugging are now released to start running user code. The developer can now examine and control the selected set of processes. At any point during the debugging session the user can change the selected set of processes, choosing to look at more or less of the ongoing parallel job.

This interface, which is essentially an agreed upon format for a table in memory that the debugger reads directly and a bit of synchronization around startup, has been widely adopted and is extremely successfully by any measure. Its limitation is that it is predicated on the notion that the set of processes, once established, is static. MPI 2 dynamic processes undermine that assumption.

2   Design Goals

This interface makes fundamental information about the identity of processes participating in MPI static or dynamic programs available to the debugger so

---

[1] Reference code and interface header files can be found in the MPICH[1] implementation or can be obtained by contacting Etnus for an up to date version of the interface specification.

that it can attach to more than just one process of the job. This is being designed within the context of the TotalView parallel debugger, but the challenges of identifying processes in a MPI 2 dynamic program are generally applicable to parallel debuggers as a class and the expectation is that this interface could be adopted by other debuggers. MPI 2 dynamic programs are those that use the dynamic process calls, MPI_COMM_SPAWN, MPI_COMM_CONNECT, MPI_COMM_ACCEPT, and MPI_COMM_JOIN defined in chapter 5 of the MPI 2 standard.[3] However, there are a few other requirements that this interface needs to satisfy.

MPI implementors work to provide the greatest possible performance available and the proposed interface must limit the impact on MPI performance.

The new interface cannot be a step backwards in terms of functionality and needs to support important parallel debugger features like subset attach.

Users need to be able to debug MPI jobs that have deadlocked and hung or that are terminated and exist only as corefiles. This means that the interface needs to allow the debugger to gather the information it needs without running the application or MPI library.

There are wide variety of different resource managers and job launchers in use. It is possible to imagine that TotalView could just interface with them. However we believe this provides no solution for 'singleton' MPI applications and would be needlessly complex for CONNECT / ACCEPT jobs that are started by multiple starters and managed as separate entities by resource managers.

Finally the new interface needs to provide the user with the information that the user will need to make sense of a parallel job. In an MPI 1 job each process is reliably and naturally named by its rank in the communicator MPI_COMM_-WORLD. MPI 2 itself does not provide for a global and stable naming scheme from the perspective of the developer who is looking at their code. MPI processes' names for one another are only understood in the context of communicators and communicator handles are purely local. To allow for comparison of results from one run to another with the same input data and program logic, the user will need a way to 'address' their MPI 2 processes in a repeatable way. This paper proposes a stable naming scheme which can be used in debuggers and other tools. The interface will ensure that sufficient information is presented to the debugger for the debugger to construct a meaningful and repeatable name for each process.

3   Design

3.1   Overview

The MPI library itself will maintain a list of processes that are part of the job. As processes are created, CONNECT to, JOIN, or are detached from the program this list of processes will be updated by the MPI library. This list is called the **proctable**. The proctable is distributed across a variable number of MPI processes.

The parallel debugger will be able to read this table from the program using the same mechanisms that it uses to perform other debugging operations.

The MPI library will change the value of a synchronization variable before making changes to the proctable data structures and will reset it afterwards. The MPI library will then call a special stub function to notify the debugger that some dynamic process event caused the proctable to change.

If the debugger is attached to the root process of the dynamic process collective then the newly created MPI processes will be temporarily held to allow the debugger the opportunity to attach to them before the end of MPI_INIT.

3.2   Proctable Elements

The proctable contains two kinds of information for each listed process. System context information is needed for the debugger to locate each listed process and potentially be able to attach to it for debugging. MPI context information is needed to uniquely and reliably name each listed process. System information for each process listed in the proctable will include things like: host name or IP address, process or task ID, program name. MPI context information for each process listed in the proctable will include: the rank of each process within its own MPI_COMM_WORLD, a unique identification for that MPI_COMM_WORLD, information about how that MPI_COMM_WORLD came to be part of this job. For the case of MPI_COMM_WORLDs created by a SPAWN operation they can be identified with the following tuple (unique ID of the MPI_COMM_WORLD of the parent root process, rank of parent root process in that MPI_COMM_WORLD, sequence of the SPAWN command among those rooted on that same process). MPI_COMM_WORLDs that are started independently and then connected together need to be given unique ids in this interface that are external to the MPI (e.g. something that is a function of the mac address, PID, and time-stamp of the launcher process).

3.3   Proctable Organization

The proctable is distributed across a set of the MPI processes, called the **directory processes**. These processes each contain a subset of the full proctable. A single MPI process may be listed in multiple directory processes; each MPI process is listed in at least one directory process.

In order to reconstruct the proctable the debugger needs to locate all the directory processes and combine their process entry information. Locating the directory processes is done through a set of processes called **meta directory** processes. Meta directory processes each contain a list of directory processes and a list of other meta directory processes. Each directory process must be listed in one meta directory process. The meta directory processes must reference one another such that they form a strongly connected graph. Starting from any meta directory process the debugger must be able to locate the full set of meta directory processes. There can be as few as one meta directory process. Meta directory processes can also be directory processes, in fact is is possible that an implementation would choose to have all directory processes also be meta directory processes. Both directory and meta directory processes are MPI processes.

Meta directory processes are separated in this interface from directory processes because meta directory processes are expected to receive and handle proctable change notification messages from other processes. MPI library implementors may decide that they don't wish to have all the directory processes assume this responsibility. In this case they can have a much smaller set of processes play the role of meta directory processes. Only this narrower set of processes needs to assume the overhead that is involved in processing proctable change messages.

The debugger always needs to be able to identify the meta directory processes. The user should not need to know which process or processes are serving as meta directory processes. So all MPI processes (including all directory processes) will have information about one meta directory process. The user can then connect TotalView to any process of a MPI job and TotalView should be able to discover all the other processes through that one meta directory process.

The intent is that $1 \leq M \leq D \ll R$ where M is the number of meta directory processes, D is the number of directory processes, and R is the number of MPI processes.

3.4   Operations on the Distributed Proctable

The primary operation that the debugger will need to do on the distributed proctable is list it out. Assuming, for example, that TotalView starts by being attached to any one of the user's MPI process. TotalView then finds the information about the meta directory process that this process references. TotalView then attaches to that meta process and gathers its information. If there are other meta directory processes TotalView walks the graph, attaching to each in turn and gathering a cumulative list of directory processes. Having gathered a full list of directory processes TotalView attaches to any of them that it has not already attached to (remember that meta directories can be directories as well). At this point TotalView is attached to the full list of directory processes and now has the full list of the users MPI processes at hand.

In order to receive notification of proctable changes TotalView will need to remain attached to all of the meta directory processes. If the user does not require notification TotalView can detach from all but one MPI process and still be able to pick up changes to the proctable when the user requests (by reattaching to the **meta directories** locating and attaching to the directories and rereading the proctable information).

MPI library operations on the proctable must be carefully designed to allow for performance at large-scales. For example, modifying the entries in a proctable should involve as few processes as possible – at most, a meta directory process, a directory process, and the processes in the collective action (SPAWN, CONNECT, ACCEPT). It is certainly preferable to involve far less than this (e.g., only the directories and the root process from a SPAWN or representative processes from each of CONNECT and ACCEPT). Since meta directory and directory processes may also be MPI processes involved in the user's application, it is also critical that the interface not require the participating processes to block on the directory process' response.

During normal startup one process might be the meta directory and one or more processes are designated directory process. Information can be propagated as needed to set this up during INIT.

During a SPAWN collective there are two examples that must be considered. If the spawning collective group includes a directory process then that directory gets the information for the newly created processes during the SPAWN collective. A change notice is sent to the meta directory. When the meta directory is able to process the notification it calls a stub routine to notify TotalView that the process table has changed.

If the spawning group does not contain a directory process then one of the group gathers data on its peers and becomes a directory process. This can be done during the SPAWN collective call. The new directory then sends notification to its meta directory process that it has assumed the new status and that a change occurred. The meta directory adds the new directory to its list and calls the stub notification routine for TotalView.

If two separately started jobs are joined with CONNECT and ACCEPT both jobs will have their own preexisting proctable structures. During the CONNECT / ACCEPT collectives the processes participating in the CONNECT / ACCEPT exchange meta directory information. Then one process on each side sends that information (the identity of the other sides meta directory processes) to its 'own' meta directory. When the meta directories each add the new peer to their list of other meta directory processes they make the entire MPI application on the other side of the CONNECT / ACCEPT operation part of the proctable.

CONNECT and ACCEPT can be called within an existing job. If this happens a new connection may be established within the set of meta directory processes but the underlying proctable remains essentially unchanged. Thinking of this as a graph operation, a new pair of vectors are added to an already connected graph but the total set of connected verticies doesn't change.

JOIN operations are almost identical to CONNECT / ACCEPT in terms of operations on the proctable.

3.5   Reading the Proctable

The program will expose one or more global loader symbols that TV will use to identify the location in memory to look at to find the information exposed by this interface.

The data will be stored in a structured way that will not depend on the program providing type information to the debugger. It can become complex for MPI vendors and users to handle the MPI library itself in such a way that the type information is preserved.

One example of an encoding that would meet the above requirement would be a simple pointer to a null terminated string. All the required information could be encoded into this string. Slightly more complex structures of pointers, integers, and strings that have better properties for efficient maintenance will likely be used.

3.6   Synchronization Between the MPI and the Debugger

During startup the processes will wait for the debugger before proceeding. This can be done using a gate variable in INIT that the debugger has to attach to trigger, or by having a barrier in INIT that the processes all need to reach before running past INIT , or using other mechanisms that achieve the same result.

Synchronization should occur at SPAWN calls if and only if the debugger is attached to the root process of the dynamic process collective. Similar synchronization should occur with CONNECT / ACCEPT , in this case however the newly related MPI processes should be held in the remote collective call, again if and only if the debugger is attached to the root process of the local collective operation.

The MPI library will declare and may check but not set a process level global variable that the debugger can set to notify the MPI process that it is being debugged.

The MPI library will maintain, on a per process level, a variable that the debugger can check to see if the proctable data-structures are being modified.

When the process needs to notify the debugger that an event has occurred it will call an agreed upon stub function. If the debugger wishes to know that this function has been called it can put a hidden breakpoint at that location. This notification will occur on one meta directory when the proctable has changed. It will occur on the root process of a dynamic process collective to notify TotalView that new processes are available and are being held so that the debugger can attach.

4   Naming Scheme for MPI 2 Proccesses in External Tools

A parallel debugger needs a way to identify the many processes being debugged to the user. Each MPI process has a handle to just one MPI_COMM_WORLD, within that MPI_COMM_WORLD each MPI process has a well defind, unique rank. In order to fully and unambiguously specify the process the user needs to have both this rank and a clear way to refer to the MPI_COMM_WORLD. For scripting and comparing the behavior or the program from one run to the next the name that the debugger gives to each MPI_COMM_WORLD should not depend on factors outside the control of the program. In section 3.2 we specified that the proctable will retain information about the MPI process that acted as the root for a newly spawned MPI_COMM_WORLD. This information can be used to construct a name for that new MPI_COMM_WORLD that is unique and descriptive of the specific sequence of SPAWN operations that lead to its creation.

5 Conclusion

The interface discussed here should be useful to any MPI library implementing MPI 2 dynamic processes and any tool designed to work with programs taking advantage of MPI 2 dynamic process features. We will be working first to proto-type the MPI library side of the interface in both Open MPI[4] and MPICH 2[5]. At the same time the parallel debugger side of the interface will be prototyped in the TotalView parallel debugger. The design will then be documented based on the experiences and lessons learned in the course of these initial prototyping efforts. This is intended to be an open interface and we welcome input from other MPI and tool developers.

References

1. Gropp, W., Lusk, E.: MPICH. http://www.mcs.anl.gov/mpi/mpich1/ (2006)
2. Cownie, J., Gropp, W.: A standard interface for debugger access to message queue information in MPI. In: Proceedings, 6th European PVM/MPI Users' Group Meeting. (1999) 51–58
3. Geist, A., Gropp, W., Huss-Lederman, S., Lumsdaine, A., Lusk, E., Saphir, W., Skjellum, T., Snir, M.: MPI-2: Extending the Message-Passing Interface. In: Euro-Par '96 Parallel Processing, Springer Verlag (1996) 128–135
4. Gabriel, E., Fagg, G.E., Bosilca, G., Angskun, T., Dongarra, J.J., Squyres, J.M., Sahay, V., Kambadur, P., Barrett, B., Lumsdaine, A., Castain, R.H., Daniel, D.J., Graham, R.L., Woodall, T.S.: Open MPI: Goals, concept, and design of a next generation MPI implementation. In: Proceedings, 11th European PVM/MPI Users' Group Meeting, Budapest, Hungary (2004) 97–104
5. Gropp, W., Lusk, E.: MPICH2. http://www.mcs.anl.gov/mpi/mpich2/ (2006)

Modeling and Verification of MPI Based Distributed Software

Igor Grudenic and Nikola Bogunovic

Faculty of Electrical Engineering and Computing, University of Zagreb, Unska 3
10000 Zagreb, Croatia
{igor.grudenic, nikola.bogunovic}@fer.hr

Abstract. Communication between processes in a distributed environment is implemented using either shared memory or message passing paradigm. The message passing paradigm is used more often due to the lesser hardware requirements. MPI is a standardized message passing API with several independent implementations. Specification and verification of distributed systems is generally a challenging task. In this paper we present a case study of specification and verification of MPI based software using abstract state machines (ASMs).

Keywords: MPI software modeling, MPI software verification, abstract state machines (ASMs).

1 Introduction

Modern scientific applications perform large number of computations in order to simulate complex natural processes or to analyze dependencies among huge data sets. Such a complex calculation can usually be efficiently split into several more or less independent calculations that can be executed in parallel.

In order to realize distributed computation scientific community relies on two most popular message passing frameworks: parallel virtual machine (PVM) [1] and message passing interface (MPI) [2]. Both frameworks support heterogeneous computing environment, synchronous and asynchronous communication and dynamic process creation. The main difference is the fact that PVM is both standard and implementation while MPI is an API definition opened for different implementations with richer set of communication primitives.

During the implementation of MPI based distributed systems some difficulties may arise as a result of the excessive synchronous or carelessly used asynchronous communications. Excessive use of synchronous communication in the complex process topologies may easily lead to numerous deadlock occurrences. The source of deadlock is usually hard to detect using plain debuggers because of deadlock's irregular occurrence pattern. Even though synchronous communication can easily lead to deadlock and reduces overall parallelism, it is very useful for coordination among processes.

B. Mohr et al. (Eds.): PVM/MPI 2006, LNCS 4192, pp. 123 – 132, 2006.

The use of asynchronous communication contributes to overall parallelism but can lead to a large number of possible process execution and communication sequences. Some of the sequences may not conform to the overall specification of the system and can be difficult to trace in the stochastic system execution.

In order to reason about complex systems featuring asynchronous communication it can be fruitful to model the system prior to its implementation. The system model should faithfully represent all the aspects of a given specification and provide a valuable insight into the complex system behavior. Also, the system model should be easily refined from the most general form to its specific implementation.

In this paper we use Abstract State Machine Language (AsmL) [3] in order to model distributed algorithm for mutual exclusion [4], which we implemented while using MPI. We also prove some of the properties of the algorithm on bounded number of states using Spec Explorer tool [5] against the AsmL model.

In Section 2 we quote previous work on modeling MPI based software. A brief description of ASMs and AsmL is given in Section 3. In Section 4 we show the modeling technique for representing distributed algorithms using AsmL and the relation of the model to the corresponding MPI implementation.

2   Previous Work

Modeling and verification of MPI based software has been practisized using the Promela language and the model checker SPIN [6]. Theorems were stated that could be used to avoid space explosions and still prove the absence of deadlock. This was accomplished by converting asynchronous communication to synchronous and then checking the synchronous communication for deadlock occurrences.

Virtual machine for the PVM system has been described in [7] using mathematical notion of ASMs. That was a pure theoretical approach to the specification of the virtual machine. The specification can be extended to specify the behavior of MPI communication because of the similarities in the two frameworks. In this paper we do not use the cited approach, but instead provide a method for specifying basic MPI communication using the AsmL language. The benefits of using AsmL lie in the existence of tools for executing the given models, and the possibilities of conformance testing using these tools.

3   Abstract State Machines

Abstract state machines [8], also known as Evolving Algebras (EA) are introduced in order to make system specification and specification refinement easier to deal with. They can be used to faithfully describe sequential [9] and parallel algorithms [10] at their natural level of abstraction, because for every such an algorithm there exists an ASM which represents it up to the elementary step.

The state S of an ASM M is a static algebra consisting of the set X (superuniverse of S), the signature Σ containing all the function names, and the interpretations of all the function names $f \in \Sigma$ where $f:X^r \to X$ (r represents the function arity). When representing computer systems all states of the ASM M share the same superuniverse and the same set of function names, while the function interpretations change in different states. The evolution of one state to another is defined by using a simple transition rule *"If Condition then Update"*. *Update* denotes a finite set of assignments of the form $f(t_1,..,t_n):=t$, where f is one of the function names in Σ with arguments $t_1,...,t_n$.

Basic ASM is defined by its initial state and the set of transition rules. There are numerous extensions to this definition suitable for modeling of different types of systems that are not further explicated here.

AsmL is the language that simplifies modeling using ASMs because it offers a richer set of constructs than the simple transition rule set, but all of these constructs eventually compile to the transition rule set. Although the elements of the AsmL language are intuitive and self explanatory, a short description of some of them is given in the rest of the paper where needed.

4 Modeling MPI Processes, a Case Study

In order to partly specify the behavior of MPI programs we present a case study in which we have designed AsmL model of an algorithm for mutual exclusion, which is afterwards implemented using MPI.

The algorithm is described in Subsection 4.1, while the procedure for modeling MPI processes is described in Subsection 4.2. In subsection 4.3 we explicate the transformation of the mutual exclusion algorithm to the AsmL model and show the guidelines for the refinement of this model towards its MPI implementation. We have verified some of the key properties of the AsmL model, as shown in subsection 4.4.

4.1 Distributed Algorithm for Mutual Exclusion

Distributed algorithm for mutual exclusion [4] is designed to order the events in distributed systems. A distributed system includes processes that communicate using asynchronous message passing. An event is characterized as either sending or receiving a message.

The goal of the algorithm is to provide exclusive access (safety property) to a resource for a given number of processes. If every process that is granted the resource eventually releases it, then every request for a resource is eventually granted (liveness property). Each process executing the algorithm maintains its local logical clock and a request queue. The value of the local clock is increased after every event and sent as a timestamp (*Tm*) within every message. The request queue is used to store all the requests for the resource that arrive from other processes. The algorithm for mutual exclusion is given by the following rules:

R1) In order to request resource, the process p_i sends the message $M=(Tm,p_i,request)$ to every other process and to its request queue.

R2) After the receipt of the $M=(Tm,p_i,request)$ the process p_j puts the message into its request queue and sends timestamped acknowledgement to p_i.

R3) When releasing the resource, the process p_i removes the message $M=(Tm, p_i, request)$ from its request queue and sends the message $M=(Tn, p_i, release resource)$ to all processes.

R4) When the process p_j receives release resource message from the process p_i it removes all the requests of the form $M=(Tm,p_i,request)$ from its request queue.

R5) Resource is granted to the process p_i when there is a request $(Tm, p_i, request)$ in its request queue ordered before (by the relation $\rightarrow$) any other request in the same queue, and the process p_i received at least one message timestamped later then Tm from all the other processes.

4.2 Modeling of MPI Semantics Using AsmL

MPI based distributed system consists of initially undetermined number of processes. The number of communicating processes may dynamically change as a result of the implemented algorithm. In order to model such an environment using ASMs, each state of the resulting ASM must contain information about all processes and the values of each process's address space. Even more, each of the processes should be uniquely identified with its process number, according to the MPI specification.

In order to implement this semantics in AsmL language we define an AsmL *process class* that represents processes in the system. Methods of the AsmL *process class* are atomic actions that can be executed by the MPI process. Every instance of the AsmL *process class* represents one MPI process. Since instance creation is possible in any phase of the parallel algorithm, in this manner it is possible to model dynamic creation of processes.

Every atomic action of the process that is implemented using the class method can be restricted by a precondition in order to define acceptable sequences of actions. When modeling the class of processes that execute the following exemplary sequence of actions:

Add(), Subtract(), Add (), Subtract()

we produce the following AsmL code, where keyword **require** denotes the precondition of the action (method):

```
class process()
var state as Boolean=true
Add()                              Subtract()
require state=true                 require state=false
  state:=false                       state:=true
```

This AsmL model represents a set of processes in which every process is allowed to execute one action (AsmL method) at a time.

Modeling interprocess communication with AsmL can be done by modeling a separate software entity that conforms to the MPI specification and mediates all of the communication. Alternative approach is to enforce MPI communication policies directly into the processes and their methods.

In this paper we use the latter approach because detailed modeling of the separate component would increase the complexity of the overall system for both implementation and execution analysis.

When modeling MPI function calls (sends and receives) of any process we create a method in the AsmL *process class* for each call. Execution sequence containing the given MPI function call is guided using method preconditions. We do not distinguish between different types of intermediate buffers (user and system buffer) because from the system state exploration point of view these two are the same. In order to provide message buffering we introduce an AsmL sequence named *messageQueue* which is used to buffer all the messages in the order in which they arrive. Every process that receives data contains *messageQueue* through which it receives all its messages. Each message is represented with an AsmL structure which contains information that identifies the sender, the tag and the useful data itself.

The representation of blocking buffered sends (MPI_SEND and MPI_BSEND) is modeled by the following AsmL method:

```
BufferedSend()
  require state=DESIRED_STATE
  receiver.messageQueue+=[Message(myId,Tag,data)]
  state:=MESSAGE RECEIVED
```

where *DESIRED_STATE* denotes the state of the system which is followed by the modeled buffered send.

The non-blocking sends can be represented with the same AsmL code because the only difference from the blocking sends is that other actions can be performed while the message is being copied from the application buffer to the intermediate buffer. This would imply that the message is certainly residing in the intermediate buffer only after the communication is successfully tested (usually with MPI_WAIT) for completion. In order to skip modeling of both unblocking send and testing for completion we can, without any loss in later simulation, assume that message is copied to the buffer when MPI_WAIT is called. In that way we must only model MPI_WAIT call as a buffered blocking send and skip the modeling of the non-blocking send call (MPI_ISEND).

In order to represent MPI_SSEND and MPI_RSEND blocking sends one must assure that the matching receives are posted before the sending process can continue its execution. This can be done by adding a special precondition to the AsmL method implementing these sends.

The AsmL representation of MPI receives is done by introducing two methods for both blocking and non-blocking receives. The first method is used to announce that there exists a posted receive, while the second method is used to actually receive data from the intermediate buffer. The announcement of the message receive is needed in order to trigger the execution of MPI_SSEND and MPI_RSEND.

All the receive message announces are stored into a set of announces (one set for each process), and the AsmL methods that implement this behavior are:

```
AnnounceMessageReceive()
  require state=DESIRED STATE
  me.announces+={Announce()}
  state:=DESIRED STATE + RECEIVE ANNOUNCED

ReceiveMessage()
  require messageQueue contains message
  (1)require STATE IN WHICH MPI WAIT IS POSTED
  (2)require DESIRED STATE + RECEIVE ANNOUNCED
  applicationBuffer=getMessage(messageQueue)
  state:=MESSAGE RECEIVED
```

After the message receive is announced, and if it is the case of blocking receive, then the only method that can be called is *ReceiveMessage* (other methods have preconditions which should restrict it from execution). The *ReceiveMessage* method for the blocking receive doesn't contain the precondition having the *(1)* label, while the same method for non-blocking receive disregards the precondition labeled *(2)* in the above code segment.

4.3 AsmL Model of the Distributed Algorithm for Mutual Exclusion

When representing distributed algorithms it is very rewarding to pick the most abstract description of states and transitions between them, and then refine it with more detailed specification. We decided to pick three basic states for each of the MPI processes. The states and possible transitions between them for one of the processes are presented in Fig. 1.

When the process is in the state *having* then it is granted the resource. Only one process can be in the state *having* at any time. The state *requesting* denotes that a process is not having the resource but it has requested it by sending appropriate messages to the other processes. After releasing the resource, but before placing new requests to obtain it again, every process may perform some calculations. These calculations should not involve the resource, and the state of the process in which these calculations may be performed is labeled *sleeping*.

We implemented every process and its internal state, which may be one of the values {*having, requesting, sleeping*}, by the following AsmL constructs:

```
enum tState
   sleeping requesting having
class process()
   var state as tState
   var clock, nOfReceivedAck as Integer
```

where every process has an internal variable *state* which may take any value from the set of the enumerated type *tState*. Besides storing the state, every process must be extended with variable that keeps tracking of the local time (local clock function). In order to implement the rule *R5)* every process should track the last timestamp it received from all other process. Since that information would increase the number of states in the whole system we introduce the counter of acknowledge messages. When a process receives acknowledge message from all the other processes than it is obvious by the rule *R2)* that it received at least one message from everyone (at least that acknowledge) that is stamped later than its request.

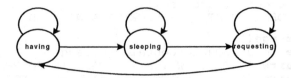

Fig. 1. Basic states of the process

There are two additional data structures that should be added to each process: one is a message queue for modeling intermediate MPI buffer, and the other is a request queue used by the mutual exclusion algorithm. We used the *messageQueue* AsmL sequence in order to represent intermediate buffer as described in subsection 4.2, and an AsmL set *requestQueue* to encode the request queue. The request queue may be described as an AsmL sequence instead of a set, but that would impose additional restriction on the algorithm in which no distinction among different request orderings is made. Both *messageQueue* and *requestQueue* are added to the process class:

```
class process()
    var messageQueue as Seq of sMessage
    var requestQueue as Set of sMessage
```

The elements of the *messageQueue* and the *requestQueue* are structures of the type *sMessage* that are used to describe messages exchanged by processes. The type *sMessage* is defined (using AsmL) as follows:

```
structure sMessage              enum tMessageType
    var clock as Integer            requestResource
    var sendingProcess as Integer   releaseResource
    var messageType as tMessageType acknowledge
```

where *clock* is used as a timestamp, *sendingProcess* indicates the sender of the message, and *messageType* determines the type of the message. There are three message types being passed by the distributed system that are identified in the enumerated type *tMessageType*. In order for every process to obtain the resource, it must send a *requestResource (REQ)* message to every other process as defined by the algorithm (rule *R1*). Upon the receipt of the *requestResource* message every process must respond with an *acknowledge (ACK)* message (rule *R2*). When the process is granted the resource and is about to release it, it notifies other processes with a *releaseResource (REL)* message (rule *R3*).

The description of distributed system implementing mutual exclusion algorithm is given by determining process execution. The execution is defined as a sequence of actions (implemented by AsmL class methods) that are guarded by preconditions. In this paper we describe AsmL methods *sendRequestResource* and *acceptMessage*. It is important to note that both methods employ only asynchronous MPI communication.

AsmL implementation of the actions sendRequestResource and acceptMessage

```
sendRequestResource()
 require state=tState.sleeping
 state:=tState.requesting
 step foreach iProcess in processes where iProcess<>me
    iProcess.messageQueue+=[NewRequest]
 requestQueue+=[NewRequest]
 clock+=1,nOfReceivedAck:=0

acceptMessage()
 require Size(messageQueue)>0 and state=tState.requesting
 messageQueue:=Tail(messageQueue)
 clock:=max(messageQueue(0).clock+1,clock+1)
```

```
match messageQueue(0).messageType
  REQ: requestQueue+={messageQueue(0)}
       sender.messageQueue+=[NewAcknowledge]
  REL: requestQueue-=request(sender)
  ACK: nOfReceivedAck+=1
```

When a process needs to enter its critical section, it executes the method *sendRequestResource*, which can be triggered only if the executing process is in the state *sleeping*. During the execution of the method new requests are placed in the message queues of all the other processes. The state and internal clock of the process are also modified. It can be observed that every message passing is not implemented by its own method (guarded by preconditions). In that way we decreased the number of methods in a class and improved readability of the model. Nevertheless, the given model can be easily extended to match the *one message - one AsmL method* translation described in the subsection 4.2. At the end of the method the number of received acknowledges is initialized.

The method *acceptMessage* is allowed to execute when intermediate buffer *messageQueue* contains at least one message. The execution of the method depends on the type of the message and is determined by the rules *R2)* and *R4)*. Regardless of the message type, the message is deleted from the intermediate buffer and the local clock of the process is updated. If the *requestResource* message is in the intermediate buffer, then it is moved to the *requestQueue*. When *acknowledge* message is received, the acknowledge counter (*nOfReceivedAck*) is incremented. Servicing the *releaseResource* message is accomplished by removal of the appropriate request from the *requestQueue*.

The given AsmL specification of the algorithm can be easily refined to the MPI implementation. The method *sendRequestResourceMPI* is a straightforward refinement of the method *sendRequestResource*.

```
void sendRequestResource()
  assert !(state==tState.sleeping);
  state=REQUESTING;
  for( iProcess=1; iProcess<=nProcesses; iProcess++)
   if(!iProcess=myRank)
     MPI_BSend(request,1,Sizeof(request),iProcess,
     TAG,MPI COMM WORLD,waitObject);
  AddToQueue(NewRequest,requestQueue);
  clock++; nOfReceivedAck=0;
```

The use of the *messageQueue* is replaced by the buffered send *MPI_BSEND* and a new function *AddToQueue* is introduced since the queue operations must be explicitly implemented. Instead of method precondition, an assertion is used in order to prevent execution of the method if the process is not in the appropriate state.

4.4 Verification of Algorithm Properties

Verification of the mutual exclusion algorithm properties is performed on the labeled transition system (LTS) which is generated from the AsmL model using Spec Explorer tool. LTS is an ordered quadruple (s_0, S, L, R) where s_0 is the initial state of the system, S is a set of all the states in the system, L is a set of actions and R is a proper subset of $S \times L \times S$. Every triple (s_1, a, s_2) denotes that there exists a state s_1 in which action a can be executed, and execution of the action a in state s_1 transforms this state to the state s_2.

The algorithm for mutual exclusion should satisfy safety and liveness properties given in the subsection 4.1.

A full verification of the given mutual exclusion algorithm is not possible using state exploration tools because the number of the states in the system is infinite. Infiniteness is caused by the constant increasing of the local clocks in the processes. In order to perform the verification we set an upper limit on the clock value and disregard all the states in which clock of any of the processes exceeds the given limit.

When checking the safety property in the Spec Explorer tool we defined the following stopping condition for state exploration in order to detect safety violation:

```
exists pi in processes where pi.state=having and
    (exists pj in processes where pi<>pj and
    pj.state=having)
```

After the breadth-first exploration of 40000 states in the system of 4 processes, the safety violation was not detected. In Fig. 2 the result of state space exploration is given for the system of two processes explored up with clocks limited to 9. The states are grouped by internal state of each of the processes. It can be observed that there is no explored state in which more that one process is granted the resource.

Checking the liveness property is usually accomplished by specification of the property in temporal logic such as CTL, and verifying the CTL formula using a model checker. The Spec Explorer tool doesn't support model checking of CTL properties, but

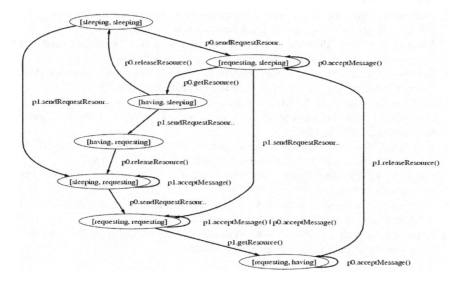

Fig. 2. Space exploration result for two processes grouped by the internal state of the process

the state groupings (Fig. 2) shows that the liveness property is satisfied. This is due to the fact that every state is reachable from any state in the finite number of steps.

5 Conclusion and Future Work

In this paper we depicted a method for modeling basic MPI constructs using AsmL language, in order to elicit convincing arguments for analyzing complex systems that way. Some pointers for refinement of the model towards the implementation are given as well. At the end of the paper we illustrated verification of safety and liveness properties using regular state space exploration tool Spec Explorer.

In our case study we have focused on the asynchronous distributed mutual exclusion algorithm to confirm the feasibility of the applied formal method. There exists a procedure for modeling MPI software using Promela language that enables verification of complex CTL specifications. We utilized a different approach, because the refinement of ASM specification is easier and the AsmL models are much simpler to work with. In the future we plan to focus on "dirtier" parts of the MPI system. We would like to provide a complete ASM description of the expected behavior of the MPI communication subsystem that would make possible conformance testing of various MPI implementations.

References

1. Sunderam, V. S.: PVM: A Framework for Parallel Distributed Computing.   In: Concurrency, Practice and Experience, December, 1990. Vol. 2,   John Wiley & Sons (1990) 315-339.
2. "MPI-2 Standard", http://www.mpi-forum.org/docs/mpi-20-html/mpi2-report.html
3. Microsoft Research: AsmL: The Abstract State Machine Language, 2002
4. Lamport, L.: Time, clocks, and the ordering of events in a distributed system. In: Communications of the ACM archive, Vol. 21 , Issue 7 (1978) 558-565
5. Campbell, C., Grieskamp, W., Nachmanson, L. et al.: Model-Based Testing of Object-Oriented Reactive Systems with Spec Explorer, Technical Report (2005)
6. Siegel, S.F., Avrunin, G. S.: Modeling wildcard-free MPI programs for verification. In: Proceedings of the tenth ACM SIGPLAN symposium on Principles and practice of parallel programming (2005) 95-106
7. Börger, E., Glässer, U.:A formal specification of the PVM architecture. In: IFIP 13th World Computer Congress, Vol. 1:, North-Holland, Amsterdam (2005) 402-409
8. Gurevich, Y.: Evolving Algebras 1993: Lipari Guide. In: Specification and Validation Methods, Oxford University Press (1993) 9-36
9. Gurevich, Y.: Sequential Abstract State Machines Capture Sequential Algorithms, In: ACM Transactions on Computational Logic (2000) 77-111
10. Blass, A., Gurevich, Y.:Abstract State Machines Capture Parallel Algorithms, In: ACM Transactions on Computational Logic (2003) 578-651

FT-MPI, Fault-Tolerant Metacomputing and Generic Name Services: A Case Study

David Dewolfs, Jan Broeckhove, Vaidy Sunderam, and Graham E. Fagg

Depts. of Math and Computer Science of the University of Antwerp (Antwerp, Belgium), Emory University (Atlanta, GA, USA) and the University of Tennessee (Knoxville, TN, USA)
{David.Dewolfs, Jan.Broeckhove}@ua.ac.be,
vss@mathcs.emory.edu, fagg@cs.utk.edu

Abstract. There is a growing interest in deploying MPI over very large numbers of heterogenous, geographically distributed resources. FT-MPI provides the fault-tolerance necessary at this scale, but presents some issues when crossing multiple administrative domains. Using the H2O metacomputing framework, we add cross-administrative domain inter-operability and "pluggability" to FT-MPI. The latter feature allows us, using proxies, to transparently replace one vulnerable module - its name service - with fault-tolerant replacements. We present an algorithm for improving performance of operations over the proxies. We evaluate its performance in a comparison using the original name service, OpenL-DAP and current Emory research project HDNS.

Keywords: FT-MPI, H2O, metacomputing, fault-tolerance, hetero-geneity.

1 Introduction

Over the course of the last ten years, clusters running some implementation of MPI have become some of the most popular supercomputing platforms. Recently, there has been a growing interest in clustering resources that feature extensive geographical distribution across multiple Administrative Domains (ADs). This raises the issue of fault-tolerance. FT-MPI [7] differs from other solutions to the fault-tolerance problem [3,4,6,10,5], in that it allows the application itself to re-store it's own state, instead of relying on automated - but potentially unscalable - solutions like global distributed checkpointing. This makes it an interesting so-lution for highly geographically distributed, heterogenous resources with a need for customized, lightweight recovery mechanisms.

However, FT-MPI is currently confined to single ADs. Also, bottlenecks and potential single points of failure (SPoFs) become an issue when deploying it over slower AD interconnects. One of the critical modules is the FT-MPI name service (NS). We have previously addressed these points [2,1] by developing a proxy-based solution which allows FT-MPI administrators to use any NS of their own choice (including any fault tolerance features available with it). Further, we use features of the H2O metacomputing framework [8] to span multiple ADs without the need for individual accounts on each system.

B. Mohr et al. (Eds.): PVM/MPI 2006, LNCS 4192, pp. 133–140, 2006.

In this paper, we focus on improving performance of operations over the proxies. We demonstrate the ability of our approach to transparently and scalably
switch between different NSs. We will also present performance test data for the
improved algorithms using different backend NSs.

2   Design Overview

2.1   Basic FT-MPI Architecture

A running FT-MPI virtual machine (VM) deploys one FT-MPI runtime per
node and a number of daemons to assist it in setting up and managing jobs: a
startup-daemon on each node (semi-critical), one or more *notifier daemons* (non
critical), and a single *naming daemon* (figure 1).

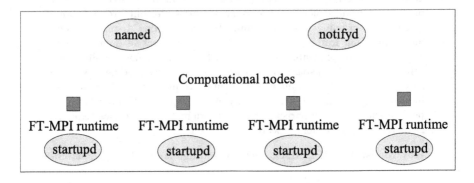

Fig. 1. a typical running FT-MPI system

Each VM needs exactly one naming daemon (however, a single NS instance
can manage multiple VMs). It provides a custom NS and serves a crucial role
in VM buildup, job startup and job recovery. Specifically, the FT-MPI runtime
uses it to keep records on VM and job membership. To ensure data consistency,
editing of records for job and task state in the NS is done by the FT-MPI runtime
of single *leader node*. The leader edits these records during the error recovery
phase to clean up job and task state. FT-MPI runtimes on other nodes are then
notified of the changes through a system of callbacks. Leaders are *elected* through
a custom call in the NS.

We note the following issues with the daemon in the currently available version
of FT-MPI: 1) it constitutes a potential SPoF, as it is highly state-retaining and
critically important for the general functioning of the VM, 2) it is also a possible
choke-point when communicating over slow AD interconnects (this issue was
recently addressed [13] and an adapted recovery algorithm should be added to
future releases of FT-MPI) and 3) it does not support features like replication
and load balancing, which would be desirable to improve scalability at very large
VM sizes. We note that many generic name servers currently available do offer
these features.

2.2   Extensions to the FT-MPI Architecture

We use proxies to bridge between the custom FT-MPI NS protocol an any generic NS, enable an operator of an FT-MPI VM to use a NS of his own choice :

- instead of directly contacting the NS, components of FT-MPI contact a proxy which resides on the gateway between the single AD and the "outside world"; this proxy acts as a "*front-end*" to the real NS, translating FT-MPI protocol calls to a format that is understood by the real, "*back-end*" name service; the front-end does not retain internal state - thus, failures can be handled through simple measures like a trivial replication scheme or a restart
- all nodes on a single AD retain an open connection to the NS front-end for that AD, and each NS front-end retains a single connection with the NS back-end (hierarchical message forwarding)
- the NS front-end is implemented as a H2O "pluglet" making it fully remotely deployable by operators on any machine that runs an H2O kernel

The setup is best illustrated by the example in figure 2.

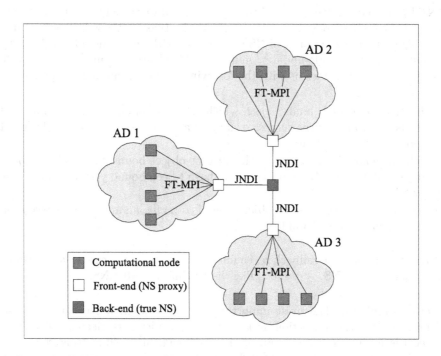

Fig. 2. An FT-MPI VM using proxies and a generic back-end NS

This approach allows FT-MPI to use one of a wide range of "off the shelf" NSs available. Many of these provide important fault-tolerance and performance features lacking from the current FT-MPI NS (load distribution, replication, checkpointing etc.).

The proxies are implemented in Java. This allowed us to use JNDI, the Java Naming and Directory Interface, which provides uniform access to a diverse set of NSs, ranging from LDAP to DNS. Any provider can make a NS "JNDI-enabled" by implementing a Service Provider Interface (SPI). All interaction with the NS is fully transparent to the user. Thus, using JNDI allows us to make access to the backend generic w.r.t. different NSs.

2.3 Concurrency, Atomicity and JNDI

FT-MPI assumes a centralized, single-threaded NS which queues all incoming requests on receive. A number of its calls resolve compound operations like increment, compare and set etc. in a single atomic call. However, 1) the new design we propose has front-ends running in parallel and accessing the back-end concurrently and 2) certain single (atomic) calls in the NS have to be resolved through multiple primitives in JNDI, requiring separate lookups and subsequent binds. This introduces the possibility for concurrency problems, e.g. race conditions.

JNDI primitives. We previously discussed a solution through the use of remote unreliable locks, composed from basic JNDI primitives[1]. We will show that it is possible to handle a majority of NS interactions without the use of said locks by exploiting the NSs single-update / multiple-callback architecture. To accomplish our goals, JNDI provides us with the following (relevant) atomic primitives:

- bind(*name,object*): binds *object*, which can contain an arbitrary number of fields to *name*; returns success or failure; appropriate exception is thrown if *name* is already bound
- rebind(*name,object*): replaces the current object bound to *name* by *object*, or acts identical to bind in case *name* hasn't been bound yet; returns success or failure
- lookup(*name*): returns the object bound to *name*; an appropriate exception is thrown in case of problems

JNDI also supports a callback mechanism, enabling us to register and "listen" for updates to the NS, very much like the original FT-MPI NS.

Leader election. The most important part of custom functionality to implement is the leader election system. Once a leader gets elected, all editing of records for job and task state is done through him, eliminating the problem of concurrency. The FT-MPI NS implements leader election as a "grab the token" type of contest. The NS provides a custom call of the general form *swap(token,old_leader,contender)* which swaps the ownership of *token* from *old_leader* to *contender* if the current owner of *token* is *old_leader*. In other words: the first contender node to get its message handled by the NS gets to swap ownership of the token (and become leader) whilst a failure message is returned to the others on all subsequent messages.

An adapted election algorithm. Given the primitives available to us, we perform leader election by implementing "grab a token" as "bind a token". For each token which is swapped during the lifetime of the VM, an object is stored in the NS with an *election_ count* keeping track of how many swaps have already been performed on it. This token is read during the initialization phase of the proxy and the counter is locally cached for later use. When an election takes place, all contender nodes send the appropriate message to their respective proxies and the following sequence of actions is performed:

1. the proxies each increase the cached leader counter for *token* by one, once - for all contenders who share the same proxy, the contest is resolved locally at the proxy
2. each proxy, for its respective local winner, attempts to bind an object under the name "<*token*>_ <counter>" - the node for which the bind succeeds is the *winner node*, all others are *loser nodes*
3. the proxy acting for the winner node rebinds *token* with the new ownership data (triggering a callback) - the winner token becomes the leader and the outcome is relayed back to the new leader node - meanwhile, the proxies handling the calls for the respective loser nodes wait for a callback on a rebind for *token*, eventually relaying the outcome to their respective loser nodes as normal
4. if something goes wrong during the winner' actions in step 3 (non-atomic), this means something is wrong with the proxy, the gateway on which it resides, or its network connection; all of these will get nodes in their respective AD into trouble and register with the FT-MPI runtime as an error - the FT-MPI runtime will then recommence the recovery procedure (including a potential new leader election) as normal

This leaves us only with the problem of compound operations: what if something goes wrong with the leader in the middle of a compound operation? JNDI only allows for atomic operations on a single object at a time. This would lead to inconsistencies in the backend. We deal with this problem by using a single state object which contains pointers to all objects involved in the compound operation. We do not directly write to the objects themselves, but to a copy, keeping the old state intact until all actions in the compound operation have been performed. When ready, a rebind of the index record turns everything over to the new state within a single operation. This may leave spurious objects in the NS, but these can easily be cleaned up by an independent garbage removal process.

Results. Advantages of this approach are 1) the ability to drastically reduce dependence on remote locks, enhancing performance by reducing the amount of JNDI calls that would normally be needed, and 2) the ability to do partial local resolution of the leader election process at the proxy, bringing down the amount of effective calls going out to the back end NS. This reduces the potential for choke-points on connections between different ADs and helps spreading load for very large, geographically dispersed VMs. Also, the number of callbacks is similarly reduced to one per proxy instead of one per node.

3   Evaluation

Setup and Experiments. To demonstrate the ability of our setup to transparently switch between multiple NS backends,we performed a comparative experiment on two nodes: one in Atlanta (Georgia), USA, the other situated in Antwerp, Belgium. The node in Atlanta is a 4 CPU 2.8 GHz Pentium 4 with 1GB memory running Mandriva Linux 2006. The node in Antwerp is a 1.90GHz Pentium 4 with 256MB memory running Suse Linux 7. This setup was used in order to simulate the conditions which the design is aimed at: geographically distributed, heterogenous resources. The node in Atlanta ran the original FT-MPI NS, OpenLDAP or HDNS depending on the test case. The node in Antwerp ran a basic client program in both cases, plus the front-end in the case of the new design.

We ran a number of performance tests comparing the original NS with two alternatives: the LDAP-based OpenLDAP using the Berkeley DB, and HDNS [12]. HDNS is a naming service initially developed for the Harness Project[11]. While developing the SPI, a completely new version of HDNS has been designed and implemented. Both of the NSs tested support distribution and a number of features like fault-tolerance and persistency, which are not available in the original FT-MPI NS.

The following experiments were performed to evaluate scalability in terms of transaction size and frequency: 1) insert and read back entries with progressively growing payloads (10-900 B, using 100 B steps from 100 to 900 B) and 2) insert

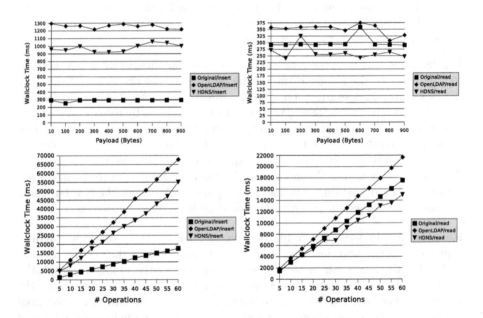

Fig. 3. Evolution of wall-clock time with increasing payload and # operations(read/write)

and read batches with a progressively growing number of equal-sized entries into the NS - measure wall-clock time for both cases. Ultimately, we want the new NS to be as scalable and stable as the original. We tested the performance of insertion and deletion without locks, allowed by the leader election mechanism described above.

Results. We note that all experiments successfully ran to conclusion and left the back-end in a consistent state. From a practical point of view, we noticed that changing between OpenLDAP and HDNS was very easily accomplished. None of these experiments required any kind of code change or recompile of either the original FT-MPI code, or the Java-code for the proxies. A few changes to a configuration file and command-line parameters suffice to change NS back-ends from one experiment to another.

Looking at the figures, we conclude that the ability to do insertion without locking (though still less efficient than the original NS) provides us with a notable performance improvement over previous experiments in which we did use locking [1], the performance gain consistently being around 40%. It should prove interesting to do further research on improving performance of compound insertion operations, bringing figures even closer to those of the original NS. We also note that HDNS performs rather well as a backbone, outperforming OpenLDAP on both insert and read operations in both experiments. On read operations it even succeeds at slightly outperforming the original NS. We are currently investigating possible reasons for this remarkable behavior. Further, both graphs show linear growth on both insert and read for both experiments, proving that our design remains scalable and stable.

4   Conclusions

In this paper, we have discussed issues concerning the deployment of FT-MPI for large scale computations on highly geographically distributed, heterogenous resources. We have shown that "vanilla" FT-MPI poses some limitations in this area due to the nature of its naming service. We have presented a design, leveraging JNDI, which address these issues by enabling operators of an FT-MPI setup to "plug in" their own name services. This feature is highly desirable as existing "off the shelf" name services often do provide numerous features for improved fault tolerance and performance.

We have discussed an algorithm which allows us to implement a leader election system without locking, and note that it is possible to minimize the amount of locking in general. This results in a significant performance gain over previous implementations, both in terms of the amount of JNDI primitives needed and the amount of data transferred over connections between multiple administrative domains. We have presented experimental results which 1) confirm the efficacy of this approach, as well as 2) show the effective ability to transparently change between different back-ends, as demonstrated by our use of both LDAP and HDNS back-ends without significant changes.

References

1. D. Dewolfs, D. Kurzyniec, V. Sunderam, J. Broeckhove, T. Dhaene, G. E. Fagg. Applicability of Generic Naming Services and Fault Tolerant Metacomputing with FT-MPI. In *Proceedings of the 12th European Parallel Virtual Machine and Message Passing Interface - Euro PVM/MPI (Springer-Verlag Berlin LNCS 3666)*, Sorrento (Naples), Italy, 2005
2. D. Kurzyniec and V. Sunderam. Combining FT-MPI with H2O: Fault-tolerant MPI across administrative boundaries. In *Proceedings of the HCW 2005-14th Heterogeneous Computing Workshop, 2005*
3. A. Agbaria, R. Friedman. Starfish: Fault-tolerant dynamic MPI programs on clusters of workstations. In *Eighth IEEE International Symposium on High Performance Distributed Computing*, 1999, pp. 31
4. A. Bouteiller, F. Cappello, T. Herault, G. Krawezik, P. Lemarinier and F. Magniette. MPICH-V2: a fault tolerant MPI for volatile nodes based on pessimistic sender based message logging. In *ACM/IEEE SC2003 Conference*, 2003, pp. 25
5. Y. Chen, K. Li, J.S. Plank. CLIP: A checkpointing tool for message-passing parallel programs. 1997. Available at http://citeseer,ist.psu.edu/chen97clip.html
6. E. Elnozahy and W. Zwaenepoel. Manetho: Transparent rollback-recovery with low overhead, limited rollback and fast output. In *IEEE Transactions on Computers, Special Issue on Fault-Tolerant Computing*, 41(5), May 1992, pp.526-531
7. G. Fagg, E. Gabriel, Z. Chen, T. Angskun, G. Bosilca, J. Pjesivac-Grbovic and J. Dongarra. Process fault-tolerance: Sematics, design and applications for high-performance computing. In *International Journal for High Performance Applications and Supercomputing*. 2004.
8. D. Kurzyniec, T. Wrzosek, D. Drzewiecki and V. Sunderam. Towards self-organising distributed computing frameworks: The H2O approach. In *Parallel Processing Letters*, 13(2), 2003, pp. 273-290
9. S. Louca, N. Neophytou, A. Lachanas and P. Eviripidou. MPI-FT: Portable fault-tolerance scheme for MPI. In *Parallel Processing Letters*, 10(4), 2000, pp. 371-382.
10. G. Stellner. CoCheck: Checkpointing and process migration for MPI. In *10th International Parallel Processing Symposium*, 1996, pp. 526-531
11. M. Migliardi and V. Sunderam. The Harness Metacomputing Framework. In *The Ninth SIAM Conference on Parallel Processing for Scientific Computing, S. Antonio*, 1999
12. D. Gorissen, P. Wendykier, D. Kurzyniec and V. Sunderam. Integrating Heterogeneous Information Services Using JNDI. In *Proceedings of the HCW 2006 - 15th Heterogeneous Computing Workshop*, Rhodes Island, Greece, April 2006
13. G. E. Fagg, T. Angskun, G. Bosilca, J. Pjesivac-Grbovic, J. Dongarra. Scalable Fault Tolerant MPI: Extending the Recovery Algorithm. In *Proceedings of the 12th European Parallel Virtual Machine and Message Passing Interface - Euro PVM/MPI (Springer-Verlag Berlin LNCS 3666)*, Sorrento (Naples), Italy, 2005, pp. 67

Scalable Fault Tolerant Protocol for Parallel Runtime Environments

Thara Angskun, Graham E. Fagg, George Bosilca,
Jelena Pješivac-Grbović, and Jack J. Dongarra

Dept. of Computer Science, 1122 Volunteer Blvd., Suite 413, The University of
Tennessee, Knoxville, TN 37996-3450, USA
{angskun, fagg, bosilca, pjesa, dongarra}@cs.utk.edu

Abstract. The number of processors embedded on high performance
computing platforms is growing daily to satisfy users desire for solving
larger and more complex problems. Parallel runtime environments have
to support and adapt to the underlying libraries and hardware which
require a high degree of scalability in dynamic environments. This paper
presents the design of a scalable and fault tolerant protocol for supporting parallel runtime environment communications. The protocol is
designed to support transmission of messages across multiple nodes with
in a self-healing topology to protect against recursive node and process
failures. A formal protocol verification has validated the protocol for
both the normal and failure cases. We have implemented multiple routing algorithms for the protocol and concluded that the variant rule-based
routing algorithm yields the best overall results for damaged and incomplete topologies .

1 Introduction

Recently, several high performance computing platforms have been installed with
more than 10,000 CPUs such as Blue-Gene/L at LLNL, BGW at IBM and
Columbia at NASA [5]. Unfortunately, as the number of components increases,
so does the probability of failure. To satisfy the dynamic requirement of such a
dynamic environment (where the available number of resources is fluctuating) a
scalable and fault-tolerance framework is needed. Many large-scale applications
are implemented on top of message passing systems for which the de-facto standard is the Message Passing Interface (MPI) [10]. MPI implementations require
support of parallel runtime environments, which are extensions of the active
operating system services, and provide necessary functionalities (such as naming resolution services) for both the message passing libraries and applications
themselves. However, currently available parallel runtime environments are either not scalable or inefficient in dynamic environments. The lack of scalable
and fault-tolerance parallel runtime environments motivates us to design and
implement such a system. A scalable and fault-tolerant communication protocol
that can be used as a basis for constructing higher level fault-tolerant parallel
runtime environment is described in this paper. The basic ability of the designed

B. Mohr et al. (Eds.): PVM/MPI 2006, LNCS 4192, pp. 141–149, 2006.

protocol is to transfer messages across multiple (multicast and broadcast rather than unicast) nodes efficiently, while protecting against recursive node or process failures.

The structure of this paper is as follows. The next section 2 discusses related work. Section 3 introduces the scalable and fault-tolerant protocol, while the section 4 presents the formal protocol verification. Experimental results are given in section 5, followed by conclusions and future work in section 6.

2   Related Work

Although there are several existing parallel runtime environments for different types of systems, they do not meet some of the major requirements for MPI implementations: scalability, portability and performance. Typically, distributed OS and single system image systems are not portable while the nature of Grid middle-wares has performance problems.

The MPICH implementation [8] uses a parallel runtime environment called Multi-purposed daemon (MPD) [3] for providing scalability and fault-tolerant through a ring topology for some operations and a tree topology for others. Runtime environments of other MPI implementations, such as Harness [1] of FT-MPI [6], Open RTE [4] of Open MPI [7] and LAM of LAM/MPI [2], do not currently provide both scalable and fault tolerance solutions for their internal communications.

The scalability and fault-tolerance issues have been addressed in several networking areas. However, those approaches could not be used or they are not efficient in the parallel runtime environments. Structured peer-to-peer networking based on distributed hash tables such as CAN [11], Chord [14], Pastry [13] and Tapestry [15] was designed for resource discovery. They are only optimized for unicast messages. Techniques used in sensor or large scale ad-hoc networking based on gossiping (or epidemic algorithm) [9] [12] mainly focus on information aggregation.

3   Scalable and Fault-Tolerant Protocol

The protocol in this paper is not aware of MPI implementation. It aims to support parallel runtime environments of various message passing implementations. However, currently work is in progress to integrate it in a fault-tolerance implementation of message passing interface called FT-MPI as well as in the modular MPI implementation called Open MPI.

The protocol is based on a k-ary sibling tree topology used to develop a self healing tree topology. The k-ary sibling tree topology is a k-ary tree, where k is number of fan-out ($k \geq 2$), and the nodes on the same level (same depth on the tree) are linked together using a ring topology. The tree is primary designed to allow scalability for broadcast and multicast operations that are typically required during MPI application startup, input redirection, control signals and termination. The ring is used to provide a well understood secondary path for

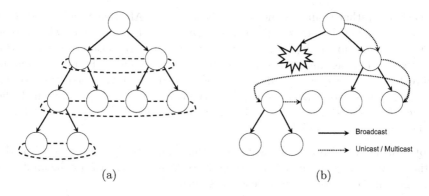

(a)   (b)

Fig. 1. (a) Binary sibling tree topology (b) Message rerouting in case of failure

transmission when the tree is damaged during failure conditions (simplest multi-path extension). In addition, typical k-ary tree only needs a single link or node failure to become bisectional, while the *k-ary* sibling tree can tolerate up to k failures. Fig. 1(a) illustrates an example of the binary (k=2) sibling tree. Each node needs to know the contact information of at most k+3 neighbors (i.e. parent, left, right and their children). The number of neighbors is kept to a minimum to reduce the state management load on each node. Both the tree and the ring topologies allow for neighbors addressing to be computed locally. Usually, we expect the k parameter to remain constant for the lifetime of the topology. The contact information of each node in some cases can be calculated locally for some tightly coupled systems or may be stored in an external directory service such as a name service of FT-MPI, a general purpose registry (GPR) of Open MPI or even a LDAP server for loosely coupled systems. The tree will automatically repair itself depending on an external recovery policy (i.e. when and how to repair it) specified by the user. The details of protocol is specified in section 3.1. The routing control of the protocol is discussed in section 3.2.

3.1 Protocol Specification

Service Specification: The goal of the protocol is to deliver messages across multiple nodes while protecting against different types of node and/or process failures. The protocol currently provides two kinds of message delivery service, which are broadcast (1 to n) and multicast (1 to m, where m $\leq$ n[1]). The broadcast service uses the k-ary tree to send messages in normal circumstance. It will use the neighbor nodes to reroute the messages in the failure cases as shown in Fig. 1(b). The multicast service treats the k-ary sibling tree as a graph. It uses best effort to deliver messages with the shortest path from a source to destinations in both normal and failure situations.

Environment Assumption: The protocol assumes that any failures are Fail-stop rather than Byzantine i.e. if a process or a node crashes, it should be

---

[1] A unicast message is a special case of multicast where m=1.

unreachable rather than pretend that it is still alive. After each failure, at least one neighbor of each node should be alive. Otherwise the k-ary tree will become bisectional, and no routing of messages between the two section of the tree will be possible. This assumption can be removed, if we allow each node to contact a directory service (considered as a stable resource) to overcome the orphan situation. The protocol also assumes that the transmission channel in which the protocol is executed can detect and recover from transmission errors (e.g. based on TCP and/or reliable UDP).

Protocol Vocabulary: There are 3 distinct kinds of messages: *hello* for the initialize message, which constructs the k-ary sibling tree; *mcast* for the multicast messages and *bcast* for the broadcast messages.

Message Format: The general message format of the protocol starts with a version number followed by a message type (i.e. the control fields *hello*, *mcast* and *bcast*). The *hello* message format consists of the above fields followed by an originator of the message indicated by *SrcID*. The *bcast* message format also contains *data* with the size DataSz. The *mcast* message consists of above mentioned fields followed by *#Dest*, *DestInd*, *DestList* and *TranList*. The *#Dest* is the number of destinations. The *DestInd* is an index, which points to the current destination in the *DestList*. The *TranList* is a transit list which contains the list of IDs of all the transit nodes in the tree to prevent looping and for back-tracking purposes.

Procedure Rules: The procedure rules can be separated into two steps: initialization and routing.

The initialization step of the procedure rules was described as follows: "Each node will register itself to the directory service (DS) and get its logical ID. It builds a logical topology and asks for the contact information of its neighbors from the DS. Once ready, it will start sending *hello* packet to its parent and its left neighbor. If the node is the right most in its level, it will also send *hello* to the left most node of the same level". After exchanging these *hello* messages, the communication channel between them will be established.

The procedure rules for routing a packet of the protocol were described as follows: "A node uses best effort to deliver messages following the shortest possible path. Sending a message procedure is dependent on the message type. If the message type is *bcast*, the node will send the message to all of its children. If a child died, it will reroute the message to all children of the child. This is done using an encapsulation technique. The node will encapsulate the broadcast message into a multicast message and send to its grandchildren. The grandchildren will decapsulate the multicast packet and continue to forward the broadcast message. However, if the message type is *mcast*, the next hop is chosen from a valid neighbor node which has the highest priority. [2] A node is said to be valid if and only if the node is not in the transit list and it is still alive. If there is no

---

[2] An implementation of the protocol may use a dynamic programming technique to improve performance by keeping the priority of neighbors for each destination in a look-up table.

possible next hop, the message will be sent back to the previous sender (i.e. back-tracking). When a node receives a message, it will first determine the header. If the message type is *hello*, it will do the initialization step. If the message type is *bcast*, it will forward to its children and handle node failure as mentioned above. If the message type is *mcast* and the node is not one of the destinations, it will add itself to the transit list and send it on to the next node. If the node is one of the destinations, but not the last one, it will remove itself from the destination list (*DestList*), decrease the destination count (*#Dest*), choose the next destination and update the destination index (*DestInd*), add itself to the transit list and send it to the next node."

Algorithm 1. Compute estimated cost

Procedure : Compute cost

1: cost $\Leftarrow$ 0 ; nextHop $\Leftarrow$ srcID
2: **while** nextHop $\neq$ destID **do**
3:    **if** myLevel $=$ destLevel **then**
4:       Choose left or right
5:    **else if** myLevel $>$ destLevel **then**
6:       nextHop $\Leftarrow$ myParentID
7:    **else**
8:       **if** $ChildID_i$ is an ancestor of destID **then**
9:          nextHop $\Leftarrow$ $ChildID_i$
10:      **else**
11:         Choose left or right, which one is closer to an ancestor of destID in myLevel
12:      **end if**
13:   **end if**
14:   cost $\Leftarrow$ cost $+1$
15: **end while**
16: return cost

Procedure : Choose left or right

1: **if** (hopLeft $\leq$ hopRight) $\wedge$ (destID $\neq$ myRightID) **then**
2:    nextHop $\Leftarrow$ myLeftID
3: **else**
4:    nextHop $\Leftarrow$ myRightID
5: **end if**

3.2   Routing Algorithm

This section discusses the routing technique used for multicast messages (which is also used by broadcast routing during failures). The goal of the routing algorithm is to find the shortest possible route in both normal and failure situations with only local knowledge stored at each node. The next hop is chosen from the highest priority node of its valid neighbors. The first algorithm (as shown in Algorithm 1) uses a rule based method to estimate a cost from the current node to the destination. The highest priority node is a neighbor which has the lowest cost (hop count). The rule is specified in such a way that a message will always go in a direction toward the destination. The second algorithm is a variant of the first algorithm, where it allows to go in a direction that does not directly route towards the destination if there is a shorter path to the destination from the current node. For example, instead of routing from left to right, it could be faster to go up a few levels, then go right and go down to the destination. The complexity of both algorithms is $O(\log_k n)$, where n is number of nodes and k is number of fan-outs. Routing with the shortest path may not be the best solution in a failure situation. The direction of the message may be changed too often such that the message is moving further from the destination. The third algorithm intends to prevent this situation by using knowledge of previously detected dead nodes from the header to compute the cost. The third method uses a graph-coloring technique of breath first search, which explores only alive neighbor nodes. However, this algorithm requires complexity $O(n + (k + 3))$, where n is number of nodes and k is number of fan-outs.

4   Protocol Verification

The main reason for the verification is to ensure that the design of the protocol did not exhibit any potential problems. The protocol has been modeled with the PROMELA [16] specification language, which is the input of the SPIN [17] verification tools. PROMELA (Process Meta Language) is a non-deterministic language, which provides a method for making abstractions of distributed system protocols. It supports dynamic creation of concurrent processes, both synchronous and asynchronous message passing via communication channels, message loss and duplicate simulation and several other features. SPIN is a model checker for asynchronous systems using an automata-theoretical. It checks for deadlocks, livelock (non-progress cycles) and non-reachable states in the entire state space. It can verify and simulate several correctness properties. If an error is found, SPIN will provide a counterexample to show a circumstance that can generate the erroneous state.

4.1   Specifying the Protocol in PROMELA

Due to the fact that the PROMELA language is based on point to point communication, there must be as many channels as nodes in order to model the broadcast system. Each node will exclusively receive messages only through this channel. They will use corresponding channel associated with the node to send messages. All the nodes will wait in a loop with the *do* repetition construct. The root node starts sending the initial messages. If a node gets a message, it will check the message type and execute portions of code corresponding to procedure rules in Section. 3.1. For simplicity reason, we use a new feature of SPIN version 4 which can include embedded C code fragments (with PROMELA's *c_code* construct) to compute node depth, neighbor IDs etc. The link failure is simulated with non-deterministic selection capability of the *if* selection construct. The SPIN verifier and simulator will randomly choose the status (up or down) of links between a node and its neighbors while the node is trying to send a message on to the next hop. In order to speed up the verification process, we reduce the size of state space by using an *atomic* construct to atomically execute its code section which represents internal computation without interleaved execution with other processes.

4.2 Verification Results

The results were conducted on a PentiumIII 550MHz, with Spin 4.2.6 on Linux. The search depth bound was 10,000 and the memory limit was 512 MB. A deadlock was discovered from the original modeling. However, after closer examination, it turns out that TCP buffer size of the communication channel in the modeling was too small. When the deadlock problem was solved, no deadlock, livelock, invalid end state, unreachable codes and assertion violation were found during verification.

5   Experimental Results

The protocol performance was evaluated in both normal and failure modes. In the case of no failure, it is obvious that the average number of hops for multicast messages decreases when the number of fan-outs increases (i.e. closer to a flat tree). On the other hand, the average number of steps to complete the message transfer for broadcast increases when the number of fan-outs increases (except that 3-ary is better than 2-ary due to more parallelism).

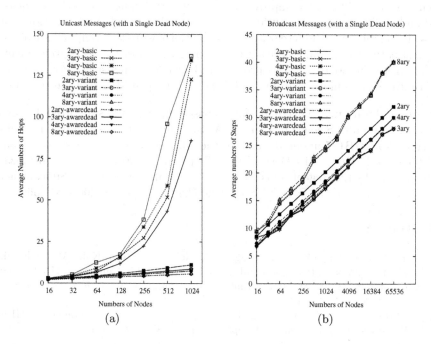

Fig. 2. Message transmission during failure situations. (a) Unicast (b) Broadcast.

During the failure mode, the dead nodes (D) are obtained from combinations of all possible nodes (N) i.e. $\binom{N}{D}$, where source node $\notin$ D. Fig 2(a) illustrates that both variant rule-based and dead node aware algorithms are scalable with unicast messages (multicast to one destination). The higher values of fan-out yields the worst performance, especially with the basic rule-based algorithm, because it has more chances to go in a direction toward a dead node. Fig 2(b) depicts that a dead node has only a small effect on the performance of a broadcast message. The results show that the basic and variant rule-based algorithms produce performance close to the dead node aware algorithm, but the rule-based algorithms are much simpler to the model e.g. a broadcast[3] with a single dead node on an AMD 2GHz machine, the simulation time of dead node aware is 15 minutes, while the basic and variant rule-based took only about 30 seconds.

6   Conclusions and Future Works

The scalable and fault tolerant protocol for parallel runtime environments was designed and developed to support runtime environments of MPI implementations. The design of the protocol has been formally proven to work under both normal and failure modes. The performance results indicate that the variant rule-based algorithm is the best choice in terms of the shortest path (and simulation computation time as well).

There are several improvements that we plan for the near future. Making the protocol aware about the underlying network topology (in both LAN and WAN environments) will greatly improve the overall performance for both broadcast and multicast message distribution. This is equivalent to adding a function cost on each possible path and integrating this function cost to the computation of the shortest path. A faster and more accurate re-routing algorithm is in development. At a longer term, we expect this protocol to be the basic message distribution of the runtime environment within the FT-MPI and Open MPI runtime systems.

Acknowledgement. This material is based upon work supported by Los "Alamos Computer Science Institute (LACSI)", funded by Rice University Subcontract No. R7B127 under Regents of the University Subcontract No. 12783-001-05 49 and "Open MPI Derived Data Type Engine Enhance and Optimization", funded by the Regents of the University of California (LANL) Subcontract No. 13877-001-05 under DoE/NNSA Prime Contract No. W-7405-ENG-36.

References

1. M. Beck, J. J. Dongarra, G. E. Fagg, G. A. Geist, P. Gray, J. Kohl, M. Migliardi, K. Moore, T. Moore, P. Papadopoulous, S. L. Scott, and V. Sunderam. HARNESS: A next generation distributed virtual machine. *Future Generation Computer Systems*, 15(5–6):571–582, 1999.
2. G. Burns, R. Daoud, and J. Vaigl. LAM: An Open Cluster Environment for MPI. In *Proceedings Supercomputing Symposium*, pages 379–386, 1994.

---

[3] 16K bcast, we model 16383 different network topologies.

3. R. Butler, W. Gropp, and E. L. Lusk. A scalable process-management environment for parallel program. In *Proceedings of the 7th European PVM/MPI User's Group Meeting on Recent Advances in Parallel Virtual Machine and Message Passing Interface*, pages 168–175, London, UK, 2000. Springer-Verlag.

4. R. H. Castain, T. S. Woodall, D. J. Daniel, J. M. Squyres, B. Barrett, and G. E. Fagg. The open run-time environment (openrte): A transparent multi-cluster environment for high-performance computing. In *Proceedings 12th European PVM/MPI User's Group Meeting on Recent Advances in Parallel Virtual Machine and Message Passing Interface*, Sorrento(Naples), Italy, September 2005. Springer-Verlag.

5. J. J. Dongarra, H. Meuer, and E. Strohmaier. TOP500 supercomputer sites. *Supercomputer*, 13(1):89–120, 1997.

6. G. E. Fagg, E. Gabriel, G. Bosilca, T. Angskun, Z. Chen, J. Pjesivac-Grbovic, K. London, and J. Dongarra. Extending the mpi specification for process fault tolerance on high performance computing systems. In *Proceedings of the International Supercomputer Conference (ICS) 2004*, Heidelberg, Germany, June 2006. Primeur.

7. E. Gabriel, G. E. Fagg, G. Bosilca, T. Angskun, J. J. Dongarra, J. M. Squyres, V. Sahay, P. Kambadur, B. Barrett, A. Lumsdaine, R. H. Castain, D. J. Daniel, R. L. Graham, and T. S. Woodall. Open MPI: Goals, concept, and design of a next generation MPI implementation. In *Proceedings 11th European PVM/MPI User's Group Meeting on Recent Advances in Parallel Virtual Machine and Message Passing Interface*, pages 97–104, Budapest, Hungary, September 2004. Springer-Verlag.

8. W. Gropp, E. Lusk, N. Doss, and A. Skjellum. A high - performance, portable implementation of MPI message passing interface standard. *Parallel Computing*, 22(6):789–828, 1996.

9. I. Gupta, R. van Renesse, and K. Birman. Scalable fault-tolerant aggregation in large process groups. In *Proceedings of The International Conference on Dependable Systems and Networks (DSN)*, pages 433–442, 2001.

10. MPI Forum. MPI: A message-passing interface standard. Technical report, 1994.

11. S. Ratnasamy, P. Francis, M. Handley, R. Karp, and S. Shenker. A scalable content addressable network. Technical Report TR-00-010, Berkeley, CA, 2000.

12. R. V. Renesse, Y. Minsky, and M. Hayden. A gossip-style failure detection service. Technical Report TR98-1687, 28, 1998.

13. A. Rowstron and P. Druschel. Pastry: Scalable, decentralized object location, and routing for large-scale peer-to-peer systems. *Lecture Notes in Computer Science*, 2218:329–350, 2001.

14. I. Stoica, R. Morris, D. Karger, F. Kaashoek, and H. Balakrishnan. Chord: A scalable Peer-To-Peer lookup service for internet applications. In *Proceedings of the 2001 ACM SIGCOMM Conference*, pages 149–160, 2001.

15. B. Y. Zhao, J. D. Kubiatowicz, and A. D. Joseph. Tapestry: An infrastructure for fault-tolerant wide-area location and routing. Technical Report UCB/CSD-01-1141, UC Berkeley, April 2001.

16. Holzmann, G.J.: Design and validation of computer protocols. Prentice Hall (1991)

17. Holzmann, G.J.: The model checker SPIN. IEEE Transactions on Software Engineering **23** (1997) 279–295.

An Intelligent Management of Fault Tolerance in Cluster Using RADICMPI*

Angelo Duarte, Dolores Rexachs, and Emilio Luque

Computer Architecture and Operating Systems Department, University Autonoma of
Barcelona, ETSE, QC/3088, Bellaterra, 08193, Barcelona, Spain
angelo@aomail.uab.es, {Dolores.Rexachs, Emilio.Luque}@uab.es

Abstract. Independence of special elements, transparency and scalability are very significant features required from the fault tolerance schemes for modern clusters of computers. In order to attend such requirements we developed the RADIC architecture (Redundant Array of Distributed Independent Checkpoints). RADIC is an architecture based on a fully distributed array of processes that collaborate in order to create a distributed fault tolerance controller. This controller works without special, central or stable elements. RADIC implements the fault tolerance activities, transparently to the user application, using a message-log rollback-recovery protocol. Using the RADIC concepts we implemented a prototype, RADICMPI, which contains some standard MPI directives and includes all functionalities of RADIC. We tested RADICMPI in a real environment by injecting failures in nodes of the cluster and monitoring the behavior of the application. Our tests confirmed the correct operation of RADICMPI and the effectiveness of the RADIC mechanism.

1 Introduction

Message-passing is a common paradigm used to create parallel algorithms. S7tandards like PVM and MPI have been largely adopted by programmers in order to implement such algorithms for executing in clusters.

The usage of cluster structures based on commodity parts; the simplicity of the message-passing paradigm; and the development of MPI implementations like LAMPI and MPICH, have contributed to the popularization of MPI. A typical user of this new class of system is interested in programming his/her algorithms without any concern about the operation of the cluster structure. Such users are typically interested in getting more performance and rarely take into consideration that some part of the cluster may fail during the application execution. Because MPI uses a fail-stop semantic, a node failure typically produces a crash into an MPI application, forcing the user to restart the execution from its beginning.

Several projects as Starfish [1], Egida [2], FT-MPI [3], MPI-FT [4], MPI/FT [5], MPICH-V [6], LAM/MPI [7], LA-MPI [8] and more recently OpenMP [9] have proposed solutions to reduce the impact of a failure by allowing the restart of the application from a point before the failure instead of its beginning. Rollback-recovery protocols [10,11] are the basis for the fault tolerance mechanism of such implementations. Nevertheless they largely differ in the way they interact with the

---

* This work was supported by the MEyC-Spain under contract TIN 2004-03388.

B. Mohr et al. (Eds.): PVM/MPI 2006, LNCS 4192, pp. 150–157, 2006.
© Springer-Verlag Berlin Heidelberg 2006

cluster architecture and with the parallel application. In order to differentiate one implementation from another, we select three features that are very important in the modern cluster and that are simultaneously linked to the behavior of the fault tolerance scheme, the cluster structure and the parallel application.

The first feature is how much the operation of the fault tolerant scheme is transparent to programmers. A non transparent scheme demands to the programmer the inclusion specific commands in the code in order to control the fault tolerance structure. Besides this scheme increases the software engineering costs, it requires changes in algorithms already coded and often demands that the programmer has a high level of knowledge about the cluster architecture in order to correctly use the available fault tolerance resources. This makes transparency a feature highly desirable for the fault tolerance schemes.

The scalability is another important feature of a fault tolerance mechanism. In order to attend such requirement, a fault tolerance scheme must operate without using central elements or global coordination between processes, because such items create a strong constraint to the scalability. So, the usage of distributed elements to build a scalable fault tolerance structure is desirable, when not mandatory. Furthermore, the scalability naturally limits the rollback-recovery protocols that may be selected to build a fault tolerance scheme.

Finally, the independence of stable resources in order to implement the fault tolerance scheme is very important for clusters based on commodity-off-the-shelf computers. Fault tolerance schemes based on a rollback-recovery protocol typically store checkpoints and message-logs in stable-storage elements which are critical and expensive in practice. Furthermore, these elements increase the final cluster cost provided that several of them must exist in order to not constraint the scalability.

Table 1. Comparison between several fault-tolerant MPI projects. (T=transparency; I= Independency of stable elements; S=Scalability)

MPI project	T	I	S	Comment
LAM/MPI	-	-	-	Uses coordinated checkpoint and requires user interference.
MPICH-V1	+	-	-	Requires central and stable elements
MPICH-V2	+	-	-	Requires central and stable elements
MPI-FT	-	-	-	Use of special MPI directives
Starfish	-	-	-	Uses global checkpoint
MPICH-V/CL	+	-	-	Requires central and stable elements. Global checkpoint
Egida	+	-	-	A framework for Rllbck. Rcovr. prot., not a entire FT system
MPI/FT	-	-	-	Centralized coordinator
FT-MPI	-	-	-	Application must manage the recovering
RADICMPI	+	+	+	Transparent/fully distributed. No stable elements required

The features we stated above (scalability, efficiency and independence of dedicated) have a straight relationship with the cluster architecture and its interaction with the parallel application in the presence of failures. **Table 1** relates these features with some recent projects about fault tolerance in MPI. We assumed all these three features as requirements that must be attended by any message-passing implementation dedicated to modern clusters and developed RADIC (Redundant Array of Distributed Independent Checkpoints) [12]. RADIC is a generic fault tolerance architecture that specifies the behavior of the cluster structure in the presence of node failures in order to assure the correct completion of a parallel application.

We created a MPI implementation RADICMPI that differs from other fault tolerant MPI implementations because it attends simultaneously to three requisites discussed above. It transparently implements a fault tolerant mechanism which does not require any central or stable elements to operate. The fault tolerance mechanism is fully distributed throughout the same nodes used by the parallel application and it does not constraint the scalability.

The remaining of this paper is organized as follows. In section 2, we explain RADICMPI and the operation of the RADIC architecture. Section 3 contains some results obtained with a prototype of our architecture. Finally, section 4 offers our concluding remarks and lists the future works.

2   RADICMPI and the RADIC Architecture

The operation of RADIC is based on an array of system processes that collaborate in order to create a distributed fault tolerance controller, or RADIC controller. Such processes are fully distributed throughout the cluster nodes in which the parallel application processes are placed. RADICMPI implements some standard MPI directives in order to transparently embed the RADIC controller in a MPI application without any modifications in the application code.

The RADIC controller performs all typical activities that any fault tolerance scheme based on rollback-recover must execute. Such set of activities differs according to the type of rollback-recovery protocol used. Nevertheless, independently of the chosen protocol, there are three elementary activities required by any fault tolerance scheme: a) collects information that will be used to recover from failures (checkpoint and message-logs); b) detects failures; and c) recovers faulty application processes assuring that the application may continue.

Although RADIC architecture was not developed having any particular rollback-recovery protocol in consideration, it should be noted that only the receiver based pessimistic message-log protocol imposes no constraint to the scalability [10]. This occurs because in such protocol no coordination or interdependency between processes is required to take checkpoints of a process or to recover a process from a failure. That was the motivation for using the receiver-based pessimist protocol in our RADICMPI implementation.

In RADIC, all the activities required by the rollback-recovery protocol are executed through the collaboration between the two set of processes of the RADIC controller: *protector* processes and *observer* processes. The total number of RADIC processes in the system is determined by the following rule: there is one *protector* for each node used by the parallel application and there is one *observer* attached to each parallel application process. The *protector* and the *observers* of a node always are connected to a *protector* in a neighbor node. **Fig. 1a** depicts a sample cluster with a set of eight application processes (P1 to P8) with their respective *observers* (O1 to O8). Each node has a *protector* process (T1 to T8). The arrow lines indicate the communication relationship between the RADIC elements.

The set *observers* attached to the application processes manage all messages delivering between the processes. Each *observer* also implements the checkpoint and message-log mechanisms for the application process to which it is attached. The set of *protectors* a) implements the failure detection; b) operates like a distributed storage mechanism for the checkpoints and c) message-logs, and also recovers faulty processes.

Observers

The first *observer* responsibility is to manage the message delivering for the application process to which it is associated. In order to perform such activity, each *observer* owns a table, namely *radictable*, which contains all information that allows it to exchange messages with the other application processes. The *radictable* has a line for each application process and each line contains the following fields: the address of the process in the cluster, a counter for the messages sent from the *observer* to another application process, a counter for the messages received by the *observer* from another application processes and the address of the *protector* of the process. **Table 2** depicts the *radictable* of the *observer* 2 at **Fig. 1**.

The *radictable* is built using the following mechanism: at the startup, every *observer* communicates its number, its node address and its *protector* node address to a leader *observer*. Using such information, the leader *observer* builds the *radictable* and transmits it to all other *observers* in the system. From this point on, each *observer* manages its own *radictable* independently. An *observer* uses the message counters of its *radictable* as identifiers for the messages exchanged between its application process and another application process. Each time that a message is successfully exchanged between two *observers*, the sender *observer* increments its sent counter in the destination line of its *radictable*. Similarly, the destination *observer* increments the message received counter regarded to the sender line in its *radictable*. Together with the *protector* address field, these counters are used for fault recovering purposes as we will explain soon.

Table 2. Structure of the *radictable* of the *observer* 2 of Fig. 1

Process rank	Process address	Sent counter	Received counter	*Protector* address
1	Addr1	3	2	Addr2
2	Addr2	0	0	Addr3
...	...	...	...	...

The second *observer* functionality is to collect the fault tolerance information (checkpoints and message-logs) from its application process and to transmit such information to its *protector*. The **Fig. 1a** depicts how *observers* and *protector* may be associated in a cluster with eight nodes, in which each *observer* is associated to only one *protector*. Another configuration could be used with, for example, several *observers* connected to one *protector* or more than one *protector* for each *observer*. The only restriction is that an *observer* must never use the *protector* in its own node. With this strategy, the RADIC controller implements the redundancy that dispenses the use of a stable-storage element in order to store the checkpoints and message-logs of a process. The mechanism works as follows.

According to the checkpoint policy, each *observer* takes checkpoints of itself and of its application process, and transmits these checkpoints to its *protector* (in a neighbor node). Similarly, in accordance with the message-log policy, each *observer* transmits the message-logs to its *protector*. Therefore, the information required to recover a process from a failure is managed by the *protector* in the neighbor node of the *observer*.

If the *protector* of an *observer* fails, the *observer* immediately fetches another *protector* in the cluster. According to this fetch algorithm, if the *protector* of an *observer* fails, the *observer* uses its *radictable* to find the address of the neighbor of its old *protector*. If such neighbor is unreachable, the *observer* runs throughout the *radictable* looking for the subsequent neighbors until a new *protector* is contacted.

The straight relationship between an process and its neighbor *protector* brings the question about what would happen if the neighbor *protector* of a faulty process fails before such faulty process is recovered, i.e., if both, the application process and its *protector* fail concurrently.

Before describing how RADIC manage this situation, we should make some considerations: firstly, node failures are rare in practice; secondly, in pessimistic message-log protocols a process recovering is independent of the state of the other application processes, therefore the recovery time of a process is, in the worst case, equal to the checkpoint interval.

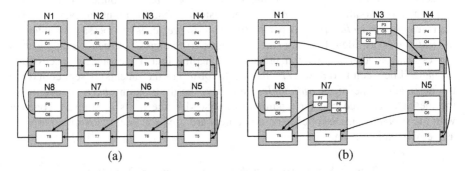

(a)   (b)

Fig. 1. a) Example the RADIC architecture in a cluster. The arrow lines indicate the relationship between *observers* (O) and *protectors* (T). b) Same cluster after a failure in the nodes N2 and N6. The pair P2/O2 recovers in N3 (the original *protector* used by the *observer* O2) and the pair P6/O6 recovers in N7 (the original *protector* used by the *observer* O6).

Basing on the first consideration, we may conclude that the probability of a concurrent failure in a node and in its *protector* is much lower that the probability of the individual failure of these nodes. Nevertheless, in order to protect the system against such combined fault, an *observer* must use more that one *protector* to store its fault tolerance information. Therefore, the checkpoints and message-logs of a process will survive to a concurrent failure in a process in one of its *protectors*. Such way of thinking could be extended if we consider that a process and its both *protectors* could concurrently fail. However, since the probability of such failure is even lower than the one in the first case, the cost of replicating the fault tolerance information in several *protectors* might not compensate the increasing in the system robustness.

Finally, the third *observer* functionality is to manage several activities related to its application process recovering. The first activity is to establish a new *protector* because a pair process/*observer* always is recovered in the node of the *protector* they were using before the failure and an *observer* may never use the *protector* of its own node. The algorithm the *observer* uses to fetch a new *protector* when it recovers is the same used when a *protector* fails. **Fig. 1b** represents the cluster after a failure in two nodes, N2 and N6. *Observers* O2 and O6 establish new *protectors* since they recover in the same node they were using before the fault.

When a process and its *observer* recover from a fault, the *observer* uses the message counter in its *radictable* to manage the message log in order to correctly get the messages its process requests. Similarly, the survivor *observers* will use the receive counters in their *radictable* to discard the repeated messages that a recovering process might send.

Protectors

The *protectors* of the RADIC controller establish a virtual neighborhood between the nodes of the cluster by means of a watchdog/heartbeat mechanism. Using such mechanism, any *protector* can detect failures in one of its neighbors. Therefore, any *protector* plays two roles related to failure detection: one as a heartbeat sender and the other as a watchdog. For example, in **Fig. 1a**, the *protector* T2 play two roles: it is the watchdog of T1 and the heartbeat sender of T3. Each watchdog *protector* also uses the heartbeat/protocol to inform to its heartbeat sender who is the *protector* to which it is sending heartbeats. So, T1 knows that T3 is the watchdog of T2; T2 knows that T4 is the watchdog of T3 and so on. This information will be used in case by the recovering mechanism explained in the next paragraphs.

To understand how the *protector* manages node failures, we suppose a failure in the node N2. In such case, the heartbeat element of T1 will detect that its watchdog neighbor (T2) is unreachable. Similarly, the watchdog element of T3 will also detect the T2 is not sending heartbeats. So, T1 and T3 start their recovering mechanisms.

T1 will use the former watchdog of T2 as its new watchdog neighbor, i.e., T1 will connect to T3. T3 in turn, will execute a greater set of activities. The first, and more simple, is to accept T1 as its new heartbeat sender. The second is to accept all *observers* that where using T2 as a *protector* (just O1 in the cluster of **Fig. 1a**). Finally, T3 will recover all application processes (together with their respective *observers*) that were in the node N2, using the stored checkpoints of such processes. After recovering, the *observers* assume the rest of the recovering activity. The **Fig. 1b** depicts the final architecture of our sample cluster after two faults: in the node N2 and in the node N6.

3 Implementation Details and Experiments

The current RADICMPI version implements a subset of MPI functions, namely: MPI_Init, MPI_Finalize, MPI_Send, MPI_Recv, MPI_Sendrecv, MPI_Wtime, MPI_Get_processor_name, MPI_Comm_rank, MPI_Comm_size, and MPI_Type_size. RADICMPI comes as a library (radicmpi.a) and a runtime environment in order to compile (radiccc) and run (radicrun) the programs.

The *observers* run as threads of the application processes. The radicc script used to compile the application code links the *observer* objects with the application objects in the link time. The *observer* main thread starts when the application process executes the MPI_Init call. The *protectors* execute as separated processes created by the radicrun script.

Our implementation was developed in C++ and tested in a homogeneous cluster of twelve 1.9GHz Athlon-XP2600 with 256MB and 40GB local disk, running Linux Fedora Core 2 with kernel 2.4.22-1.2199-8, interconnected by a 100BaseT Ethernet switch. The checkpoints were performed using the BLCR (Berkeley Labs Checkpoint/Restart) library version 0.4.2.

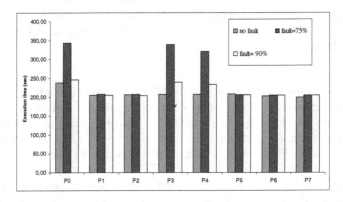

Fig. 2. Execution of our matrix-multiplication program with faults injected in different execution times

We conducted the tests using a ping-pong program to evaluate the correct operating of our implementation and a master-worker matrix multiplication program for performance studies. The **Fig. 2** depicts the example of one execution of the matrix-multiplication with faults injected in the node of the process P3 at 75% and 90% of the total execution time without failures for a situation with checkpoints of 60 seconds. RADICMPI manage the recover of P3 in the node of the process P4. This explains the enlargement of the execution time of such processes.

4   Conclusions

In this paper we described RADICMPI, an implementation used to efficiently manage the fault tolerance in clusters. RADICMPI dispenses from the use of central stable-storages by replicating the critical information throughout the cluster nodes used by the parallel application. Such strategy reduces the cost and simultaneously increases the efficiency of the cluster.

RADICMPI works as a library that implements some standard MPI directives and transparently embeds the fault tolerance activities of the RADIC architecture in such directives. The great advantage of this mechanism is that a programmer just has to recompile her/his application code with the RADICMPI library in order to execute using the RADIC architecture.

The future works are oriented to increase the number of MPI directives of RADICMPI, specially the MPI non-blocking functions. We also will work in the performance improvement of the implementation, taking special attention to reduce the impact of the rollback-recovery protocol mechanism.

References

1. A.M. Agbaria and R. Friedman. Starfish: fault-tolerant dynamic MPI programs on clusters of workstations. In Proceedings of 8th International Symposium on High Performance Distributed Computing, 167–176, August 1999.
2. S. Rao, L. Alvisi, and H. Vin. Egida: An extensible toolkit for low-overhead fault-tolerance. In Proceedings of IEEE Fault-Tolerant Computing Symposium (FTCS-29), Madison. USA. June 1999.
3. G. Fagg and J. Dongarra. FT-MPI: Fault tolerant MPI, supporting dynamic applications in a dynamic world. In Euro PVM/MPI User's Group Meeting 2000, pages 346–353, Berlin, Germany, 2000. Springer-Verilag.
4. S. Louca, N. Neophytou, A. Lachanas, and P. Evripidou. MPI-FT: Portable fault tolerance scheme for MPI. Parallel Processing Letters, 10(4):371–382, 2000.
5. R. Batchu, J. Neelamegam, Z. Cui, M. Beddhua, A. Skjellum, Y. Dandass, and M. Apte, MPI/FT: Architecture and taxonomies for fault-tolerant, message-passing middleware for performance portable parallel computing. In Proceedings of the 1st IEEE International Symposium of Cluster Computing and the Grid, Melbourne, Australia, 2001
6. G. Bosilca, A. Bouteiller, F. Cappello, S. Djilali, G. Fedak, C. Germain, T. Herault, P. Lemarinier, O. Lodygensky, F. Magniette, V. Neri, and A. Selikhov. MPICH-V: Toward a scalable fault tolerant MPI for volatile nodes. In Proceedings of SuperComputing 2002 (SC2002), November 2002.
7. S. Sankaran, J.M. Squyres, B. Barrett, A. Lumsdaine, J. Duell, P. Hargrove, and E. Roman. The LAM/MPI checkpoint/restart framework: System-initiated checkpointing. In Proceedings of LACSI Symposium, Sante Fe, New Mexico, USA, October 2003.
8. R.T. Aulwes, D.J. Daniel, N.N. Desai, R.L. Graham, L.D. Risinger, M.A. Taylor, T.S.Woodall, and M.W. Sukalski. Architecture of LA-MPI, a network-fault-tolerant MPI. In Proceedings of 18th International Parallel and Distributed Processing Symposium. IEEE, April 2004.
9. E. Gabriel, G.E. Fagg, G. Bosilca, T. Angskun, Jack J. Dongarra, J.M. Squyres, V. Sahay, Prabhanjan Kambadur, Brian Barrett, A. Lumsdaine, R.H. Castain, D.J. Daniel, R.L. Graham, and T.S. Woodall. Open MPI: Goals, concept, and design of a next generation MPI implementation. In Proceedings, 11th European PVM/MPI Users' Group Meeting, pages 97–104, Budapest, Hungary, September 2004.
10. E.N. Elnozahy, L. Alvisi, Y.M. Wang and D.B. Johnson. *A Survey of Rollback-Recovery Protocols in Message-Passing Systems*. ACM Computer Survey, 34(3):375–408, 2002.
11. S. Kalaiselvi and V. Rajaraman. *A Survey of Checkpointing Algorithms for Parallel and Distributed Computers*. SADHANA:Academic Proceedings in Engineering Sciences, Vol. 25, Part 5, 489-510, Bangalore, India, October 2000.
12. A. Duarte, D. Rexachs and E. Luque. A distributed scheme for fault-tolerance in large Clusters of Workstations. In Proceedings of Parrallel Computer 2005 (Parco2005). September 13-16, 2005. Málaga. Spain. (in press)

Extended mpiJava for Distributed Checkpointing and Recovery

Emilio Hernández, Yudith Cardinale, and Wilmer Pereira

Universidad Simón Bolívar,
Departamento de Computación y Tecnología de la Información,
Apartado 89000, Caracas 1080-A, Venezuela
{emilio, yudith, wpereira}@ldc.usb.ve

Abstract. In this paper we describe an *mpiJava* extension that implements a parallel checkpointing/recovery service. This checkpointing/recovery facility is transparent to applications, i.e. no instrumentation is needed. We use a distributed approach for taking the checkpoints, which means that the processes take their local checkpoints independently. This approach reduces communication between processes and there is not need for a central server for checkpoint storage. We present some experiments which suggest that the benefits of this extended MPI functionality do not have a significant performance penalty as a side effect, apart from the well-known penalties related to the local checkpoint generation.

1 Introduction

Parallel checkpointing algorithms are important in parallel environments where long-term parallel processes are executed. These algorithms can be classified into two main groups:coordinated and uncoordinated, according to the approach used to coordinate the capture of local checkpoints [1], i.e. the state of single nodes. In the coordinated approach, it is necessary to synchronize all processes in order to produce a consistent (normally centralized) global checkpoint. In the uncoordinated approach, the processes take their local checkpoints independently. As a consequence, some of the checkpoints may not belong to any consistent global checkpoint. In order to reduce the number of useless checkpoints, the processes may exchange information about their checkpointing activities. This information is mainly piggy-backed on the messages sent between processes. Under this scheme, processes can take forced checkpoints to avoid orphan messages, and log in-transit messages. The algorithms based on communication-induced protocols are called quasi-synchronous [2].

In a previous work [3], we proposed a distributed checkpointing protocol as a combination of the protocols described in [4] and [5]. The protocol described in [4] logs in-transit messages in order to be able to re-send them during the recovery process. On the other hand, the protocol described in [5], focuses on avoiding orphan messages by taking checkpoints that have not been scheduled previously (forced checkpoints) based on a communication-induced checkpointing protocol. Those protocols provide fault-tolerance in asynchronous systems, and assume

B. Mohr et al. (Eds.): PVM/MPI 2006, LNCS 4192, pp. 158–165, 2006.

that each ordered pair of processes is connected by a reliable, directed logical channel whose transmission delays are unpredictable but finite.

In this paper we describe an extension to the *mpiJava* functionality that implements the combined checkpointing protocol. This quasi-synchronous protocol was implemented on the JNI wrappers, at each point-to-point communication method (send/receive), without changing the *mpiJava* API. It means that checkpointing and recovery facilities are transparent to applications. That is, the extended *mpiJava* is used without making any changes to the application code. Additionally, we present a case study in which we implement this algorithm. We show the results of some experiments that suggest that the benefits of this extended MPI functionality does not imply a significant performance penalty.

There are several works that implement parallel checkpointing/recovery or migration schemes in combination with communication libraries such as PVM and MPI. Most of these proposals use a coordinated approach [6,7,8,9]. MPICH-V2 [10] is a fault tolerant MPICH version that implements an uncoordinated protocol. However, a centralized checkpoint server is used to control logging messages. The algorithm we implemented does not need a centralized component because each node decides independently whether a forced checkpoint is needed.

2   Extended mpiJava

We have extended *mpiJava* functionality to include the control information needed to decide when forced checkpoints have to be taken and when to log messages. We added procedures "send_protocol" and "receive_protocol" (related to the combined protocol) at each point-to-point communication method without changing their interfaces. It means that the extended *mpiJava* preserves the same *mpiJava* API. Specifically, "send_protocol" and "receive_protocol" were added at `Comm` package, where all point-to-point communication methods are defined. We implemented a new package called `Ckpt` to define and manage data structures of the combined protocol (see Figure 1). In this package there are two important procedures, `initialize` and `take_checkpoint`, which are described below.

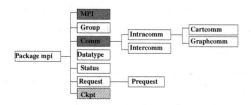

Fig. 1. Principal classes of the extended mpiJava

The `initialize` procedure is executed before any "send" or "receive". It initializes the data structures used by the checkpointing protocol. This procedure is called by MPI.Init, which is the first *mpiJava* method invoked by each **process**.

The `take_checkpoint` procedure takes the checkpoints, either normal or forced. It first increases the checkpoint number corresponding to the current process (its logical clock). Then, for each message, it checks whether the message is not an in-transit one, in which case the message is suppressed from the message log. Finally, the current state should be saved. A copy of the checkpoint number array and a copy of the log are also saved in stable storage.

The `mpiJava_send` procedure is the wrapper of `mpi_send` routine. The "send_protocol" registers the event "send" and executes the actual mpi_send routine. The current message is logged in the v_log, while procedure take_checkpoint, as explained above, identifies the messages that may become in-transit during recovery.

```
Procedure mpiJava_send(m, dest_rank)
known_received[dest_rank][dest_rank] := +1;
sent_to[dest_rank] := true;
append (m, known_received[dest_rank][dest_rank], dest_rank) to v_log
// Send message m to process dest_rank
actual_mpi_send(m, dest_rank, greater, ckpt, taken, known_received);
end procedure
```

The `mpiJava_receive` procedure is the wrapper of the `mpi_receive` routine. The "receive_protocol" executes the actual mpi_receive routine and takes forced checkpoints (if necessary) to avoid orphan messages during a potential recovery. `ckpt[source_rank]` is the logical clock of process with `rank= source_rank` when message m was sent. This procedure is more complex because it actually checks the control information and has to consider several cases. A detailed description of this procedure follows:

```
Procedure mpiJava_receive (m, source_rank)
// Receives message m from process source_rank
actual_mpi_receive(source_rank, m, r_greater, r_ckpt, r_taken, r_known_received);
// Evaluating control information to avoid orphan messages
if (∃ rank : (sent_to[rank] ∧ r_greater[rank]) ∧ (ckpt[source_rank] > ckpt[my_rank]) ∨
            (r_ckpt[my_rank] = ckpt[my_rank] ∧ r_taken[my_rank]))
    then take_checkpoint // forced checkpoint
end_if
switch
    case ckpt[source_rank] > ckpt[my_rank] do
        ckpt[my_rank] := ckpt[source_rank]; greater[my_rank] := false;
        forall rank ≠ my_rank do greater[rank] := r_greater[rank] end_do;
    end case
    case ckpt[source_rank] = ckpt[my_rank] do
        forall rank do greater[rank] := greater[rank] ∧ r_greater[rank] end_do;
    end case
    // case ckpt[source_rank] < ckpt[my_rank] do nothing
end_switch;
forall rank ≠ my_rank do
    switch
        case r_ckpt[rank] > ckpt[rank] then
            ckpt[rank] := r_ckpt[rank]; taken[rank] := r_taken[rank];
        end case
        case r_ckpt[rank] = ckpt[rank] then
            taken[rank] := taken[rank] ∨ r_taken[rank]
        end case
        // case r_ckpt[rank] < ckpt[rank] do nothing
    end_switch;
end_do;
JNI_deliver(m); // gives message m to the application
known_received[my_rank, source_rank] := known_received[my_rank, source_rank] + 1;
```

forall $(rank\_x, rank\_y)$ **do** $known\_received[rank\_x, rank\_y] :=$
    $max(known\_received[rank\_x, rank\_y], r\_known\_received[rank\_x, rank\_y]);$
end_do
end procedure

We added a new method named `InitAndRecoverNative` at the `MPI` package, which should be called during the recovery process. `InitAndRecoverNative` is similar to `InitNative` except that the first one calls method `ReadCheckpoint`, which is in charge of re-sending all in-transit messages stored in checkpoint files. Method `ReadCheckpoint` is part of the package `Ckpt`.

3   A Case Study: Checkpointing Service on a Java-Based Grid Platform

In this section we describe the checkpointing service implemented on SUMA/G[1] [11], a distributed platform that transparently executes both sequential and parallel Java programs on remote machines. It extends the Java execution model to be used on Globus-based grid platforms. The Execution Agents of SUMA/G are JVMs that can be deployed in Globus worker nodes. These agents provide checkpointing services by using an extended JVM that is able to capture local checkpoints (currently, we use the extended JVM described in [12]). There are two key components: `SUMAgCkpMonitor` and `SUMAgRecover`. The Execution Agent starts the application, as well as the `SUMAgClassloader` and the thread `SUMAgCkpMonitor` in the extended JVM. If a fault occurs, the `SUMAgRecover` is invoked for restoring the application execution from its last consistent global checkpoint.

Figure 2 shows the interactions between `SUMAgCkptMonitor` and the extended *mpiJava* wrappers in a node. This scheme is followed for all processes of the parallel application. `SUMAgCkpMonitor` periodically takes checkpoint, in an asynchronous way (step 1 in figure 2). Every time an asynchronous checkpoint is taken, `SUMAgCkptMonitor` calls method `take_checkpoint` to update and save control information (steps 2 and 3).

For each "send" or "receive" communication call executed by the application (step 4), the extended *mpiJava* wrappers update the control information (i.e, data structures used by the combined checkpointing protocol), as shown in step 5 in figure 2. In case of a "receive" call, the `receive_protocol` decides if a forced checkpoint is needed in order to avoid orphan messages (step 6). In this case, `take_checkpoint` method saves the control information and the state of application threads. To save state of application threads it is necessary to make an upcall to `SUMAgCkpMonitor` (steps 7 and 8). After `send_protocol` or `receive_protocol` is executed, the actual mpi_send or mpi_receive routine is executed (step 9).

3.1   SUMA/G Recovery Process

If a failure occurs in the platform while an application is running, an exception is caught by the `SUMAg Proxy`. SUMA/G launches the recovery algorithm

---

[1] http://suma.ldc.usb.ve

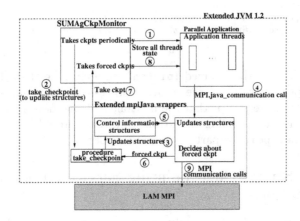

Fig. 2. Interaction between SUMA/G, Extended JVM and Extended *mpiJava* during normal execution

described in [3] which determines the last consistent global checkpoint (step 1 in Figure 3). When a new execution platform is identified, a recovery process is initiated on each node of the parallel application. A `SUMAgRecover` thread is initiated on each node, one per process in the parallel application (step 2). Each `SUMAgRecover` reads the state of threads from checkpoint files and restores threads execution (step 3), executes method `InitAndRecoverNative` (step 4) to re-start MPI and re-send all in-transit messages (step 5). Eventually, each process will receive corresponding in-transit messages (step 6) and the execution will continue normally.

4   Experimental Results

We evaluated the overhead produced by the extended *mpiJava* wrappers on two parallel programs. We used a modest platform to make the measurements, specifically several small clusters of 143 MHz SUN Ultra 1 workstations running Solaris 7, with 64 MB Ram, connected through a 10Mbps switched Ethernet LAN. The first program, called "Pi_Number", is simple and calculates π. The processes keep the set of digits obtained during the execution, so the checkpoint size increases during the execution. The second program is a kernel of a real application, called "Acoustic_Par", which solves an acoustic wave propagation model on a homogeneous, two dimensional medium. Its checkpoint size is constant during the execution.

We measured the total execution time invoking (T_{ckp}) and without invoking (T_{nockp}) the checkpointing service. The net checkpointing overhead and checkpointing overhead percentage are given by O_{ckp} and $O\%_{ckp}$.

$$O_{ckp} = (T_{ckp} - T_{nockp}) \qquad O\%_{ckp} = \frac{(T_{ckp} - T_{nockp})}{T_{nockp}} * 100$$

In table 1 each row represents a single execution. The checkpoints were taken every 2 minutes. The overhead ($O\%_{ckp}$) exhibited when the checkpoint service

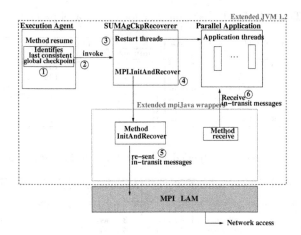

Fig. 3. Interaction between SUMA/G, Extended JVM and Extended *mpiJava* during recovery process

is active increases with the number of processors. This is mainly due to the checkpoint calls and the distributed checkpointing protocol. Note that the T_{nockp} for the "Pi_Number" application increases as the number of processes increases. This is due to the fact that all processes roughly carry out the same amount of work, regardless of the number of processes. Even though this is not a typical parallel program, this example can help measure the checkpointing overhead. On the other hand, the execution time of "Acoustic_Par" reduces as the number of processors is increased.

In these experiments the higher checkpointing overhead is 6.11%. However, we have measured how much of that overhead is due to the extended *mpiJava*. Tables 2 and 3 show the measurements taken from the master node P_0 of each application. $T_{ThrMonitor}$ is the time taken by SUMAgCkpMonitor to save the state of the threads by calling extended JVM facilities. $T_{save}(F_{general})$ is the time, also taken by SUMAgCkpMonitor, spent on saving the checkpoint information (such as number of threads, name of main thread, thread names, etc.). The time to save the control information is $T_{actmpijava}$, while $T_{save}(\overline{M_{intransit}})$ represents the total time to log all in-transit messages in stable storage. T_{send} and $T_{receive}$ represent the accumulated time of "send_protocol" and "receive_protocol" procedures respectively during the execution. Table 2 also shows the average size of in-transit messages $(\overline{M_{intransit}})$ and total number of sent (M_{sent}) and received $(M_{received})$ messages. In case of "PI_number", P_0 does not send any message, thus we show T_{send} in $P_i \neq P_0$ only as a reference. The total overhead of extended *mpiJava* wrappers is denoted as $O_{mpiJavaExt}$ and is calculated as follows:

$$O_{mpiJavaExt} = T_{actmpijava} + T_{save}(\overline{M_{intransit}}) + T_{send} + T_{receive}$$

Last column of table 3 shows $O_{mpiJavaExt}$ and its percent from total checkpointing overhead (O_{ckp}). Note that the overhead of the extended *mpiJava* only represents 15%.

Table 1. Total execution time for the parallel programs

App name	# of proc	$T_{nockp}(1)$	$T_{ckp}(2)$	Checkpoints			O_{ckp} (min)	$O\%_{ckp}$
				#	min size (KB)	max size (KB)		
	2	6.74 m	7.1 m	3	1	131	0.36	5.1%
Pi_Number	4	7.31 m	7.77 m	3	1	139	0.48	5.92%
	6	7.68 m	8.18 m	4	1	142	0.5	6.11%
	2	5.45 m	5.51 m	2	1	1	0.06	1.1%
Acoustic_Par	4	3.56 m	3.65 m	2	1	1	0.09	2.47%
	8	2.64 m	2.71 m	2	1	1	0.07	2.58%

Table 2. Overhead of Extended *mpiJava* taken from master node (P_0)

App name	# of proc.	# of ckpts	$T_{ThrMonitor}$	T_{save} ($F_{general}$)	$T_{actmpijava}$	$M_{intransit}$	$T_{save}(M_{intransit})$
	2	3	274 msec	42 msec	3 msec	220B	40 msec
Pi_Number	4	3	392 msec	42 msec	3 msec	255B	42 msec
	6	4	405 msec	42 msec	4 msec	410B	45 msec
	2		43 msec				
Acoustic_Par	4	2	73 msec	8 msec	2 msec	3KB	7 msec
	8		54 msec				

Table 3. Overhead of Extended *mpiJava* taken from master node (P_0) (cont)

App name	# of proc.	M_{sent}	T_{send}	$M_{received}$	$T_{receive}$	O_{ckp}	$O_{mpiJavaExt}$ ($O\%_{mpiJavaExt}$)
	2		0.8 msec	2000	1.5 msec	360 sec	44.5 sec (12%)
Pi_Number	4	0	(on $P_i \neq P_0$)	4000	2.9 msec	480 sec	47.9 sec (10%)
	6			6000	4.6 msec	500 sec	53.6 sec (11%)
	2	140	0.01 msec	102	0.07 msec	60 sec	9.08 sec (15%)
Acoustic_Par	4	280	0.08 msec	204	0.15 msec	90 sec	9.23 sec (10%)
	8	350	0.08 msec	274	0.09 msec	70 sec	9.17 sec (13%)

5   Conclusions

We present an implementation of an uncoordinated checkpointing/recovery protocol on *mpiJava*. The implemented checkpointing/recovery facility is transparent to applications. We use a distributed approach for taking the checkpoints, which means that the processes take their local checkpoints independently. This approach reduces communication between processes and a central server for checkpoint storage is not needed. This checkpointing/recovery facility does not need any instrumentation of the source code but it needs an extended JVM for local checkpoint generation.

We have tested this extended *mpiJava* functionality and the results suggest that this approach does not involve a significant performance overhead, especially when compared with the overhead of taking the local checkpoints. However, experiments with big parallel applications are important to corroborate these results. We expect more parallel *mpiJava* applications to become available in the public domain for continuing to test the checkpointing facility presented. In the future we are going to implement this facility on the rest of the *mpiJava* functions, such as the broadcast.

References

1. Elnozahy, E.N., Alvisi, L., Wang, Y.M., Johnson, D.B.: A survey of rollback-recovery protocols in message-passing systems. ACM Computing Surveys **34** (2002) 375–408
2. Manivannan, D., Singhal, M.: Quasi-Synchronous Checkpointing: Models, Characterization, and Classification. IEEE Transactions on Parallel and Distributed Systems **10** (1999) 703 – 713
3. Cardinale, Y., Hernández, E.: Parallel Checkpointing Facility in a Metasystem. In: Proceedings of Parallel Computing Conference (PARCO'01), Naples, Italy (2001)
4. Mostefaoui, A., Raynal, M.: Efficient message logging for uncoordinated checkpointing protocols. Technical Report 1018, Institut de recherche en informatique et systemes aleatoires (IRISA) (1996)
5. Helary, J., Mostefaoui, A., R. Netzer, Raynal, M.: Communication-based prevention of useless checkpoints in distributed computations. Technical Report 1105, Institut de recherche en informatique et systemes aleatoires (IRISA) (1997)
6. Stellner, G.: Cocheck: Checkpointing and process migration for MPI. In: 10th International Parallel Processing Symposium. (1996)
7. Sankaran, S., Squyres, J.M., Barrett, B., Lumsdaine, A., Duell, J., Hargrove, P., Roman, E.: The Lam/Mpi Checkpoint/Restart Framework: System-Initiated Checkpointing. International Journal of High Performance Computing Applications **4** (2005) 479–493
8. Zhang, Y., Xue, R., Wong, D., Zheng, W.: A Checkpointing/Recovery System for MPI Applications on Cluster of IA-64 Computers. In: ICPP 2005 Workshops. International Conference Workshops. (2005) 320–327
9. N. Woo, H. Y. Yeom, T.P.: MPICH-GF: Transparent Checkpointing and Rollback-Recovery for Grid-enabled MPI Processes. IEICE Transactions on Information and Systems, Special Section on Hardware/Software Support for High Performance Scientific and Engineering Computing **E87-D** (2004) 1820–1828
10. Bouteiller, A., Cappello, F., Herault, T., Krawezik, G., Lemarinier, P., Magniette, F.: MPICH-V2: a Fault Tolerant MPI for Volatile Nodes based on Pessimistic Sender Based Message Logging. In: Proceedings of High Performance Networking and Computing (SC2003). (2003)
11. Cardinale, Y., Hernández, E.: Parallel Checkpointing on a Grid-enabled Java Platform. Lecture Notes in Computer Science **3470** (2005) 741 – 750
12. Bouchenak, S.: Making Java applications mobile or persistent. In: Proceedings of 6th USENIX Conference on Object-Oriented Technologies and Systems. (2001)

Running PVM Applications on Multidomain Clusters

Franco Frattolillo

Research Centre on Software Technology
Department of Engineering, University of Sannio, Italy
frattolillo@unisannio.it

Abstract. ePVM has been developed to enable PVM applications to run across multidomain clusters made up of computing nodes belonging to non-routable private networks, but connected to the Internet through publicly addressable IP front-end nodes. However, ePVM cannot relieve programmers of the classic burden tied to the problems related to the deployment of PVM runtime libraries and program executables among computational resources belonging to distinct administrative domains. This paper presents *ehPVM*, a lightweight software infrastructure that enables programmers to deploy ePVM applications among computational resources belonging to multidomain clusters. Thus, ePVM programmers can run their applications without having to use grid software toolkits or resources providers whose configurations usually result in being tedious and time-consuming activities.

1 Introduction

ePVM [1,2] is an extension of PVM [3] purposely developed to enable PVM applications to run across multidomain clusters made up of computing nodes belonging to non-routable private networks, but connected to the Internet through publicly addressable IP front-end nodes. In particular, ePVM enables programmers to build "extended virtual machines" (EVMs) made up of sets of clusters. Each cluster can be a set of interconnected computing nodes provided with private IP addresses and hidden behind a publicly addressable IP front-end node. During computation, it is managed as a normal PVM virtual machine where a master *pvmd* daemon is started on the front-end node, whereas slave *pvmds* are started on all the other nodes of the cluster. However, the front-end node is also provided with a specific ePVM daemon, called *epvmd*, which allows the cluster's nodes to interact with the nodes of all other clusters of the EVM, thus creating a same communication space not restricted to the scope of the PVM daemons belonging to a single cluster, but extended to all the tasks and daemons running within the EVM. In fact, ePVM enables both publicly addressable IP nodes and those ones hidden behind publicly addressable IP front-end nodes of the clusters in the EVM to be directly referred to as hosts. This means that the hosts belonging to an EVM, even though interconnected by a two-levels physical network, can run PVM tasks as in a single, flat distributed computing platform.

B. Mohr et al. (Eds.): PVM/MPI 2006, LNCS 4192, pp. 166–173, 2006.

ePVM has also taken advantage of some marginal implementation improvements [4], and has been provided with a parallel file system, called ePIOUS [5], which represents the optimized porting of PIOUS [6] under ePVM. The porting has been developed taking into account the two-levels physical network topology characterizing the cluster grids built by ePVM. As a consequence, it can exploit the basic ePVM ideas and architecture to provide ePIOUS with a file caching service that can speed up file accesses across clusters.

However, despite its capability to run large-scale applications on heterogeneous cluster grids, the ePVM system cannot relieve programmers of the usual burden of using parallel programming systems for high performance applications, such as, for example, MPI [7]. In fact, such burden is essentially determined by the problems related to the deployment of PVM runtime libraries and program executables among computational resources belonging to distinct administrative domains. Such problems are further complicated by the heterogeneity of multidomain environments, which are usually based on diverse hardware platforms, operating systems, access policies, and middleware systems [8].

This paper presents a lightweight software infrastructure, called *ehPVM*, which enables programmers to deploy ePVM applications among computational resources belonging to multidomain clusters. In particular, *ehPVM* implements simple services that make such resources, which can be also hidden from the Internet, available to dynamically load and run ePVM applications. Thus, ePVM programmers can run their applications without having to use grid software toolkits or resources providers whose configurations usually result in being tedious and time-consuming activities [8,9].

The paper is structured as follows. Section 2 describes *ehPVM*. Section 3 describes how ePVM applications can be deployed by exploiting *ehPVM*. Section 4 reports conclusion remarks.

2 The *ehPVM* Software Infrastructure

ehPVM is a flexible and lightweight Java software infrastructure able to support the deployment of ePVM applications among computational resources belonging to clusters interconnected by heterogeneous multidomain, non-routable networks. It has been designed as a customizable collection of components, and it can aggregate or cross-access varied computational resources spanning distinct administrative domains, networks and institutions.

ehPVM has been developed in Java in order to achieve two main goals:

- to gain maximum compatibility across heterogeneous computing resources;
- to provide fine-grained access control mechanisms that allow users to utilize computing resources without compromising their integrity.

However, *ehPVM* also implements an actual support for resource-dependent functionalities as well as provides features that allow the hiding of heterogeneity from programmers.

ehPVM can build and dynamically reconfigure a "minimal" metacomputer on which ePVM applications can be deployed and run. Such a metacomputer

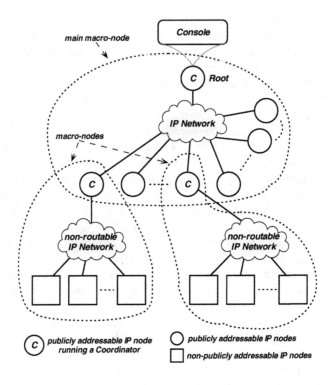

Fig. 1. The organization of the metacomputer

can harness computing resources directly available on the Internet as well as those belonging to multidomain, non-routable private networks, i.e. computing nodes not provided with public IP addresses, but connected to the Internet through publicly addressable IP front-end nodes. Therefore, even though all the computing nodes building the metacomputer result in being actually arranged according to a hierarchical physical network topology consisting of two levels (the level of the publicly addressable IP nodes and the level of the not publicly addressable IP nodes belonging to non-routable private networks), they virtually appear as arranged according to a flat network topology (see Figure 1).

Computing nodes can be PCs, workstations or units of parallel systems interconnected by heterogeneous or dedicated networks. However, *ehPVM* supplies all the basic services that enable programmers to deploy ePVM applications on the underlying physical computational and network resources in a transparent way. To this end, the metacomputer results in being abstractly composed of computational *nodes* interconnected by a flat virtual network and unambiguously identified by integer values automatically assigned at the metacomputer start-up.

The *nodes* hidden from the Internet or connected by dedicated, fast networks can be grouped in *macro-nodes*, which thus abstractly appear as single, more powerful, virtual computing units. In fact, the metacomputer is assumed to be

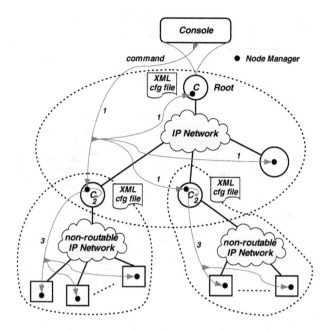

Fig. 2. The scheme of a command execution

made up of at least one *macro-node*, called the *main macro-node*, which groups all the publicly addressable IP *nodes* taking part in the metacomputer.

Each *node* maintains status information about the runtime organization of the metacomputer, such as information about the identity and liveness of the other *nodes*. To this end, the hierarchical physical organization of the metacomputer allows each *node* to keep and update only information about the configuration of the *macro-node* which it belongs to, thus promoting scalability, since the updating information has not to be exchanged among all the *nodes* of the metacomputer.

Each *macro-node* is managed by a special task, called *Coordinator* (C), which:

- runs onto the publicly addressable IP *node* of each non-routable, private network interconnecting other non-directly addressable IP *nodes*;
- creates the *macro-node* by activating *nodes* within the private network that it manages;
- takes charge of updating the status information of each *node* grouped by the *macro-node*;
- monitors the liveness of *nodes* to dynamically change the configuration of the *macro-node*;
- carries out the automatic "garbage collection" of the crashed *nodes* in the *macro-node*.
- acts as a gateway for system communications towards the *nodes* belonging to the *macro-node*.

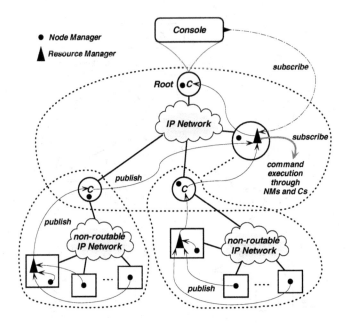

Fig. 3. The publish/subscribe information service

The metacomputer is controlled by the *Coordinator* of the *main macro-node*, called *Root*, which is directly interfaced with the user through the *Console*, by which the configuration of the metacomputer can be dynamically managed (see Figure 1). To this end, a *node* wanting to make its computing power available to *ehPVM* has to run a server, called *Node Manager* (NM), which takes charge of interacting with the *Console* as well as running further software components that enable the *node* to participate in the metacomputer (see Figure 2).

The *Console* can only interact with the NMs running on publicly addressable IP *nodes*, i.e. the *nodes* belonging to the *main macro-node*. Therefore, when a NM hosted by a *node* running a *Coordinator* receives a command from the *Console*, such as, for example, a "configuration" command, it has to forward it to the NMs of the *nodes* grouped by its *macro-node*. To this end, the NM contacts the *Coordinator* of its *macro-node*, which exploits the configuration information stored in a specific XML file to forward the received command to the NMs of the *nodes* inside the *macro-node*. In fact, information contained in the XML file is initially provided by the administrator of the *nodes* grouped by the *macro-node*, and then dynamically updated by the *Coordinator* (see Figure 2).

ehPVM also provides a simple "publish/subscribe" information service implemented by two servers: the *Resource Manager* (RM) and the NM (see Figure 3). In particular, each *macro-node* has an RM, which can be allocated onto one of the *nodes* belonging to the *macro-node*. A RM is periodically contacted by the NMs of the *nodes* belonging to the *macro-node* and wanting to publish information about the CPU power and its utilization or the available memory or the communication performance. Information is collected by the RM and made then available to the "subscribers", which are the *Coordinator*s belonging to the metacomputer. Thus, each *Coordinator* can know the maximum computing/communication power made available by its *macro-node*. Furthermore, this information is also made available to the *Root*, which can thus know the power of all the *macro-node*s making up the metacomputer. This allows the user to know the globally available computing power and reserve a part of it by issuing a subscription request to the *Console*. Then, the *Console* can ask the RM of the *main macro-node* for selecting and reserving only the required computing resources. The result of this process is an XML file containing system information to create a metacomputer without having to consult anew the RM.

3   Deploying ePVM Applications

ehPVM implements an architecture that can host dynamically loaded software components. To this end, it supplies a Java runtime environment in the form of a component container. In fact, components can be uploaded by users who have appropriate permissions, while the administrators of the *node*s aggregated by *ehPVM* retain complete and fine-grained control over the computing and network resources they manage.

*Coordinator*s, NMs and RMs are all implemented as services directly run by the Java Virtual Machine (JVM) at the *ehPVM* start-up. In particular, each NM may request the JVM to link a dynamic native library. More precisely, a NM can transparently resolve and link at runtime the appropriate version of a library precompiled for different platforms by following a four-steps scheme. In particular, the NM:

1. obtains the library path referred to the code repository;
2. resolves the actual Java "resource" name by combining information about the library path, requested library name, and detected platform type;
3. stages the resource to a temporary local file;
4. loads the library from that file.

To this end, NMs assume an automatic platform detection type based on the classification adopted by PVM [3]. Furthermore, *ehPVM* assumes that the loaded resources may originate from an arbitrary URL.

To run an ePVM application, the only files that must be present at the *node*s aggregated by *ehPVM* are: the ePVM daemons, the ePVM runtime library, and the application itself [1,2] (see Figure 4). In fact, ePVM has not external dependencies except for the standard C library, and this makes precompiled versions of these files portable within a single platform type. Therefore, to set up the ePVM runtime, the *Console* requests NMs to load the ePVM daemons and runtime library. Then, NMs take charge of staging and starting up the appropriate platform-specific versions of the ePVM daemons, fetching them from a network code repository. To this end, it is worth noting that the execution of the "set up" command requested by the *Console* is managed by *ehPVM* as a normal command, according to what shown in Figure 2. This means that only

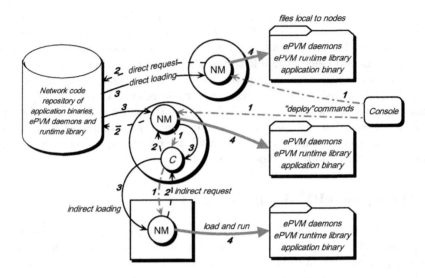

Fig. 4. The deployment of an ePVM application

the NMs belonging to the publicly addressable IP *nodes* can directly load the
ePVM runtime, whereas the NMs running on *nodes* inside *macro-nodes* can load
the runtime through their *Coordinators*.

It is worth noting that *node* administrators have only to authorize the deployment, but they have not to configure the ePVM runtime, since NMs can stage
and start up the appropriate platform-specific versions of the ePVM daemons
and runtime library. Furthermore, users are allowed to execute ePVM applications on computing nodes where they have not a login account but instead, only
restricted *ehPVM* access. As a consequence, the setup procedure implemented
by *ehPVM* hides heterogeneity from users as well as releases *node* administrators from the responsibility to install or configure additional software except for
the *ehPVM* support itself.

The ePVM runtime setup is followed by the application deployment (see Figure 4). To this end, the application binary has to be stored in a specific directory,
designated by the ePVM daemons, of the file system of each *node*. Therefore,
the *Console* can ask NMs for loading the application binary from a URL code
base that the user may specify along with the executable name. As illustrated in
Figure 4, the execution request coming from the *Console* is handled by publicly
IP addressable NMs, which use their already described capabilities to fetch the
platform-specific versions of the application and save them in files local to *nodes*.
Then, these NMs ask their *Coordinators* for forwarding the execution request to
NMs running on the *nodes* inside *macro-nodes*, thus enabling the deployment of
the application on all the *nodes* aggregated by the metacomputer.

After the deployment of the application, each NM can directly invoke the
ePVM daemons to launch the application from the local file. However, to provide protection against malicious code, NMs can assess whether the deployed

application can be trusted. To this end, NMs can base their decisions upon the code source and/or user who requested the execution, according to the basic Java security model. In particular, NMs can restrict the code source to designated places that can be specified as URL base paths and/or verify code signatures and/or use flexible authentication mechanisms to determine user identity.

Finally, it is worth noting that the functionalities implemented by *ehPVM* are mainly related to the application and environment start-up, and thus have only a minimal effect on the ePVM performance already documented in [1,2].

4   Conclusions

This paper has presented *ehPVM*, a lightweight software infrastructure that enables programmers to easily deploy ePVM applications among computational resources belonging to multidomain clusters. *ehPVM* supplies flexible services that can make resources hidden from the Internet available to dynamically load and run ePVM applications. Thus, ePVM can take advantage of important functionalities that can be attained with no changes to its implementation and with only marginally lowered performance, as demonstrated by preliminary tests.

References

1. Frattolillo, F.: A PVM extension to exploit cluster grids. In: Procs of the 11th EuroPVM/MPI Conference. Volume 3241 of LNCS., Budapest, Hungary (2004) 362–369
2. Frattolillo, F.: Running large-scale applications on cluster grids. Int'l Journal of High Performance Computing Applications **19**(2) (2005) 157–172
3. Geist, A., Beguelin, A., et al.: PVM: Parallel Virtual Machine. A Users' Guide and Tutorial for Networked Parallel Computing. MIT Press (1994)
4. Petrone, M., Zarrelli, R.: Utilizing PVM in a multidomain clusters environment. In: Procs of the 12th EuroPVM/MPI Conference. Volume 3666 of LNCS., Sorrento, Italy (2005) 241–249
5. Frattolillo, F., D'Onofrio, S.: An efficient parallel file system for cluster grids. In: Procs of the 12th EuroPVM/MPI Conference. Volume 3666 of LNCS., Sorrento, Italy (2005) 94–101
6. Moyer, S.A., Sunderam, V.S.: PIOUS: a scalable parallel I/O system for distributed computing environments. In: Procs of the Scalable High-Performance Computing Conference. (1994) 71–78
7. M.P.I. Forum: MPI-2: Extensions to the Message-Passing Interface, http://www.mpi-forum.org/docs/mpi-20.ps. (1997)
8. Kurzyniec, D., Hwang, P.and Sunderam, V.: Failure resilient heterogeneous parallel computing across multidomain clusters. Int'l Journal of High Performance Computing Applications **19**(2) (2005) 143–155
9. Schopf, J.M., Nitzberg, B.: Grids: The top 10 questions. Scientific Programming, special issue on Grid Computing **10**(2) (2002) 103–111

Reliable Orchestration of Distributed MPI-Applications in a UNICORE-Based Grid with MetaMPICH and MetaScheduling

Boris Bierbaum[1], Carsten Clauss[1], Thomas Eickermann[2],
Lidia Kirtchakova[2], Arnold Krechel[3], Stephan Springstubbe[3],
Oliver Wäldrich[3], and Wolfgang Ziegler[3]

[1] Chair for Operating Systems, RWTH Aachen University,
52056 Aachen, Germany
{boris, carsten}@lfbs.RWTH-Aachen.DE
[2] Central Institute for Applied Mathematics, Research Centre Jülich,
52425 Jülich, Germany
{th.eickermann, l.kirtchakova}@fz-juelich.de
[3] Fraunhofer Institute SCAI
53754 Sankt Augustin, Germany
{arnold.krechel, stephan.springstubbe, oliver.waeldrich,
wolfgang.ziegler}@scai.fraunhofer.de

Abstract. Running large MPI-applications with resource demands exceeding the local site's cluster capacity could be distributed across a number of clusters in a Grid instead, to satisfy the demand. However, there are a number of drawbacks limiting the applicability of this approach: communication paths between compute nodes of different clusters usually provide lower bandwidth and higher latency than the cluster internal ones, MPI libraries use dedicated I/O-nodes for inter-cluster communication which become a bottleneck, missing tools for co-ordinating the availability of the different clusters across different administrative domains is another issue. To make the Grid approach efficient several prerequisites must be in place: an implementation of MPI providing high-performance communication mechanisms across the borders of clusters, a network connection with high bandwidth and low latency dedicated to the application, compute nodes made available to the application exclusively, and finally a Grid middleware glueing together everything. In this paper we present work recently completed in the VIOLA project: MetaMPICH, user controlled QoS of clusters and interconnecting network, a MetaScheduling Service and the UNICORE integration.

Keywords: MetaMPICH, Grid, Co-allocation, UNICORE, Network QoS.

1 Introduction

1.1 The VIOLA Project

The work presented here is carried out in the context of VIOLA [1] (Vertically Integrated Optical testbed for Large Applications), a co-operative project with a

B. Mohr et al. (Eds.): PVM/MPI 2006, LNCS 4192, pp. 174–183, 2006.

consortium of 12 partners from German research labs, universities and industry, lead by DFN, the German NREN. VIOLA is funded by the German federal ministry of education and research BMBF. The project has set up an optical testbed in North-Rhine-Westfalia with an extension to Bavaria. Main objectives are evaluation and testing of advanced networking equipment and technologies in a close-to-production environment and development of software for user-driven dynamical provisioning of network bandwidth and quality-of-service. A set of initially four applications with high communication demands has been selected to provide real-life requirements and to stress-test the network. Three of them are performing distributed simulations with MPI-based codes on the currently five Linux- and Solaris-based clusters in the testbed. The clusters are attached to the testbed with multiple Gigabit-Ethernet adapters, in most cases one adapter in each node of the cluster. The clusters are interconnected over the testbed via 10 Gigabit-Ethernet. Given this complex and bandwidth-rich environment, it is obvious that a scalabe high-performance MPI-implementation with wide-area support is a prerequisite for efficient use of the testbed. Also, Grid middleware is required to orchestrate the various resources and to provide reliable, secure and seamless access to them. For the former, the RWTH has extended their Metacomputing MPI-implementation MetaMPICH [12], for the latter we have integrated a MetaScheduling Service into the Grid system UNICORE [15]. In VIOLA the MetaScheduling Service does not orchestrate Web Services as the applications are not wrapped in services and the orchestration is made for a synchronous start. However, as described in Section 6 future versions of the MetaScheduling service will also support workflows - or choreography - of two or more Web Services.

1.2   Related Work

Besides MetaMPICH, there are other MPI implementations enabling the coupling of compute resources over wide-area networks, most notably PACX-MPI [4], MPICH/Madeleine [3], and MPICH-G2 [10]. The features that differentiate MetaMPICH from some or all of these approaches are the startup mechanism using a single configuration file, the choice between two different methods to couple clusters (*routers* and *multidevice*), and the fact that it is not tied to a specific grid system. There are also a number of approaches for co-allocation of resources like KOALA, CSF, GridWay or products like MP Synergy or Moab. However, most of them are not providing advance reservation and neither reliable SLAs nor co-allocation of compute resources with the interconnecting network guaranteeing a user-requested QoS. In the UNICORE Plus project [7] a proof-of-concept implementation of a MetaScheduling Service based on a proprietary negotiation protocol has been implemented. It supports PACX-MPI as MPI library, the the CCS (Computing Center Software) [14] as reservation system for advance reservation of compute resources. Our system uses some of the ideas of the UNICORE Plus development (e.g. making the functionality accessible via a UNICORE client plugin), but is besides that a completely independent design and implementation. Related projects based on optical Grid testbeds are e.g. the

Japanese g-lambda project or the Polish CLUSTERIX project. These projects use different middleware and have a different focus, but co-operation has been launched to exchange developments made and to work on interoperability.

1.3   Remainder of the Paper

The remainder of the paper is organised as follows. In Section 2 we present the MetaMPICH library developed at RWTH. The VIOLA MetaScheduling environment is described in Section 3, followed by the description of the MetaMPICH integration in both the MetaScheduling Service and the UNICORE system in Section 4. Experiences made are discussed along a use case we present in Section 5. An overview about further developments for the MetaScheduling environment and the MetaMPICH library in Section 6 concludes the paper.

2   MetaMPICH

Based on MPICH1 [9], MetaMPICH was originally developed to couple MPPs from different vendors in the Gigabit Testbed West project [8]. Since those systems internally had very fast networks, but only dedicated I/O-nodes for external communication, a router-based communication architecture was chosen, as depicted in the left part of Figure 1. We call each of those coupled systems a *meta host* in the context of meta computing.

In a second stage, MetaMPICH was extended to support the emerging class of PC-based cluster systems with high-performance interconnects like SCI [17]. MetaMPICH has been optimised for coupling such clusters, as published in [12]. One advantage of the router approach was the possibility to couple an arbitrary number of cluster nodes via fast, dedicated external connections to achieve higher scalability and higher bisectional bandwidth between the systems.

However, when coupling clusters via the high-bandwidth VIOLA network, the communication performance between clusters is limited by the speed with which the I/O-nodes can handle the traffic. Since all compute nodes within the VIOLA project are connected to the VIOLA WAN, it becomes preferable to let the application processes communicate directly with each other. To enable this, we implemented a new architecture for MetaMPICH.

The result is shown in the right part of Figure 1. It is called the *multidevice* architecture, because it enables the usage of multiple MPICH communication drivers (called *devices*) side-by-side. Note that every node of system A can send data directly to every node of system B and vice versa. That way, this approach allows to run large applications that benefit from the dedicated internal cluster networks and from the connecting high-performance optical network at the same time. Nevertheless, in order not to lose the flexibility of a router driven communication, which is the only choice in some environments, MetaMPICH also supports setups combining router-based and multidevice coupling.

The needs of the VIOLA project also led to several other improvements of MetaMPICH: Support for Myrinet was added by integrating code from the MPICH-GM distribution. The device for TCP/IP communication, ch_usock,

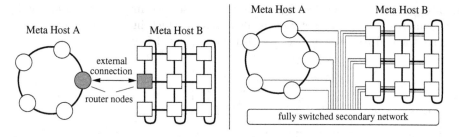

Fig. 1. Communication Architectures of MetaMPICH

was made *instantiable* to be able to use it for cluster-internal communication as well as for coupling clusters at the same time. The syntax of the *meta configuration file* [13], with which a coupled system is configured, was extended to support several new requirements, e.g. automated startup of server processes for remote parallel I/O.

3   VIOLA MetaScheduling Environment

The MetaScheduling Service (MSS) has beed developed to ensure that all resources necessary for executing the distributed applications are available. The MSS receives the information on resources needed for an application from the UNICORE client via an agreement proposal [2] containing the specification of resources and QoS. The MSS then starts the negotiation process with the local Resource Management Systems (RMS) of these resources, where the compute resources are managed by the local scheduling systems and the network resources by the ARGON (Allocation and Reservation in Grid-enabled Optic Networks) system. Due to the heterogenous nature of the employed RMS we used a set of adapters to suport the MSS during the negotiation process by providing a stable interface to the different RMS (see Fig. 2). The negotiation process consits of four main phases:

1. querying the local RMS for free slots to execute the application within a preview period
2. determining a common time slot
3. if such a time slot exists, perform a reservation request of this slot on behalf of the user;
   otherwise restart the query with a later start time of the preview period
4. check whether the reservation was made for the correct time slot on all systems, if yes, we are done;
   otherwise restart the query with a later start time of the preview period.

The successful negotiation and reservation is sent back as agreement to the UNICORE client which then continues processing the job as usual. Once the job starts at the negotiated starting time the MSS collects the IP addresses of the compute nodes finally allocated by the local RMS. The IP addresses are used to generate the meta-configuration file as described in Section 4 below and are communicated to the network RMS ARGON which in turn is then able to manage the end-to-end connections with the requested QoS.

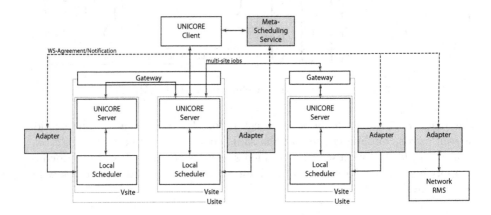

Fig. 2. The meta-scheduling architecture

4   Grid Middleware Integration

4.1   Integration into the MetaScheduling Service

In this section we describe how the MSS supports the preparation of the runtime environment for MetaMPICH applications. As mentioned in Section 3 an important functionality of the MSS is the coordinated allocation of multiple resources (e.g. compute and network) at different sites. Compute reservations mainly consist of the number of required nodes at a compute site, the duration, an executable and the start time for the reservation. Network reservations specify a network service, which consists of a set of point-to-point connections, the bandwidth for each connection, the duration and a start time of the service. Further on, each connection is specified by two connection endpoints (source and destination), where each endpoint is associated with one compute site in the VIOLA network, respectively with the router that represents this compute site in the network.

The process of negotiating the execution time for a MetaMPICH application allocating the required resources was already described in Section 3. After submission of all reservations the MSS continuously monitors the partial reservations. When all reservations entered the state running (active), the MSS queries the compute RMS in order to determine which nodes (IP addresses of the nodes) where finally assigned to a reservation. This information is collected and aggregated by the MSS, and is then published as runtime configuration to all subsystems.

Publishing this runtime configuration to the network RMS ARGON comprises the completion of the reservation data in terms of which compute nodes (list of

IP addresses) belong to each connection endpoint defined in the reservation. This address data is used at the network layer to create access control lists (ACL) at the routers in order to enable the compute nodes belonging to a MetaMPICH job to communicate with each other using the QoS level specified in the network reservation. Therefore the ARGON system implements a bind functionality that allows completing reservation data of existing reservations at runtime.

The runtime configuration data is used in a different way for compute resources. Here an XML configuration file is created on each cluster, which contains the nodes (a list of IP addresses) that belong to a reservation for every site. This configuration file then is used together with the job description submitted by UNICORE to generate a MetaMPICH configuration during the startup process of the application, as further described in Section 4.2.

4.2   UNICORE Integration

As the underlying Grid software, UNICORE is responsible for several tasks during the lifetime of a MetaMPICH application:

- providing a user interface for specification of the Meta-Job,
- interacting with the MetaScheduling Service to allocate the requested resources,
- management of the job: start and monitor the sub-jobs on the individual clusters, retrieve and present the job output.

The first and second task are performed by the UNICORE user-client by means of a Metacomputing-plugin, developed specifically for this purpose. The third is one of the core UNICORE server responsibilities. Managing MetaMPICH jobs did not require changes of this server, but just some user-level wrapper scripts for starting the MPI-application.

The Metacomputing-plugin is an extension of UNICORE's graphical user-client, that lets the user specify the MetaMPICH job in a convenient way: In a main panel, the user specifies the duration and favored start time of the job and selects the clusters, on which the individual sub-jobs shall run. In a communication matrix, the number of MetaMPICH router-pairs (which defaults to 0) and the required bandwidth between each pair of clusters can be specified. Then for each sub-job, the user enters the executable, the number of MPI-tasks and various other optional configuration parameters in a separate form.

The job description entered via the plugin is sent along with the job encoded in XML. At job startup time, when the actually allocated nodes are known, the MetaMPICH configuration-file is created on each of the participating clusters, based on the XML job description and the IP-addresses provided by the MetaScheduling system.

The startup of the MetaMPICH application is also different from the standard way, where a single execution of 'mpirun' will start the sub-jobs on all clusters via ssh. In the UNICORE integrated version, each MetaMPICH sub-job is represented as a UNICORE sub-job and is started individually by the local scheduling system of the cluster. The advantages are that no ssh-logins between the clusters are required and that the sub-jobs can be monitored individually by UNICORE.

5   Use Case: Distributed Algebraic Multi Grid Solver

Solving huge, linear, sparse systems of equations is an important subtask in many simulation codes, e.g. of computational fluid mechanics, structural mechanics or semiconductor device simulation [16], [5]. Typically, the efficiency of a such simulation code is restricted by the efficiency of the linear solver used. Algebraic multigrid methods provide a well established, state-of-the-art solver technology for wide classes of applications. They are optimal since they turn out to be numerically scalable : the time for solving a problem in a certain class grows only linearly with the problem size. The AMG solver technology has been made available in the VIOLA Grid for all simulation codes where the solver of a linear system is a numerical bottleneck. We have shown that the VIOLA Grid is suitable for these industrial simulation codes.

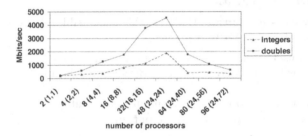

Fig. 3. Throughput for data redistribution on two clusters. The throughput for integers is lower than for the doubles as more messages of smaller length have to be sent for the integer data.

Especially, when considering problem sizes which are expected to become relevant in the nearest future. The parallelization of algebraic multigrid methods requires various communication patterns and therefore is a real challenge for the network and the communication software. When starting the solver process, huge amounts of data describing the problem to be solved have to be distributed to the computing MPI processes. The most important factor here is the transfer rate of the network. After having distributed the problem data, a hierarchy of coarse and fine grids has to be calculated. In this setup phase the network latencies become more important. The reason is an increased number of messages which are at the same time of much smaller length than in the redistribution phase. For the same reason the network latencies become the most important in the solution phase.

An existing parallel, MPI-based algebraic multigrid code has been ported to the VIOLA Grid using MetaMPICH. It could be demonstrated that the throughput of the network can be exploited for the program phase dealing with the redistribution of the problem data, provided that enough processors are used. In addition, the timings show that in the Compute Grid VIOLA the redistribution of a sparse matrix problem is always an option: typically, the time of the redistribution of a larger problem is strictly less than the time for solving it.

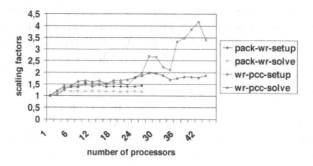

Fig. 4. Scaling test using two clusters

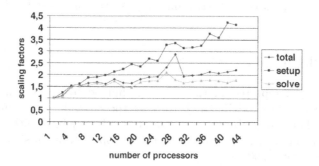

Fig. 5. Scaling test using three clusters

Fig. 4 and 5 show the results of scaling tests for AMG when connecting two and three PC-Clusters respectively using MetaMPICH. The PC-Clusters are up to 80 km apart. For a scaling test the problem size per processor is kept fixed. Scaling factors are defined by dividing the time of the larger problem on the larger number of processors by the time used for the smaller problem on one processor. Therefore, scaling factors of one are ideal, whereas scaling factors of n imply a restriction to a factor of 1/n of the maximum efficiencies that can be anticipated for the larger problem. The scaling factors (Fig. 5) for the total time and the solving time are very satisfactory. The scaling factors for the setup however indicate that restriction of communication in the setup will have to be investigated further [11]. The total scaling factors are not bogged down by the scaling factors of the setup as the time for the solution phase is dominating.

6   Future Perspectives

Methods improving the interoperability of MetaMPICH are currently in development, namely a new device that can use other MPI implementations for communication inside a meta host and support for the Interoperable MPI (IMPI) standard. To improve application performance, the implementation of optimized collective communication operations for wide-area networks and support for MPI process topologies are planned. The results of the VIOLA project wrt orchestration of services including support for user-driven dynamical provisioning of network bandwidth and quality of-service will be adopted and made available on a European level in the EU funded LUCIFER project. While the current version of the VIOLA Grid testbed expects the user to describe the resource demands of his application using the UNICORE client and do a pre-selection of resources satisfying this demand, we are working in several other projects to have applications providing this information. E.g. together with the Swiss EPFL an interface for a resource Broker responsible for generating a candidate set of potentially suitable resources to run an application has been defined and is currently implemented [6]. Furthermore, the MetaScheduling Service will be made available for GT4 or gLite based Grid environments in the near future. It will then support reliable orchestration and reservation across Grids based on different middleware.

Acknowledgements

Some of the work reported in this paper is funded by the German Federal Ministry of Education and Research through the VIOLA project under grant #01AK605L. This paper also includes work carried out jointly within the Core-GRID Network of Excellence funded by the European Commission's IST programme under grant #004265.

References

1. VIOLA – Vertically Integrated Optical Testbed for Large Application in DFN, 2006. 29 March 2006 <http://www.viola-testbed.de/>.
2. A. Andrieux and K. Czajkowski and A. Dan and K. Keahey and H. Ludwig and T. Nakata and J. Pruyne and J. Rofrano and S. Tuecke, and M. Xu.   WS-Agreement - Web Services Agreement Specification, 2006. 19 April 2006 <https://forge.gridforum.org/projects/graap-wg/document/WS-AgreementSpecificationDraft.doc/en/31>.
3. O. Aumage and G. Mercier. MPICH/MADIII: a Cluster of Clusters Enabled MPI Implementation. In *Proceedings of the 3rd IEEE/ACM International Symposium on Cluster Computing and the Grid (CCGrid 2003)*, pages 26–36, Tokyo, Japan, 2003.
4. T. Beisel, E. Gabriel, M. Resch, and R. Keller. Distributed Computing in a Heterogeneous Computing Environment. In *Recent Advances in PVM and MPI – Lecture Notes in Computer Science*, pages 180–187, 1998.
5. T. Clees. AMG Strategies for PDE Systems with Applications in Industrial Semiconductor Simulation., 2005. Fraunhofer Series n Information and Communication Technology, vol. 6. Fraunhofer SCAI, Sankt Augustin, Germany.
6. K. Cristiano, R. Gruber, V. Keller, P. Kuonnen, S. Maffioletti, N. Nellari, M.-C. Sawley, M. Spada, T.-M. Tran, O. Wäldrich, Ph. Wieder, and W. Ziegler. Integration of ISS into the VIOLA Meta-scheduling Environment. In *Proc. of the 2nd CoreGRID Integration Workshop*, volume 4 of *CoreGRID Series*, pages 47–54. Springer, 2006. To appear.

7. D.Erwin (eds). UNICOREi Plus Final Report. Technical report, Research Center Jülich, Germany, 2003. ISBN 3-00-011592-7.

8. T. Eickermann, R. Völpel, and P. Wunderling. Gigabit Testbed West – Final Report. Technical report, Research Center Jülich, Germany, 2000.

9. W. Gropp, E. Lusk, and A. Skjellum. A High-Performance, Portable Implementation of the MPI Message Passing Interface Standard. *Parallel Computing*, 22(6):789–828, 1996.

10. N. Karonis, B. Toonen, and I. Foster. MPICH-G2: A Grid-Enabled Implementation of the Message Passing Interface. *Journal of Parallel and Distributed Computing (JPDC)*, 63(5):551–563, 2003.

11. A. Krechel and K. Stüben. Parallel algebraic multigrid based on subdomain blocking. *Parallel Computing*, 27:1009–1031, 2001.

12. M. Pöppe, S. Schuch, and T. Bemmerl. A Message Passing Interface Library for Inhomogeneous Coupled Clusters. In *Proc. of CAC Workshop at IPDPS'03*, 2003.

13. M. Pöppe, S. Schuch, R. Finocchiaro, C. Clauss, and J. Worringen. MP-MPICH User Documentation and Technical Notes., 2005. Aachen: Lehrstuhl für Betriebssysteme, RWTH Aachen.

14. F. Ramme, T. Romke, and K. Kremer. A Distributed Computing Center Software for the Efficient Use of Parallel Computer Systems. In *Proc. of HPCN Europe, Int. Conf. on High-Performance Computing and Networking 1994*, volume 797 of *Lecture Notes in Computer Science*, pages 129–136. Springer, 1994.

15. A. Streit, D. Erwin, Th. Lippert, D. Mallmann, R. Menday, M. Rambadt, M. Riedel, M. Romberg, B. Schuller, and Ph. Wieder. UNICORE - From Project Results to Production Grids. In *L. Gandinetti (Edt.), Grid Computing: The new Frontiers of High Performance Processing, , Advances in Parallel Computing 14*, Elsevier, 2005.

16. K. Stüben. A Review of algebraic multigrid. *Comp. Appl. Math.*, 128:281–309, 2001.

17. J. Worringen. SCI-MPICH: The Second Generation. In *Proceedings of SCI-Europe 2000 (Conference Stream of Euro-Par 2000)*, pages 11–20, Munich, Germany, 2000.

The New Multidevice Architecture of MetaMPICH in the Context of Other Approaches to Grid-Enabled MPI

Boris Bierbaum, Carsten Clauss, Martin Pöppe,
Stefan Lankes, and Thomas Bemmerl

Chair for Operating Systems, RWTH Aachen University, Germany
mp-mpich@lfbs.rwth-aachen.de
http://www.lfbs.rwth-aachen.de

Abstract. MetaMPICH is an MPI implementation which allows the coupling of different computing resources to form a heterogeneous computing system called a *meta computer*. Such a coupled system may consist of multiple compute clusters, MPPs, and SMP servers, using different network technologies like Ethernet, SCI, and Myrinet. There are several other MPI libraries with similar goals available. We present the three most important of them and contrast their features and abilities to one another and to MetaMPICH. We especially highlight the recent advances made to MetaMPICH, namely the development of the new *multidevice* architecture for building a meta computer.

Keywords: MPI, Grid Computing, Meta Computing, MPICH, Multidevice Architecture.

1 Introduction

Since several years ago, there is a strong trend in the scientific computing community towards distributed high performance computing. By coupling multiple computing resources to a single meta computer, a higher degree of parallelism can be achieved and thereby larger problem instances can be solved. Such a coupled system is heterogeneous by nature, e.g. regarding administration policies, network technologies, and communication performance between different processes. This heterogeneity leads to specific issues which do not have to be dealt with if homogeneous systems are used. As MPI is currently the most important API for implementing parallel programs, several MPI libraries have been developed which try to address these issues in an efficient and user-friendly manner.

In this paper, we compare and contrast four different MPI implementations that support heterogeneous systems, one of which, *MetaMPICH*, is being developed at the Chair for Operating Systems, RWTH Aachen University. As they are under active development and have already been used for running large-scale parallel applications, we chose MPICH-G2 [1], PACX-MPI [2], and MPICH/Madeleine [3] as our examples besides MetaMPICH. Several other projects are of interest in this context, e.g. MagPIe [4], IMPI [5], and STAMPI [6], but they are not covered in this paper.

B. Mohr et al. (Eds.): PVM/MPI 2006, LNCS 4192, pp. 184–193, 2006.

We compare the MPI implementations with regard to several topics that are of high importance for heterogeneous systems: the method by which the coupling is done, approaches to support different architectures and networks, methods of grid integration, and the way in which the system's heterogeneity is exploited to support efficient data communication. To keep this paper focused, we concentrate on the important features of each implementation and do not provide extensive benchmark results. A detailed benchmark comparison between PACX-MPI and MPICH/Madeleine can be found in [7].

This paper is organized as follows: In Section 2, we describe MPICH-G2, PACX-MPI, and MPICH/Madeleine with regard to the above mentioned topics. In Section 3, we present MetaMPICH, especially the new features that have recently been added to it. Section 4 describes experiences with the usage of MetaMPICH and Section 5 concludes the paper.

2 Approaches to Meta Computing

For the scope of this paper, we view a meta computer as being composed of coupled entities called *meta hosts*, which may be clusters, MPPs, or SMP servers. Figure 1 shows the three possible methods for coupling: using dedicated *router* processes communicating over an additional interlinking network (a), *all-to-all* connectivity between application processes (b), or connecting meta hosts via *gateway* nodes which are part of two coupled entities (c). A description of the advantages of each coupling method can be found in Section 3.1, Section 3.2, and Section 2.3, respectively.

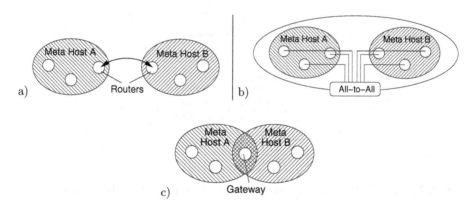

Fig. 1. Coupling Methods for Heterogeneous Network Architectures

We have identified the following approaches to support multiple networking technologies inside a meta computer (see Figure 2): Building the support into the driver layer of the MPI implementation (a), using an external communication library providing access to multiple networks (b), and using an additional layer on top of other MPI libraries (d and e). For those MPI implementations which are

186   B. Bierbaum et al.

based on MPICH [8] (like MPICH-G2, MPICH/Madeleine, and MetaMPICH), the first approach means to have multiple ADI (Abstract Device Interface) *devices* built into the library, each of which supports the message transfer via a different underlying mechanism. This is what we call a *multidevice* architecture.

For grid integration, a Grid-enabled MPI implementation can either be tied to a specific grid system or provide its own method to configure a meta computer, which then has to be integrated with a resource management system.

Two different methods for efficient communication in a heterogeneous system may be provided: *transparent*, optimizing as much as possible without breaking the source code compatibility with other MPI implementations, or *intransparent*, by providing additional means for the developer to adapt the application to the meta computer topology while sacrificing compatibility.

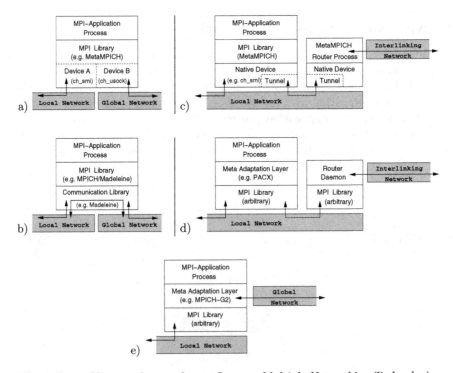

Fig. 2. Layer View on Approaches to Support Multiple Networking Technologies

2.1 MPICH-G2

MPICH-G2 is designed to couple parallel systems using arbitrary MPI libraries over wide-area-networks via the *Globus Toolkit*. Globus is responsible for tasks like authentication and process creation on all participating machines. MPICH-G2 in itself is implemented as an ADI device (the *globus2* device), which provides processes with all-to-all connectivity via the Globus system.

By relying on other MPI implementations for communication inside of a part of the coupled system (see Figure 2e), MPICH-G2 does not have to support network technologies directly. Thus, much flexibility is gained at the expense of some communication overhead. A lot of work has been performed to implement topology-aware collective communication operations based on a multilevel tree model [9] (transparent exploitation of heterogeneity). The description of the system regarding this model is also made available to an MPI application programmer via communicator attributes to assist in (intransparently) optimizing the application for a given topology.

2.2 PACX-MPI

PACX-MPI is not a self-contained MPI implementation, but rather a tool for adding meta computing capability to existing MPI libraries. Thereby, the PACX-MPI library is located between the user application and the native cluster MPI implementation. In case of local communication, the message is just handed over to the underlying MPI library. But, in case of inter-cluster communication, the message is forwarded to dedicated router daemons (Figure 2d). These routers, residing on two special I/O-nodes of each participating cluster (one for outgoing and one for incoming messages), can then send the message to the respective remote cluster.

Since PACX-MPI is utilizing an underlying vendor MPI library, its approach is slightly similar to MPICH-G2 in this context. But unlike MPICH-G2, inter-cluster connections are explicitly established between the router daemons only and not between all participating nodes. By exploiting the often well-adjusted vendor MPI implementation for local communication and by taking the connection between meta hosts into account as a possible bottleneck, PACX-MPI transparently provides optimized collective communication operations.

2.3 MPICH/Madeleine

In contrast to MPICH-G2 and PACX-MPI, MPICH/Madeleine is not used to couple underlying MPI libraries. Instead, it relies on the *Madeleine* communication library, which provides connectivity between processes running on different parts of the meta computer (Figure 2b). Therefore, for a cluster to be used in an MPICH/Madeleine system, the cluster interconnect must be supported by Madeleine. The glue between the upper layers of MPICH and the Madeleine library is the `ch_mad` ADI device. The developers of MPICH/Madeleine chose not to follow a multidevice approach like that described in Section 3.2, because they assumed that it would require a lot of integration work to make multiple devices coexist, a claim that is repeated in [7].

For coupling meta hosts, MPICH/Madeleine provides support for the all-to-all and the gateway approach. The latter avoids using a slow network between clusters, but the gateway approach is not suitable for wide-area distributed computing, because there must be one node which is part of two meta hosts. The system is configured via configuration files with a very simple syntax. Unlike MPICH-G2, MPICH/Madeleine is not tied to a specific grid system. A means to intransparently exploit the heterogeneity of the meta computer is provided by an additional array of pre-defined communicators.

3   MetaMPICH

MetaMPICH is part of MP-MPICH, which is an ongoing development project at our chair. In its original architecture [10], it allowed to couple MPPs and later clusters of PCs to meta computers via router processes (Figure 1a). This feature is described in Section 3.1. It is now complemented by the new multidevice architecture which is detailed in Section 3.2.

One core concept of MetaMPICH is to have a single, central configuration entity for the whole meta computer, the so-called *meta configuration file* [11]. A part of an exemplary meta configuration file is shown in Figure 3. With this file, it is possible to configure the system in a very detailed way. Unlike the configuration files of MPICH/Madeleine, which contain only host names (a shortcoming criticized in [7]), a meta configuration file differentiates between hosts and network interfaces and can easily handle multiple NICs of the same type in the same machine. This way of configuring the system gives users a maximum of control over the configuration and allows to precisely match the configuration on the underlying hardware.

```
NUMHOSTS 2                          # meta computer with 2 meta hosts
OPTIONS
SECONDARY_DEVICE ch_usock (          # secondary device configuration
  PORTBASE=2200,
  NETMASK=192.168.2.0/24            # use 2nd TCP/IP net
)
...
METAHOST metahost_a {                # start of meta host definition
   type = ch_smi;                    # meta host is an SCI cluster
   nodes = node01(134.130.62.12,     # node with multiple NICs for
                 192.168.2.100), ... # connection between meta hosts
   ...
}
...
CONNECTIONS
PAIR metahost_a metahost_b 0 -        # connected via secondary device
...                                   # (0 router connections)
```

Fig. 3. Part of a meta configuration file

Like MPICH-G2 and MPICH/Madeleine, MetaMPICH provides MPI application programmers with a tool to intransparently optimize their code for the underlying heterogeneous system via the predefined MPI communicator MPI_COMM_LOCAL, which contains all application processes running on the same meta host [12].

MetaMPICH does not enforce the usage of a specific grid system (like MPICH-G2). In fact, it can be used without grid middleware at all, if the services provided by such software are not needed, e.g. for ad-hoc-coupling of local compute resources, something which MPICH/Madeleine is also especially suited for.

3.1   The Router-Based Architecture

The first version of MetaMPICH implemented a router architecture for coupling the meta hosts, because the target platforms were MPP systems with dedicated I/O-nodes for external communication. These nodes were used to run the router processes (Figure 2c). When MetaMPICH was extended to support clusters of PCs later on, the router architecture still proved to be suitable. These clusters had high-performance interconnects and were part of a Fast Ethernet LAN. As the router connection over this LAN was a communication bottleneck when coupling clusters (see Section 4), the emerging Gigabit Ethernet technology was used to couple dedicated I/O-nodes of the clusters via router processes.

Routers can be used to build distributed meta computers with meta hosts hidden behind a firewall, because the concept eliminates the need to give all nodes of a cluster system access to an external network. Those issues regarding the coupling of private networks via gateways and proxies are also discussed in [13] concerning PACX-MPI and MPICH-G2. The MetaMPICH implementation of the router concept provides bundling of network interfaces (similar to PACX-MPI) to increase bandwidth between meta hosts as well as static load balancing among multiple router connections between two meta hosts to offload the I/O-nodes. In contrast to PACX-MPI, a single router process is responsible for incoming and outgoing messages, but those can be handled by different threads, if desired.

3.2   The Multidevice Architecture

Recently, we complemented the router-based architecture of MetaMPICH by implementing the multidevice approach shown in Figure 2a, which provides users of MetaMPICH with a new and unique way for coupling compute resources. The first step was to enable *build configurations* with multiple devices: Whereas an MPICH installation can contain a single device only and thus can communicate just over a single data transfer mechanism (e.g. one network technology), MetaMPICH can now be built with multiple devices linked together. The network to be used can then be chosen when starting the application. Moreover, MetaMPICH supports a plugin mechanism to dynamically load the needed device at startup time.

The next step of development was to extend the idea of *choosing* one of multiple devices at startup time to the possibility of *using* several devices simultaneously at runtime. Although the original MPICH already includes rudimentary support for multiple devices, up to now, this feature has merely been used to implement special-purpose devices, e.g. for loop-back functionality.

However, our approach is in fact the implementation of two coexistent and *in-dependent* communication devices on each meta host. While the *primary* device is cluster-specific, enabling each meta host to benefit from its internal high-speed network, the *secondary* device couples the meta hosts. E.g., if two SCI clusters are coupled via Ethernet, the ch_smi device [14] provides intra-cluster connectivity, whereas the ch_usock device (communication via sockets) is responsible for communication between processes running on different clusters (see Figure 3). To enable this, some work had to be performed to really separate the devices from one another, because originally they were developed under the assumption to be the only device in use.

MPICH devices expect runtime configuration information (e.g. process ranks) in the form of command line parameters. These are constructed by some startup mechanism (e.g. shell script or grid system) or must be supplied by the user when starting each process individually. In multidevice configurations, the command line parameters are passed to the primary device, whereas the parameters for the secondary device are internally constructed from the meta configuration file. They are then passed to the secondary device as "artificial" command line parameters, so that a device can use the same form of parameter passing when being used as primary or secondary device.

We were cautious to not break the compatibility to MPICH, so that devices written for MPICH can still be integrated into MetaMPICH to support new network architectures. The success of this attempt is proved by the fact that afterwards we were easily able to support the deployment of Myrinet clusters in the VIOLA network (see Section 4) by taking the code from the MPICH-GM distribution from Myricom[1].

By making use of a secondary device, it is possible to build routerless meta computers, i.e. configurations in which all meta hosts communicate via the secondary device. Additionally, to give the users as much flexibility as possible, mixed configurations can be set up, in which some meta hosts are coupled via router processes and some via a secondary device. To our knowledge, there has not been any grid-enabled MPI implementation offering these combination of coupling methods yet. If there is no router connection defined between two meta hosts, they are automatically connected via the secondary device, if one has been configured before (see Figure 3). To be able to use the ch_usock device as a secondary device for meta hosts which also use it internally, this device has been made *instantiable*, i.e. multiple instances of it can coexist and be used concurrently.

4   Grid Integration and Performance

Using MetaMPICH's configuration file for building a meta computer has two main advantages: It provides fine-grained control over the details of the system and it is independent of a specific grid management system. The drawback is that some work has to be performed to integrate MetaMPICH with a grid system,

---

[1] http://www.myri.com.

which can then automatically configure a meta computer and prevent the users from having to write their own meta configuration files.

This work was done for the VIOLA project [15], whose goal is to implement and operate a computational grid based on various computing centers in Germany, which are linked by a high-speed optical WAN. To enable a transparent view on the grid resources for the applications, the VIOLA grid uses the UNI-CORE middleware [16]. This middleware allows the user to submit computing jobs without the need to configure the jobs individually for the target systems. In order to not only use a remote resource separately (e.g. a remote cluster), but also to distribute large applications across all the grid resources (e.g. the local and various remote clusters), MetaMPICH and an additional Meta-Scheduler [17] have been integrated into the UNICORE framework.

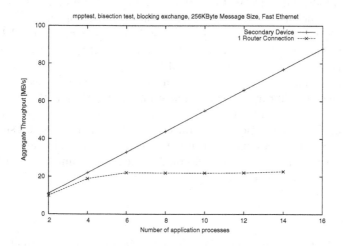

Fig. 4. Scalability of different Connections between Meta Hosts

While being used for running distributed parallel applications on the VIOLA systems, the multidevice architecture showed its main advantage over the router approach: Establishing router connections over a high-bandwidth network (e.g. the VIOLA WAN) is like introducing an artificial bottleneck into the meta computer, because of the limited speed with which the router nodes can handle the traffic. Because the VIOLA systems have an all-to-all connectivity to each other via the optical network, a secondary device can be used instead to couple them. Our performance measurements showed that in this case the available bandwidth between the meta hosts scaled linearly with the number of application processes, because the fast external VIOLA network was never saturated.

In general, such a situation arises when the network between the meta hosts can sustain a much higher bandwidth than can be utilized by a single router connection over that network. Figure 4 shows this for a Fast Ethernet connection between two SCI clusters. We measured the aggregate throughput between the clusters with *mpptest* [18] by letting pairs of processes exchange data, with one process running on each cluster. Additionally to providing lower bandwidth, the router processes utilize one processor on each cluster which is then lost for running application processes. Of course, the available bandwidth between the meta hosts can be increased with multiple router connections, but then, even less processors can run application processes.

5   Conclusion and Outloook

In this paper, we presented the features of MetaMPICH in the context of other MPI implementations with similar goals. MetaMPICH offers two methods of coupling clusters and single machines: the first is based on router processes and the second on a multidevice architecture. Both coupling methods may be combined side-by-side in a meta system containing three or more meta hosts. Compared to the other MPI implementations we mentioned, MetaMPICH is the only one that supports heterogeneity by integrating multiple ADI devices. The new multidevice architecture showed its scalability when running parallel applications on a high-bandwidth optical WAN.

Currently under development for MetaMPICH is a new device that utilizes an underlying MPI library for communication, giving MetaMPICH capabilities similar to MPICH-G2 and PACX-MPI in that area. To improve application performance, the implementation of optimized collective operations and support for the MPI process topology mechanism is planned. Because this paper is focused on features, we plan to publish a thorough benchmark comparison of the MPI implementations described herein.

References

1. Karonis N., Toonen B., and Foster I.: MPICH-G2: A Grid-Enabled Implementation of the Message Passing Interface. Journal of Parallel and Distributed Computing, Vol. 63, No. 5, pp. 551–563, May 2003
2. Beisel T., Gabriel E., Resch M., and Keller R.: Distributed Computing in a Heterogeneous Computing Environment. Proc. of the 5th European PVM/MPI Users' Group Meeting, pp. 180–187, Liverpool, UK, September 1998
3. Aumage O. and Mercier G.: MPICH/MADIII: a Cluster of Clusters Enabled MPI Implementation. Proc. of the 3rd IEEE/ACM International Symposium on Cluster Computing and the Grid, pp. 26–36, Tokyo, Japan, May 2003
4. Kielmann T., Hofmann R., Bal H., Plaat A., and Bhoedjang R.: MagPIe: MPI's Collective Communication Operations for Clustered Wide Area Systems. Proc. of the 7th ACM SIGPLAN Symposium on Principles and Practice of Parallel Programming, pp. 131-140, Atlanta, Georgia, May 1999
5. George W., Hagedorn J., and Devaney J.: IMPI: Making MPI Interoperable. Journal of Research of the National Institute of Standards and Technology, Vol. 105, pp. 343–428, 2000
6. Kimura T. and Takemiya H.: Local Area Metacomputing for Multidisciplinary Problems: A Case Study for Fluid/Structure Coupled Simulation. Proc. of the 12th International Conference on Supercomputing, pp. 149–156, Melbourne, Australia, 1998

7. Balkanski D., Trams M., and Rehm W.: Heterogeneous Computing With MPICH/Madeleine and PACX MPI: a Critical Comparison. Chemnitzer Informatik-Berichte, CSR-03-04, December 2003
8. Gropp W., Lusk E., Doss N., and Skjellum A.: A High-Performance, Portable Implementation of the MPI Message Passing Interface Standard. Parallel Computing, Vol. 22, No. 6, pp. 789–828, September 1996
9. Karonis N., De Supinski B., Foster I., Gropp W., Lusk E., and Bresnahan J.: Exploiting Hierarchy in Parallel Computer Networks to Optimize Collective Operation Performance. Proc. of the 14th IEEE International Parallel and Distributed Processing Symposium, pp. 377–384, Cancun, Mexico, May 2000
10. Pöppe M., Schuch S., and Bemmerl T.: A Message Passing Interface Library for Inhomogeneous Coupled Clusters. Proc. of the 17th IEEE International Parallel and Distributed Processing Symposium, p. 199, Nice, France, April 2003
11. Pöppe M., Schuch S., Finocchiaro R., Clauss C., Worringen J.: MP-MPICH User Documentation and Technical Notes. Aachen: Lehrstuhl für Betriebssysteme, RWTH Aachen, 2005
12. Clauss C., Pöppe M., and Bemmerl T.: Optimising MPI Applications for Heterogeneous Coupled Clusters with MetaMPICH. Proc. of the IEEE International Conference on Parallel Computing in Electrical Engineering, pp. 7–12, Dresden, Germany, September 2004
13. Müller M, Hess M. and Gabriel E.: Grid enabled MPI solutions for Clusters. Proc. of the IEEE International Symposium on Cluster Computing and the Grid, pp. 18–25, Tokyo, Japan, May 2003
14. Worringen J.: SCI-MPICH: The Second Generation. Proc. of SCI-Europe 2000 (Conference Stream of Euro-Par 2000), pp. 11–20, Munich, Germany, 2000
15. Bierbaum B., Clauss C., Eickermann T., Kirtchakova L., Krechel A., Springstubbe S., Wäldrich O., and Ziegler W.: Reliable Orchestration of distributed MPI-Applications in a UNICORE-based Grid with MetaMPICH and MetaScheduling. Proc. of the 13th European PVM/MPI Users' Group Meeting, Bonn, Germany, September 2006
16. Erwin D. and Snelling D.: UNICORE: A Grid Computing Environment. Journal of Lecture Notes in Computer Science, Volume 2150, pp. 825ff, 2001
17. Grund H. and Ziegler W.: Resource Management in an Optical Grid Testbed. Journal of ERCIM News No. 59, October 2004
18. Gropp W. and Lusk E.: Reproducible Measurements of MPI Performance Characteristics. Proc. of the 6th European PVM/MPI Users' Group Meeting, pp. 11–18, Barcelona, Spain, September 1999

Using an Enterprise Grid for Execution of MPI Parallel Applications – A Case Study

Adam K.L. Wong and Andrzej M. Goscinski

School of Engineering and Information Technology, Deakin University
Geelong, Vic 3216, Australia
{aklwong, ang}@deakin.edu.au

Abstract. An enterprise has not only a single cluster but a set of geographically distributed clusters – they could be used to form an enterprise grid. In this paper we show based on our case study that enterprise grids could be efficiently used as parallel computers to carry out high-performance computing.

1 Introduction

Many parallel programs can benefit from running even on a small cluster [4, 12], which can be commonly found in many enterprises. If an enterprise owns many of such clusters, it could be efficient to connect them together via a fast network to form a multi-cluster system even though they are geographically dispersed. This kind of systems, generally referred as an enterprise grid [3], can potentially act as a parallel computer to carry out high-performance computing.

Communication latency and the heterogeneous nature of enterprise grids make high performance executions of parallel programs a challenging task. Heterogeneity of an enterprise grid can come from both computers and networks. First, an individual cluster may be made up of heterogeneous computers (with different processors and memories) – intra-cluster heterogeneity; and even if different clusters of an enterprise grid are internally homogeneous they could differ from each others – inter-cluster heterogeneity. The mismatch of processing speeds in different computers and different clusters may cause process coordination problems and therefore it affects the execution performance of the programs. Second, the LAN connecting computers in one cluster may be different from that of another. Besides, the clusters of the enterprise grid are usually connected by a much slower WAN and therefore communication bottlenecks usually occur at the inter-cluster communications, making achieving any speedup of parallel programs impossible. The improved networking technology in recent years has lowered the gap in the communication speed of LANs and WANs, which could lead to execution performance gains.

The aim of this paper is to report on the outcome of our study into the execution performance of MPI parallel applications on an enterprise grid. First, we study whether it is feasible to employ geographically distributed computer clusters of an enterprise to carry out high-performance parallel computation. Second, we show how to improve the execution performance of parallel programs on heterogeneous clusters

B. Mohr et al. (Eds.): PVM/MPI 2006, LNCS 4192, pp. 194 – 201, 2006.

of the enterprise by a simple technique of dynamic load balancing using the source initiative strategy.

2 Related Work

The study into parallel processing on enterprise grids can be generally classified into two categories and the major aspect distinguishing them is co-allocation – the simultaneous allocation of computers of different clusters for running a single parallel job. The first category investigates parallel job scheduling without co-allocation which is targeted to provide a high throughput for multi-cluster systems [5]. The second category investigates parallel job execution with co-allocation on multiple clusters, which is targeted to provide a high performance execution of parallel applications [2, 6]. Our research belongs to the latter category.

[6] provides a detailed simulation carried out to study the influence of inter-cluster communication on the performance of co-allocation based execution. It shows a bandwidth-centric job communication model for different jobs scheduling policies with co-allocation. However, the execution performance of parallel programs, has not been addressed in their study.

The execution performance study of eight parallel applications running on wide-area clusters is shown in [2]. The measurement had taken into account the latency and bandwidth differences between local- and wide-area networks. It had also presented several optimization techniques for the selected applications running on wide-area clusters. The parallel applications were implemented restrictively in Orca [1] and executed with the support of fast broadcasting communication provided by the Orca runtime system. In contrast, our study employs MPI and computer clusters running with a commodity-based operating system such as Linux.

3 Enterprise Grid Setup

We set up an enterprise grid, the Deakin Enterprise Grid (DEG), comprising two PC clusters located on the two different campuses of our School. Each cluster has 16 PCs connected by a 100Mbps Ethernet. However, the two clusters have different processors speeds and memory sizes, which allow us to study inter-cluster heterogeneity. The Geelong computers are based on a 350MHz Pentium Processor and 383Mb RAM whereas the Melbourne computers are based on a 800MHz Pentium Processor and 512Mb RAM. These two clusters located 80km apart are linked together via an ATM-based MAN (155Mbps microwave link). Currently, all the computers are running the RedHat Linux operating system. LAM/MPI [7] is used to support the development and execution of MPI parallel applications.

4 MPI Applications

We selected for our experiments a set of MPI parallel applications from the domains of both parallel benchmarks and real world parallel applications. These applications are the representative programs in the areas of computational science, engineering and bioinformatics. They are characterized by a wide range of computation and communication attributes of parallel programs.

4.1 NAS Parallel Benchmarks

The NAS Parallel Benchmarks (NPB) suite [8] has been widely used to objectively measure and compare the performance of parallel computer systems. It consists of a set of eight programs derived from computational fluid dynamics codes, of which the NAS 2.4 version can be used as MPI applications and run on computer clusters. They can be categorized into computation- or communication-bound.

We have found that out of the eight NAS programs, EP (Embarrassingly Parallel), LU (LU solver), BT (Block Tridiagonal solver) and MG (MultiGrid) can represent a broad range of communication patterns of parallel applications that can commonly be found in many real world applications. Thus, these four programs are selected for our experiments. Since the NAS benchmark programs are designed to study parallel computation with static load balancing only, these programs must be compiled for a specific grid size and number of processes when a benchmarking is to be carried out.

4.2 Parallel FastDNSml

Maximum likelihood is a useful but highly computation-intensive technique that can analyze relationships among genes and DNA sequences of phylogenetic trees. One well established and commonly used code for maximum likelihood phylogenetic inference is the fastDNAml program. There is also its MPI parallel implementation [10], which we used in our study.

The parallel fastDNAml consists of a set of processes. A master which constructs the trees to be evaluated, workers which do the evaluation and branch length optimization, and a foreman which manages all the workers. This programming paradigm can support a program-level dynamic load balancing in clusters.

4.3 MPI-Povray

Ray-tracing is a compute-intensive rendering technique that has been widely used in many areas such as computer graphics and games, film animations and automobile engineering. MPI-Povray [11] is a parallelized version of POVRAY [9] – the most popular ray-tracer. It is functioned by distributing works amongst a number of processing elements and the communication between the processing elements is achieved by the MPI message passing. Basically, the application follows the master/worker parallel programming paradigm where a master process divides an image of a trace into smaller blocks and assigns the blocks to many worker processes.

4.4 Classification of Program Attributes of the Selected Applications

The execution behavior of the selected parallel applications is determined by a number of program attributes including computation, communication, memory and topology. For example, the computation attribute of the programs is governed by the program size; the communication attribute depends on the communication volume and pattern; the memory attribute affects the size of memory spaces allocated from computers; and the topology attribute defines the size (process number) and the structure (process connections defined by communication volume and communication pattern) of the programs. These attributes of the four selected NAS parallel benchmarks, parallel fastDNAml and MPI-Povray are shown in Table 1.

Table 1. Attributes of the selected applications [12]

Program	Attributes			
	Computation	Comm. Volume	Comm. Pattern	Topology
EP	Computation-Bound	Negligible	Point-to-point	Any
LU	Computation-Bound	Medium	Point-to-point	Power-of-2
BT	Communication-Bound	High	Collective	Square-of-n
MG	Communication-Bound	High	Collective	Power-of-2
fastDNAml	Computation-Bound	Low	Point-to-point	Any
MPI-Povray	Computation-Bound	Low	Point-to-point	Any

5   Execution Measurements

We studied two dimensions of heterogeneity in enterprise grids: inter-cluster network and inter-cluster computers, and their influence on the execution performance of parallel applications. In the experiments, the processes of a parallel application were co-allocated (one process per computer) on the two DEG clusters to gain a higher level of parallelism and by this to hopefully achieve better execution performance. Clusters C_G^n (Geelong) and C_M^n (Melbourne) had n computers in each cluster.

The first set of experiments was carried out using the four selected NPB programs. The objectives were to demonstrate the feasibility of high-performance computing on enterprise grids, and to study the influence of the program's computation and communication characteristics on its execution through the use of benchmarks.

The second set of experiments was carried out using two real parallel applications: fastDNAml and MPI-Povray. The objectives were to support our claim that high-performance computing can be practically conducted on enterprise grids. Also, in contrast to a static load balancing approach adopted in the NAS programs, a simple but effective program-level dynamic load balancing approach adopted in these two applications was studied to provide better execution performance of the programs.

5.1   Execution of NAS Parallel Benchmarks on DEG

We choose the problem size of class B for the NAS programs to make sure that the execution memory requirement can be met without memory swapping. We could not satisfy fully the topology requirement. EP, LU and MG can run effectively with a topology of a power-of-2 number of processes. However, BT performs best when it runs with a square number of processes. We carried out the experiments using a power-of-2 number of computers for each selected NAS program learning from our experiments that the loss of performance for executing BT with a power-of-2 number of computers will not distort the experiment outcomes significantly.

The first experiment aimed to profile the inter-cluster communication overhead of parallel programs with various communication requirements. Different co-allocation ratios of 16 computers in DEG (C_G^n and C_M^{16-n} where $n = 0, 2,...16$) were used and the execution time of the programs was measured. The results presented in Table 2 show that the performance of computation-bound parallel programs: EP and LU, executed on the enterprise grid is comparable to that achieved when executing on a single-cluster, which however is bounded by the slowest cluster, C_G. It is visible that unless all computers of the fast cluster could be allocated, using computers of both clusters (co-allocation) influences the execution performance only slightly.

Table 2. Executions of NAS Programs with different co-allocation ratios in the DEG

Co-allocation Ratio		Execution Time of NAS Programs (min)			
No. of Computers in C_G^{n}	No. of Computers in C_M^{16-n}	EP	LU	BT	MG
16	0	37.3	41.8	47.1	37.9
14	2	37.4	41.9	48.9	45.5
12	4	37.3	42.2	59.8	47.9
10	6	37.4	42.4	58.7	53.9
8	8	37.7	44.5	61.3	61.0
6	10	37.3	45.2	61.7	47.7
4	12	37.8	41.8	62.3	57.7
2	14	37.2	40.7	49.6	43.4
0	16	16.6	27.9	33.3	30.0

For communication-bound parallel programs: BT and MG, inter-cluster communication overhead is the highest when computers are evenly allocated in C_G and C_M. In this case it is preferable to use either of the two clusters (hopefully the faster one) rather than to exploit co-allocation. Inter-communication negatively influences the execution performance of the enterprise grid.

However, even in the case of the user who cannot receive the requested number of 16 computers of the fast cluster but only received n, and used also $16 - n$ much cheaper computers of the slower cluster, there is evident improvement of the execution performance. This is better visible in the case of the user who can afford computers of the faster cluster. In summary, co-allocation works well in these cases.

The second experiment aimed to compare the execution performance of the NAS parallel programs on the single- and multi-clusters of DEG. For the case of the grid, we evenly allocated computers from C_G and C_M since it represents the worst case situation (with highest inter-cluster communication overhead).

To demonstrate how to use clusters effectively in an enterprise grid, we compared the execution time of the NAS parallel programs running on C_G^{n} and C_M^{n} against $C_G^{n/2} + C_M^{n/2}$; and on C_G^{n} and C_M^{n} against $C_g^{n} + C_m^{n}$ where n = 2, 4,...16. $C_G^{n/2} + C_M^{n/2}$ represents the situation where neither of C_G nor C_M has all n requested computers available for such parallel program execution and therefore symmetric co-allocation is used. We studied as whether the execution performance of a parallel program running on $C_G^{n/2} + C_M^{n/2}$ is at least as good as that of running on the slowest single-cluster, C_G^{n}. On the other hand, $C_G^{n} + C_M^{n}$ represents the situation when there are more than n requested computers (we assume 2n) available in DEG and thus they can be used to increase the level of parallelism. We studied as whether the execution performance of a parallel program with higher level of parallelism can be improved despite the inter-cluster communication cost. The result of the comparisons is presented in Figures 1(a, b, c and d).

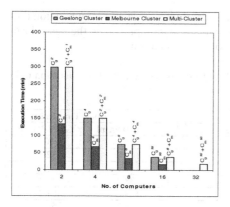

Fig. 1a. Improvement of EP

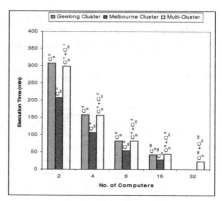

Fig. 1b. Improvement of LU

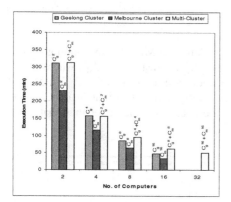

Fig. 1c. Improvement of BT

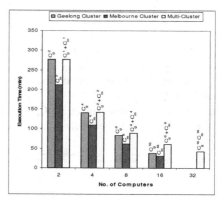

Fig. 1d. Improvement of MG

For computation-bound programs: EP and LU, their execution performance on $C_G^{n/2} + C_M^{n/2}$, is comparable to that of on the slow cluster, C_G^n, and their execution performance on $C_G^n + C_M^n$ is almost two times better than that of on C_G^n. For communication-bound programs: BT and MG, although their execution performance on $C_G^{n/2} + C_M^{n/2}$ is worse than that of on C_G^n and C_M^n especially for big n, their execution performance on $C_G^n + C_M^n$ is still better than that of on C_G^n and C_M^n.

5.2 Execution of Parallel FastDNAml and MPI-Povray on DEG

We adopted a problem size of 50 taxa of genes (a sample data file used in [10]) for Parallel fastDNAml and a problem size of 50 frames animation of 640x480 pixels image for MPI-Povray. Both Parallel fastDNAml and MPI-Povray are considered computation-bound parallel programs – they share communication characteristics with LU. They also support a program-level source initiative load balancing mechanism provided by the master/slave parallel programming paradigm.

Table 3 shows the execution performances of Parallel fastDNAml and MPI-Porvay using different co-allocation ratios of 16 computers in DEG (C_G^n and C_M^{16-n} where $n = 0, 2,...16$). The results show the execution performance gained by balancing the workloads on computers of C_G and C_M, where a computer of C_M is over two times more powerful than that of C_G, outweighs the added inter-cluster communication cost. The table also shows that increasing the number of computers of the fast cluster improves the overall execution performance of the application.

Table 3. Executions of parallel fastDNAml and MPI-Povray with different co-allocation ratios in the DEG

Co-allocation ratio		Execution Time (min)	
No. of Computers in C_g^n	No. of Computers in C_m^{16-n}	fastDNAml	MPI-Povray
16	0	173.3	110.3
14	2	160.8	92.3
12	4	147.9	77.7
10	6	138.9	69.4
8	8	129.2	61.2
6	10	122.2	54.6
4	12	114.6	50.1
2	14	108.6	45.7
0	16	105.3	41.2

We compared the execution time of fastDNAml and MPI-Povray running on C_G^n and C_M^n against $C_G^{n/2} + C_M^{n/2}$; and on C_G^n and C_M^n against $C_G^n + C_M^n$, where n = 2, 4,...16. Fig. 2 and Fig. 3 show that the execution performances of both fastDNAml and MPI-Porvray on $C_G^{n/2} + C_M^{n/2}$ are significantly better than that of on C_G^n; and their execution performances on $C_G^n + C_M^n$ are over two time better than that of on C_G^n and significantly better than that of on C_M^n.

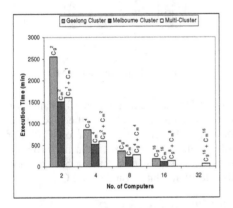

Fig. 2. Improvement of fastDNAml   **Fig. 3.** Improvement of MPI-Povray

An analysis of Table 3, Fig. 2 and Fig. 3 also shows that adding any available number of computers k, where $k < n$, of the fast cluster to the slow cluster that originally supports a parallel application by $(n - k)$ computers, where $n \leq 16$, can improve the execution performance of the application. Thus, there is some tradeoff for the user of the slow cluster to pay the cost of using some computers of the fast cluster to improve the execution performance. The same is experienced by the user of the fast cluster who could improve the execution performance by increasing the level of parallelism and importing services of computers of the slower cluster.

6 Conclusions

The results of the study into the feasibility of using enterprise grids to execute parallel applications and improve their overall execution performance demonstrate that computer co-allocation on multiple clusters for high performance is feasible and sound. For inter-cluster network heterogeneity, our experiments on the NAS parallel benchmarks have demonstrated that not only computation-bound parallel programs: EP and LU can work well in such an environment, but communication-bound parallel programs: BT and MG can also benefit from increased level of parallelism (executed on more computers) and thus improve the execution performance despite the inter-cluster communication cost. For inter-cluster computers heterogeneity, our experiments on the Parallel fastDNAml and MPI-Povray have demonstrated that the execution improvement gained through balancing the workloads on computers of heterogeneous clusters can outweigh the inter-cluster communication cost.

References

1. H.E. Bal, M.F. Kaashoek, and A.S. Tanenbaum.   Orca: A Language for Parallel Programming of Distributed Systems. *IEEE Transactions on Software Engineering*, Vol. 18, No. 3, pp. 190-205, 1992.
2. H.E. Bal, A. Plaat, M.G. Bakker, P. Dozy and R.F.H. Hofman. Optimizing Parallel Applications for Wide-Area Clusters. In *Proceedings of the 12th International Parallel Processing Symposium*, pp. 784-790, Orlando, Apr., 1998.
3. Enterprise Grid Alliance. URL: http://www.gridalliance.org, accessed: Sept. 2005.
4. A.M. Goscinski and A.K.L. Wong. Performance Evaluation of the Concurrent Execution of NAS Parallel Benchmarks with BYTE Sequential Benchmarks on a Cluster. In *Proceedings of the 11th International Conference on Parallel and Distributed Systems (ICPADS 2005)*, Fukuoka, Japan, Jul., pp. 313-319, 2005.
5. L. He, S.A. Jarvis, D.P. Spooner, X. Chen and G.R. Nudd. Dynamic Scheduling of Parallel Jobs with QoS Demands in Multiclusters and Grids. In *Proceedings of the 5th International Workshop on Grid Computing*, pp. 402-409, 2004.
6. W.M. Jones, L.W. Pang, D. Stanzione and W.B. Ligon III. Characterization of Bandwidth-aware Meta-schedulers for Co-allocating Jobs in a Mini-grid. *The Journal of Supercomputing*. Vol. 34, No. 2, pp. 135-163, 2005.
7. LAM/MPI. URL: http://www.lam-mpi.org, accessed: Jul. 2005.
8. NAS Benchmarks. URL: http://www.nas.nasa.gov/Software/NPB/, accessed: Jul. 2005.
9. The POVRAY Homepage. Persistence of Vision Ray-tracer. URL: http:// www. povray.org, accessed: Jul. 2005.
10. C.A. Stewart, D. Hart, D.K. Berry, G.J. Olsen, E.A. Wernert and W. Fischer. Parallel Implmentation and Performance of fastDNAml – A Program for Maximum Likelihood Phylogenetic Inference. In *Proceedings of the 2001 ACM/IEEE Conference on Supercomputing (CDROM)*, Nov., pp. 20-20, 2001.
11. L. Verrall. MPI-Povray: Distributed Povray Using MPI Message Passing. URL: http://www.verrall.demon.co.uk/mpipov, accessed: Jul. 2005.
12. A.K.L. Wong and A.M. Goscinski. Execution Environments and Benchmarks for the Study of Applications' Scheduling on Clusters. *Lecturer Notes in Computer Science – Distributed and Parallel Computing*, Springer, No. 3719, pp. 204-213, 2005.

Self-adaptive Hints for Collective I/O

Joachim Worringen

C&C Research Laboratories, NEC Europe Ltd.
Rathausallee 10, D-53757 Sankt Augustin, Germany
mpi@ccrl-nece.de
http://www.ccrl-nece.de

Abstract. The processing of MPI-IO operations can be controlled via
the MPI API using *file hints*, which are passed to the MPI library as *MPI
info objects*. A file hint can affect how the MPI library accesses the file on
the file system level, it can set buffer sizes, turn special optimizations on
and off or whatever parameters the MPI implementation provides. However, experience shows that file hints are rarely used for reasons that will
be discussed in the paper. We present a new approach which dynamically
determines the optimal setting for file hints related to collective MPI-IO
operations. The chosen settings adapt to the actual file access pattern,
the topology of the MPI processes and the available memory resources
and consider the characteristics of the underlying file system. We evaluate our approach which has been implemented in MPI/SX, NEC's MPI
implementation for the SX series of vector supercomputers.

Keywords: MPI-IO, collective operations, self-adaptation, file hints.

1 Introduction

MPI-IO [3] is the part of the MPI-2 standard that defines an API for parallel
file access from within an MPI application. The two most relevant API elements
that MPI-IO offers, and which are not offered by the standard APIs for file I/O
such as POSIX, are *collective I/O* and the notion of *file views*, which allow the
flexible definition of non-contiguous file accesses. These two features simplify
the I/O-related code of the application, while at the same time offer significant
potential for optimization by the MPI library.

A few fundamental techniques for such optimizations exist and are widely
used. However, the parameters of these techniques have static default values
which in many cases adversely impacts the performance due to the resulting
access patterns on the file system level. Although most of these parameters can
be set by the MPI application via *file hints* passed to the MPI library as MPI_Info
objects, experience shows that file hints are rarely ever used. The reasons for this
are manifold: firstly, the algorithms and thus the effect of these parameters are
often complex and beyond the understanding of the typical application or library
programmer[1]. Next to this, the fact that these parameters are not performance

---

[1] Not in the sense that he wouldn't be able to understand, but rather unwilling to
take care.

B. Mohr et al. (Eds.): PVM/MPI 2006, LNCS 4192, pp. 202–211, 2006.

portable, not necessarily portable between different MPI implementations and their effects being highly dependent on the characteristics of the I/O system used make it very unattractive to the MPI user to care at all for their possible settings. Instead, we consider it the job of the MPI library to determine the optimal settings for these parameters automatically.

In this paper, we present an adaptive and dynamic approach for the refinement of these optimization parameters that have been implemented in MPI/SX, NEC's MPI implementation for the SX series of supercomputers. It provides flexible algorithms and aims to set related parameters to values which generate access patterns optimally matching with the performance characteristics and access semantics of the underlying file system. We discuss the current status of optimizations for collective and non-contiguous MPI-IO operations in the next section, followed by a more detailed discussion of our motivation in section 2. This leads us to the presentation of the chosen optimization approaches in section 4. Section 5 presents results of performance measurements which compare the new technique with the standard approach.

2   Motivation

MPI/SX is NEC's MPI-2 implementation for the SX series of vector computers. It includes MPI-IO based on the ROMIO MPI-IO library, but has been adapted and tuned since then. We observed performance below the expected values for different access patterns that are used in *b_eff_io* [8] on a large SX-8 system with NEC's GFS file systems. A chosen file hint could improve the performance for some access patterns, but lead to decreases in other patterns. We recognized that such manual optimizations are not reasonable for the MPI user, and decided to let the MPI library care for this.

For the initial approach as presented in this paper, we chose the SX vector computer together with the GFS file system as the target platform. NEC GFS[2] [6] is a file system optimized for large-scale SANs (Storage Area Networks) based on Fibre Channel networks. It provides NFS client access, but uses *third-party transfer* to optimize large block transfers. This means that file access below 64KiB will be processed via the standard NFS v3 protocol, while for larger blocks, the client only receives the block numbers from the server and uses this information to transfer the file data directly via the Fibre Channel interconnect.

2.1   Access Semantics

The fact that a GFS file system is visible as a NFS file system has the advantage that no special software is required on the client side for file system access. NFS client software is installed by default on all operating systems used in a HPC computing center. GFS client file system drivers that uses third-party transfer for increased performance is available for Linux and SUPER-UX and can be used on systems that are connected to the Fibre Channel SAN.

---

[2] Not related to the GFS file system provided by RedHat Linux.

However, NFS is not a POSIX compliant file system due to its non-coherent client side caching and other limitations of the stateless server architecture. Considerable efforts are required to make sure that MPI-IO on top of NFS complies with the semantics defined in the MPI-IO standard when processes from multiple nodes access the same file: each read or write access needs to be locked with NFS block size granularity (4KiB), and buffers have to be flushed explicitly via a fcntl() call.

2.2   Performance Characteristics

To get the baseline performance of a typical SX node accessing a GFS file system on POSIX-level, we ran the *iozone* benchmark [5] with a single process for access sizes from 4KiB to 128MiB. The performance for initial reads peaks at about 500MiB/s (close to the hardware limit) with a block size of 16MiB, and for the same access size at 340MiB/s for writes. However, when having to lock and synchronize each access as described in the previous section, this block size increases to 128MiB.

Only for read accesses, caching increases the performance by about 20%. For 50% of the peak bandwidth, the access size has to be about 2MiB for uncoordinated access, but nearly 8MiB for locked and synchronized access. For access sizes below 64KiB, the bandwidth is about 10MiB/s without any caching effects.

2.3   Scaling Characteristics

With MPI-IO, it is typical that all processes of an application access *the same file* concurrently. However, most file systems are not optimized for this access pattern. Instead, they assume that many processes access many *different* files at the same time. Therefore, and together with the problem of the NFS access semantics in general, we need to be aware of how the GFS file system behaves in this respect. It is also relevant to know how the bandwidth scales with the number of processes and/or nodes. We measured the scaling characteristics using the *b_eff_io* benchmark and cite the accumulated bandwidth with an access size of 64MiB (S_{chunk}) for the access types *separate* where each process operates on its own file (label *multiple files*) and *segmented* where all processes access disjoint locations in the same file (label *single file*) in Figure 1. We scaled the number of nodes from 1 to 32, and tested with a single MPI process per node (label *1 proc*) and with 8 MPI processes per node (label *8 proc*). Please note that we need to lock and synchronize file access in this test.

It shows that for the access type *separate*, we only need enough concurrent processes to saturate the hardware at 500MB/s. However, with all processes accessing a single file, the necessary locking and flushing adds overhead which limits the peak bandwidth to 380MB/s. The bandwidth even decreases for a large number of processes (here at 128 processes) due to the contention at the lock server.

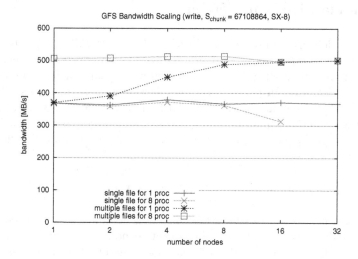

Fig. 1. Scaling of file access bandwidth of a single GFS file system for access types *separate* and *segmented*

3   Related Work

The basic optimization technique for collective read and write access in MPI-IO is *two-phase I/O* [9], which is similar in concept to *collective buffering* [4]. It is efficient for non-contiguous file views where it transforms each process' non-contiguous I/O request (described by an arbitrary number of *(offset, length)* tuples) into a small number of contiguous I/O requests which are performed by an adjustable number of *aggregator processes (AP)*, which are chose from the group of MPI processes that perform the collective access. After having read a contiguous range of the file, these *APs* exchange data with all other processes to provide them with the data from the file corresponding to the original request. As this technique transforms a potentially large number of fine-grained concurrent file accesses into a small number of ordered coarse-grained file accesses plus data exchange via MPI messages, it provides substantial performance improvements. Self-adaptive optimization has been applied to determining the optimal algorithm for collective communication in MPI [11,1], although these approaches are not in practical use. In contrast, the adaptivity of numerical software libraries [2] has proved to be a very successful approach, and is critical to achieve high performance on different platforms. We are aware of one MPI-IO implementation [7] which contains similar optimizations. These optimizations can be controlled by file hints i.e. to differentiate between *sparse* and *dense* collective file access. However, self-adaptive means of setting these or other file hints are not described. Also, this MPI-IO implementation is targeted exclusively at IBM's GPFS file system.

4   Self-adaptive Hint Setting

The most widely used implementation of MPI-IO is ROMIO [10], which is actively maintained and is part of current MPI implementations, both open-source and vendor-specific. MPI/SX uses a modified and optimized ROMIO as a base for the MPI-IO part. In this section, we discuss how the file hints related to collective I/O (defined in the MPI standard and available in ROMIO) influence the way the collective MPI-IO call is processed, and present our approach to dynamically adapt the parameters related to these hints for optimized performance of the current access.

4.1   Number of Processes for Two-Phase I/O

The file hint cb_procs[3] can be set to an integer value to indicate how many *APs* should be accessing the file for collective non-contiguous file access via two-phase I/O. The default value of this hint is the number of processes in the appendant communicator, and the user will rarely ever think of setting this to a different value.

We have observed that a static setting of this value is not optimal. For small access sizes in a multi-node environment, it is better to let few processes perform a few, but larger I/O operations to reduce the relative locking overhead and exploit the higher bandwidth for larger access sizes and MPI messages. As a simple measure, we specify a minimal I/O size for each *AP* which reduces the number of *APs* for small access sizes. However, we also have to consider concurrency and thus do not reduce the number of *APs* too much. For accesses that are large enough, we use at most 4 *APs* per node.

Finally, the access ranges of the *APs* are aligned to the NFS block size to avoid lock contention.

4.2   Placement of Aggregator Processes

A very complex file hint is cb_config_list. It is designed to allow the individual placement of *APs* on specific nodes which may e.g. have special I/O systems or faster I/O facilities than other nodes. The nodes are referenced by their MPI names which makes it hardly applicable in batch processing environments where the names of the execution nodes are not known in advance.

For this reason, and because the GFS file system is always configured symmetrically (all nodes of a system have identical connections toward the storage systems), we chose to not support this option. Instead, we assign the task of being an *AP* to the MPI processes in a round-robin manner over the nodes. This is closely related to the setting of the cb_procs file hint as it is done right after having determined the required number of *APs* for the current I/O operation.

---

[3] The name of this file hint in ROMIO is cb_nodes, but was changed in MPI/SX as it counts processes and not nodes.

4.3   Decision Between Collective and Individual I/O

Even if the user calls a collective MPI-IO function, the MPI library does not automatically use the optimized collective file access routines described above. Instead, it decides between the different I/O strategies based on the actual access pattern. The default optimization approach in ROMIO is to check if any two access ranges overlap. If yes, two-phase I/O is chosen; if no, each process performs the file access individually. This decision process can be controlled by setting the file hint cb_read for read access respectively cb_write for write access to one of the three values *enable*, *disable* or *automatic*. The default content is *automatic* which behaves as described. Statically setting these file hints to either *enable* or *disable* would require that the user has exact knowledge of the access pattern generated by the application. Choosing the wrong content can significantly reduce the I/O performance.

For similar reasons as described in section 4.1, we consider a static setting to be inappropriate anyway. But also the automatic setting based on the algorithm described above is not optimal as there are situations where the I/O accesses of all processes do not overlap but two-phase I/O still has advantages. In MPI/SX, a NFS file is locked with NFS block size granularity (4KiB) to ensure full consistency for multi-node operation. Individual file accesses can not be coordinated, and if file adjacent accesses are not aligned with the NFS block size, processes have to wait on locks to become available. When collectively accessing the file with two-phase I/O, the file accesses of the *APs* can properly be aligned to avoid any waiting time. Next to this, also for non-overlapping access ranges, two-phase I/O allows to avoid overloading the file system with many small requests as discussed above.

Of course, two-phase I/O is not always the best approach. If the access ranges are not adjacent, but are separated by *gaps*, two-phase I/O will become more ineffective with increasing gap size in relation to the effective data size. Therefore, we perform a more detailed analysis of the access pattern than simply testing for any overlaps. Our analysis returns information on the ratio between gaps and data in the collective access range. The decision between collective and individual I/O is made based upon a fixed threshold for this ratio.

4.4   Buffer Size for Collective Buffering

A very important factor for the performance is the size of file buffers required for techniques like two-phase I/O. The size of the file buffer determines the block size of the file access, and this has a strong correlation with the bandwidth as shown in Section 2.2. Looking at the GFS baseline performance, it would be possible to use the minimum of 128MB and the actual access range as buffer size. However, it is not known to the MPI library if this amount of memory is available on the node. If the node starts paging out memory because of the allocation of file buffers, the application performance effectively decreases. On the other hand, fixing the default buffer size at some lower bound will waste performance in many cases in which a larger buffer size would not harm the application performance because sufficient memory resources are available. As discussed before, the possibility to let the user set the buffer size is of little practical value.

Instead, we again implemented a dynamic approach. Because two-phase I/O is (for large requests) processed in multiple iterations, and typical applications perform multiple I/O operations during their execution, we iteratively determine a suitable buffer size. Our approach starts with the buffer size at a lower bound (2MB in our case), and measures the effective bandwidth of the iteration of the two-phase I/O. It then increases the buffer size for the next iteration, and compares the bandwidth achieved with the increased buffer size with the bandwidth of the previous iteration. If the bandwidth increased, it continues to increase the buffer size to an upper bound. If the bandwidth stayed about the same, the buffer size is left unchanged. Otherwise, it goes back to the previous buffer size.

We keep the state of this dynamic buffer size determination persistent between I/O calls, together with information on the I/O call itself. As soon as we find the optimal buffer size, we stick with it as long as the size of the I/O calls stays about the same. A significant change of the size of the I/O call or the installation of a new file view indicates that the application enters a new phase, and therefore the file buffer size calibration is triggered again. Next to this, we recalibrate the buffer size after a fixed number of collective I/O calls to ensure continued optimization by adapting to possibly changed conditions.

5   Performance Evaluation

This section shows the results of some typical test cases where we compare the static default setting of the parameters with the self-adaptive setting of MPI/SX.

5.1   Scaling Collective Write Operations

Figure 1 indicates that the accumulated file system performance decreases when a single file is accessed by an increasing number of processes concurrently. In this test case, we let a varying number of processes store a fixed amount of data in a file. We compare three approaches: each process serves as an AP and accesses the same amount of data in the file (called *symmetric*, default in ROMIO); the number of APs is set to the number of nodes in use via an explicit file hint (*explicit*); finally, we let MPI/SX perform the scheduling adaptively (*adaptive*). We show the relative performance differences for fixed file sizes of 50MiB and 5GiB between the *adaptive* and *symmetric* respectively *explicit* setting of the cb_procs in Fig. 1 for different process counts and 1 or 8 processes per node (*ppn*). It shows that the adaptive setting is sometimes only slightly worse (nearly within the measuring noise), but offers significant performance advantages for certain cases (here for large process counts) without any user interaction.

Table 1. Relative performance differences of *adaptive* vs. *symmetric* and *explicit* setting of the cb_procs file hint for 1 process per node (*1 ppn*) and 8 processes per node (*8 ppn*). Negative differences indicate performance degradation through adaptivity.

	adaptive vs. symmetric				adaptive vs. explicit			
	filesize 50 MiB		filesize 5 GiB		filesize 50 MiB		filesize 5 GiB	
processes	1 ppn	8 ppn	1 ppn	8 ppn	1 ppn	8 ppn	1 ppn	8 ppn
4	4.0%	n/a	0.5%	n/a	-2.7%	n/a	23.2%	n/a
8	-1.5%	-4.2%	11.3%	4.1%	5.5%	12.6%	-22.1%	missing
32	n/a	39.0%	n/a	-11.0%	n/a	24.8%	n/a	-12.3%
64	n/a	-2.8%	n/a	123.6%	n/a	0.6%	n/a	106.7%

5.2   Collective Write of Adjacent Blocks

The access pattern used in this test case is a call to MPI_File_write_all where the 16 processes write blocks of 1500 bytes to adjacent, non-overlapping offsets. The static algorithm would thus choose to let each process perform its individual write operations. The adaptive algorithm detects that the accesses are non-overlapping and adjacent (in this case, without any gaps), and chooses two-phase I/O. Because of the small data size, all I/O is performed by one process without any lock contention. The distribution of lock acquisition latencies for both techniques is shown in Table 2.

Table 2. Key values for latency of lock operations (milliseconds)

variant	minimum	maximum	average	99th quantile
static	2.154	208.493	22.376	160.597
adaptive	2.277	5.024	3.113	5.024

The distribution of the latencies for the static algorithm show that significant lock contention happens, while the adaptive approach delivers latencies within a very narrow range of 3ms. Due to this effects, the bandwidth for adaptive algorithm increases by a factor of 8 for write access and by a factor of 3 for read access. To achieve the same behaviour with ROMIO, the user needs to provide the file hints cb_nodes=1 and cb_write=yes. This setting, however, leads to significant performance penalties for many other access patterns.

5.3   Collective Large Block Access

To document the advantage of the adaptive buffer size setting, we compare the bandwidth for collective non-contiguous file access as measured by the *b_eff_io* benchmark with a fixed buffer size of 8MiB and the self-adaptive buffer size. The initial buffer size is set to 2MiB and was observed to typically increase to 32MiB. For very large accesses, it sometimes increases further (the maximum buffer size is set to 128MiB). Table 3 illustrates the performance gains of 50%

and more achieved for these large-block accesses (64MiB) for a single process and for 64 processes. The remaining parameters are identical for both test, i.e. a single *AP* was running on each node.

Table 3. Performance improvement for large block transfers with adaptive buffer sizing (total bandwidth in MB/s for *scattered* access in *b_eff_io* with 64MiB block size)

variant	single process	64 process on 8 nodes
static	204.4	205.6
adaptive	298.9	354.4

5.4   Choice of I/O Technique

A customer code (which can not be disclosed) running on an NEC SX-8 installation with GFS has a 3-dimensional regular data decomposition and writes this distributed data into a single file. The I/O was originally tunneled through a single process using MPI communication and native I/O operations. It was then changed to perform parallel I/O using MPI-IO. However, the chosen approach was still very naive and resulted in repeated calls of `MPI_File_write_all()` for a contiguous file view. This means, the data access of the processes in the file is not overlapped. We evaluated a test case with 16 processes running on 2 nodes. Without this optimization, the test case required about 7.5 minutes to complete as all processes effectively performed individual I/O operations which caused contention at the lock server. With the optimized decision between collective and individual I/O described in Section 4.3, the completion time was below 2 minutes because the I/O technique that was then used (collective buffering) avoids lock contention and performs less file accesses with larger blocks.

6   Conclusions

We have shown that it is possible to replace the cumbersome manual setting of file hints with self-adaptive algorithms to optimize collective I/O. By analyzing the actual access pattern more thoroughly, and taking into account the process topology and file system characteristics, these algorithms are able to optimize the basic techniques for collective I/O for most of the tested access patterns achieving the same or better performance than can be achieved by setting file hints manually, which requires detailed knowledge of the inner workings of MPI-IO. For testing and special benchmarking purposes, the user can always deactivate the self-adaptivity of the file hints by setting them to a fixed value.

The presented ideas are valid for other MPI-IO implementations as well. Generally, we consider self-adaptivity an important quality of an MPI implementation. Therefore, we aim at providing self-adaptive algorithms for other hints as well, like for non-collective I/O or one-sided communication. This will make the sweet spot for performance cover not only specific communication and application scenarios, but all situations that an MPI application can create.

Acknowledgments

We are grateful to the *High Performance Computing Center Stuttgart* (HLRS) for letting us perform the required benchmarks on their NEC SX-8 system and Holger Berger of NEC HPCE for his support. All data management, analysis and plotting was done with perfbase [12].

References

1. X. Y. Ahmad Faraj and D. Lowenthal. STARMPI: Self Tuned Adaptive Routines for MPI Collective Operations. In *Proceedings of the 20th ACM International Conference on Supercomputing*, Queensland, Australia, June 2006.
2. J. Dongarra and V. Eijkhout. Self-Adapting Numerical Software and Automatic Tuning of Heuristics. In *Proceedings of the International Conference on Computational Science*, number 2660 in LNCS. Springer, June 2003.
3. Message-Passing Interface Forum. *MPI-2.0: Extensions to the Message-Passing Interface*, chapter 9. MPI Forum, June 1997.
4. B. Nitzberg and V. Lo. Collective buffering: Improving parallel I/O performance. In *Proceedings of the Sixth IEEE International Symposium on High Performance Distributed Computing*, pages 148–157. IEEE Computer Society Press, August 1997.
5. W. D. Norcott and D. Capps. *Iozone Filesystem Benchmark.* http://www.iozone.org, March 2006.
6. A. Ohtani, H. Aono, and H. Tomaru. A File Sharing Method for Storage Area Network and its Performance Verification. *NEC Research & Development*, 44(1):85–90, January 2003.
7. J.-P. Prost, R. Treumann, R. Hedges, B. Jia, and A. Koniges. MPI-IO/GPFS, an Optimized Implementation of MPI-IO on top of GPFS. In *Proceedings of the 2001 ACM/IEEE conference on Supercomputing*, Denver, November 2001. Association for Computing Machinery.
8. R. Rabenseifner, A. E. Koniges, J.-P. Prost, and R. Hedges. *The Parallel Effective I/O Bandwidth Benchmark: b_eff_io*, chapter 4. Kogan Page Ltd., 2004.
9. R. Thakur, W. Gropp, and E. Lusk. Data sieving and collective I/O in ROMIO. In *Proceedings of the Seventh Symposium on the Frontiers of Massively Parallel Computation*, pages 182–189. IEEE Computer Society Press, 1999.
10. R. Thakur, W. Gropp, and E. Lusk. On Implementing MPI-IO Portably and with High Performance. In *Proceedings of the Sixth Workshop on Input/Output in Parallel and Distributed Systems*, pages 23–32, May 1999.
11. S. S. Vadhiyar, G. E. Fagg, and J. Dongarra. Automatically Tuned Collective Communications. In *Proceedings of the 2000 ACM/IEEE conference on Supercomputing*, Dallas, USA, November 2000.
12. J. Worringen. Experiment Management and Analysis with perfbase. In *Proceedings of the IEEE International Conference on Cluster Computing*, Boston, September 2005. IEEE Computer Society Press.

Exploiting Shared Memory to Improve Parallel I/O Performance

Andrew B. Hastings[1] and Alok Choudhary[2]

[1] Sun Microsystems, Inc.
andrew.hastings@sun.com
[2] Northwestern University
choudhar@ece.northwestern.edu

Abstract. We explore several methods utilizing system-wide shared memory to improve the performance of MPI-IO, particularly for non-contiguous file access. We introduce an abstraction called the *datatype iterator* that permits efficient, dynamic generation of (offset, length) pairs for a given MPI derived datatype. Combining datatype iterators with overlapped I/O and computation, we demonstrate how a shared memory MPI implementation can utilize more than 90% of the available disk bandwidth (in some cases representing a 5× performance improvement over existing methods) even for extreme cases of non-contiguous datatypes. We generalize our results to suggest possible parallel I/O performance improvements on systems without global shared memory.

Keywords: Parallel I/O, shared memory, datatype iterator, non-contiguous access, MPI-IO.

1 Introduction

The rich MPI derived datatype facility can describe arbitrary regions of in-memory and in-file data. Via this facility, an application using MPI-IO may issue I/O operations that are non-contiguous in memory and/or in a file[1]. Previous work has explored optimizing these operations via data sieving[2], the two-phase collective optimization[2], list I/O[3], and datatype I/O[4], primarily in the context of commodity clusters. The first two optimizations are broadly available through the open source ROMIO implementation distributed as part of MPICH2[5].

In a data sieving read, each process repeats this cycle: read the next large contiguous chunk of file data into a working buffer and extract the pieces needed by the MPI read operation. Each cycle typically transfers a subset of the data covered by the associated datatypes. Writes are similar, except locks serialize access to each chunk, and a read-modify-write may be necessary for each chunk. (ROMIO uses data sieving for non-interleaved collective I/O operations.)

In the two-phase collective optimization, a subset of processes are designated *aggregators*. All processes construct lists of (offset, length) pairs (*flattening* their memory and file datatypes), and send flattened file datatypes to the aggregators.

B. Mohr et al. (Eds.): PVM/MPI 2006, LNCS 4192, pp. 212–221, 2006.

For each chunk all processes use flattened memory datatypes to transfer an appropriate subset of their data to or from the aggregators via MPI messages, and the aggregrators use the flattened file datatypes to transfer the data to or from the working buffer. The messages serve not only to transfer the data, but also to synchronize the processes and ensure that I/O is complete before accessing the working buffer.

In list I/O, the MPI-IO implementation flattens the memory and file datatypes to lists of (offset, length) pairs. These lists are then communicated (concurrently with the data on a write) to a new list I/O interface in the filesystem, where it may apply techniques such as data sieving to optimize the operation.

Since the memory and file lists can be quite long for non-contiguous datatypes, datatype I/O replaces the lists with two compact datatype representations extracted from the MPI derived datatypes specified by the MPI-IO call. As in list I/O, these compact representations are communicated (concurrently with the data on a write) to a new datatype I/O filesystem interface.

In this work we investigate algorithms for optimizing MPI-IO in the context of a shared memory computer. As participants in the DARPA High Productivity Computer Systems initiative[6], researchers at Sun Microsystems, Inc., have been exploring the use of shared memory in petascale computer systems[7]. Exploiting shared memory in the MPI-IO implementation offers an opportunity to improve I/O performance without altering how applications express I/O operations.

2   Exploiting Shared Memory

In a global shared memory system the filesystem typically has direct access to both user memory and I/O devices, and it is efficient to transfer data independently from control information. Further, the data transfer step in the two-phase collective optimization can be performed via shared memory, bypassing the packing, copying, and unpacking operations usually required to implement MPI messages. We extended ROMIO with new I/O methods that exploit shared memory.

mmap. In a shared memory system it is relatively efficient to use the POSIX *mmap* operation to map a file directly into the address space of several processes. In contrast, mmap on a cluster might require additional bookkeeping and data transfer overhead to provide distributed shared memory. In the mmap I/O method, processes must synchronize initially to compute and set the new file length, but then can proceed independently using a data sieving-like algorithm: map a file chunk, then use the flattened datatypes to copy data into or out of the mapped chunk. The advantage over data sieving is that pages of the file are shared in memory; thus I/O transfers happen only once. The disadvantage is that memory management hardware limitations require existing file contents to be read before each write operation, even when overwriting an entire page.

Collective shared data. In the ROMIO two-phase collective, processes exchange data with aggregators via MPI messages. In the *collective shared data* method, we arrange for aggregators to have direct access to every process's address space.[1] Each aggregator copies data between its working buffer and appropriate application memory locations in other processes without costly MPI messages. The locations in application memory and the working buffer are identified via the flattened datatypes. In contrast to the message-based collective, the processes need only synchronize at the very start and very end of the MPI-IO operation, no matter how many cycles through the working buffer are needed to complete the operation.

Collective shared buffer with flattened datatypes. In the *collective shared buffer* methods, we arrange for each process to have direct access to every aggregator's working buffer. Each process uses the flattened datatypes to copy its own application data to or from working buffers in the appropriate aggregators. With a single working buffer per aggregator it is necessary for all processes to synchronize before beginning their copy operations (to wait for the read or write of the previous chunk to complete) and after their copy completes (to notify the aggregators of copy completion and that it is safe to initiate the read or write of the next chunk). We eliminated one synchronization step by splitting each working buffer into multiple sub-buffers and performing I/O asynchronously. On each cycle of a write operation, for example, the aggregators: (1) wait for I/O to complete on the next sub-buffer, (2) synchronize with all processes, and (3) initiate I/O on the just-completed sub-buffer. We measured a 40-90% performance improvement for our collective shared buffer algorithms on the FLASH I/O benchmark (Section 3.3) by enabling sub-buffering.

Collective shared buffer with dynamic offset/length generation. This method replaces the flattened datatypes with dynamic (offset, length) generation. A problem with flattening is that the entire list must be generated before any actual I/O can begin. Also, the flattened list may be large and thus compete for space in the processor cache with the application data being transferred.

We introduced a new abstract data type called a *datatype iterator*, representing a cursor into a specific MPI datatype. The function dtc_next advances the cursor to the next contiguous block in the associated datatype and returns the (offset, length) for that block. dtc_extent_tell and dtc_size_tell return the extent or size within the datatype corresponding to the current cursor position. dtc_extent_seek and dtc_size_seek position the cursor to a specific extent or size within the datatype.

The datatype iterator concept is similar in some ways to the *segments* used to transfer a datatype subset (*partial processing*) in MPICH2's dataloops, although segments seem not to have been applied to MPI-IO[8]. Datatype iterators are also similar to the flattening-on-the-fly technique (like dataloops but with added optimizations useful for vector processors) of listless I/O, which specializes the MPI pack/unpack interfaces to perform partial processing[9]. However, our datatype iterator interface appears to be unique: it allows a data transfer where

---

[1] Our experimental implementation uses *shmget*/*shmat* to attach a single shared memory region to every process. Each process (or aggregator, for collective shared buffer) allocates its application data (or working buffer) in a contiguous subset of the region.

both source and destination buffers are non-contiguous, and factors out separate *seek* and *tell* operations while still supporting partial processing.

The key data structure in the datatype iterator implementation is a stack with depth equal to the maximum nesting level of the derived datatype. Each stack element tracks the current position within the corresponding nested derived datatype. The basic algorithm for advancing the cursor descends the derived datatype tree to look ahead to the next contiguous block of bytes. If the lookahead is contiguous to the current accumulated block, add it and continue; otherwise, remember the lookahead for the next call and return the accumulated block.

The main processing loop for the collective shared buffer with dynamic offset/length generation algorithm is similar to that with flattened datatypes (including use of asynchronous I/O). However, no flattening is necessary before starting the main loop; instead, each process constructs datatype iterators for its own file and memory datatypes. The core of the "copy data" step for a write operation (the pseudocode below) demonstrates the power of datatype iterators. The code copies data directly from the (possibly non-contiguous) application buffer to the (possibly non-contiguous) destination locations in the shared working buffer without the need to pack and/or unpack data in an intermediate buffer (in contrast to the *direct_pack_ff* technique of [10]). Further, a contiguous datatype is not a special case: the code works efficiently for both contiguous and non-contiguous datatypes.

```
while (file_off + file_len <= end_off) {
                  // Entire file block still fits in current chunk
  while (file_len >= mem_len) {   // Mem block fits in file block
    src = app_buf + mem_off;
    memcpy(dest, src, mem_len);        // Copy remaining mem block
    file_off += mem_len;
    file_len -= mem_len;
    dest += mem_len;
    (mem_off, mem_len) = dtc_next(mem_dtc);      // Next mem block
  }
  while (mem_len >= file_len) {   // File block fits in mem block
    dest = temp_buf + file_off - start_off;
    memcpy(dest, src, file_len);       // Copy remaining file block
    mem_off += file_len;
    mem_len -= file_len;
    src += file_len;
    (file_off, file_len) = dtc_next(file_dtc); // Next file block
    if (file_off + file_len > end_off)
      break;
  }
}          // Elided: post-loop handling of tail end of file block
```

Another illustrative paradigm is the pseudocode to position the memory and file datatype cursors to match the start of the current file chunk:

```
dtc_extent_seek(file_dtc, start_off);
file_size = dtc_size_tell(file_dtc);
dtc_size_seek(mem_dtc, file_size);
```

The use of datatype iterators considerably simplifies the implementation of the two-phase collective I/O method. Using lines of code as a proxy for complexity, we compared our two collective shared-buffer implementations. The list-based routine required 952 lines (with 1210 lines of supporting functions), while the datatype iterator-based routine required 358 lines (with 617 lines of supporting functions, including the datatype iterator implementation), an overall savings of 62% for datatype iterators. It seems reasonable to expect similar savings for a non-shared-memory-based two-phase collective algorithm.

3   Performance Evaluation

To evaluate the performance of our new shared memory-based I/O methods against other existing methods, we ran three MPI-IO benchmarks. Our benchmark hardware is a Sun Fire™ 6800 server with 24 processors at 1.2 GHz and 96 GBytes of RAM. Four Sun StorEdge™ T3 disk arrays are connected via four dedicated 1 Gbit Fibrechannel host adapters. We used the Sun StorageTek™ QFS 4.5 filesystem[11] and the Solaris™ 9 operating system. The filesystem is configured with metadata on one disk array and data striped across the remaining three disk arrays using a 512 MByte disk allocation unit per array. Aggregate peak read or write bandwidth to the three data arrays does not exceed 300 MBytes/second. Some of the benchmark problem sizes are small enough to fit within the RAM cache of the disk arrays, so we explicitly disabled this cache for a fairer comparison to larger problems that do not fit in cache.

QFS offers both *buffered* I/O (caches file blocks in system memory, then copies or maps them to/from user memory) and *direct* I/O (host adapter copies file blocks directly between user space and disk array). Direct I/O usually delivers higher performance than buffered for writes and for reads of data not already present in buffer cache (lower bookkeeping overhead and one fewer copy operation) but requires user code to align file offsets. List I/O, datatype I/O, and mmap are by their nature restricted to buffered I/O. As a baseline, we measured both buffered and direct I/O results for data sieving and the ROMIO two-phase collective I/O methods, but only direct I/O for the remaining, higher-performing methods. QFS does not support datatype I/O; results for other buffered I/O methods suggest datatype I/O would have similar performance to those methods. In some cases we omit list I/O or data sieving results because their performance was so poor that the corresponding runs took too long to complete.

We required an MPI implementation that included ROMIO's implementation of MPI-IO and supported a shared memory transport on our test platform. LAM 7.1.1[12] seemed to be the only available implementation meeting both requirements at the time of our experiments. We implemented datatype iterators directly inside LAM, avoiding the public MPI interfaces for inspecting datatypes. We upgraded ROMIO to version 1.2.4 with additional flattening code from version 2005-06-09[5]. All code was compiled for a 64-bit execution model.

For each method, we picked one set of tuning parameters (primarily working buffer size, number of sub-buffers, and number of aggregators), chosen to obtain the best results across the selected range of problem types and sizes. The collective shared buffer with dynamic generation method required the least aggregate working buffer space of the collective methods and seemed least sensitive to parameter changes.

Each reported result is the average over three runs. Time constraints prevented us from flushing the filesystem buffer cache between runs, and one benchmark pre-reads file contents into the buffer cache before beginning measurements. Therefore, the results for I/O methods utilizing buffered reads include time to access the buffer cache but not time to transfer data from disk.

3.1 ROMIO 3D Block Test

The ROMIO 3D block test (coll_perf.c), included in the ROMIO test suite, measures bandwidth to a 600×600×600 array of integers stored in an 824 MByte file. Each process uses a contiguous memory datatype, but the portion of the array file accessed by each process is determined by a block distribution (MPI_DISTRIBUTE_BLOCK).

Figure 1 shows our results. When the number of processes is not an integer's cube, data is distributed unevenly among the processes, accounting for several zig-zags in the graphs. As expected, buffered methods outperform direct on reads, but suffer from cache management overhead and extra copying on writes. Among direct methods, the collective shared buffer with dynamic generation method achieves the best read and write performance with little sensitivity to the uneven data distribution. Data sieving has the poorest direct I/O performance; repeated access to the same disk block causes extra disk seeks and must be serialized for writes.

3.2 Tile Reader Benchmark

The tile reader benchmark[13] implements tiled access to a two-dimensional dense dataset. A *tile* represents an individual display unit; displays are arranged in an array to collectively present a large image to a human viewer. Each tile is 1024×768 pixels with 24 bits per pixel; the tiles overlap by 128 pixels vertically and 270 pixels horizontally to improve edge merging. Each process reads its corresponding tile from the file to a contiguous memory buffer.

Figure 2 presents results for array sizes from 2×2 to 6×4 with corresponding file sizes 7 to 37 MBytes. (The number of processes is the product of the two dimensions.) Buffered methods again benefit from a warm filesystem buffer cache to outperform direct methods. Among direct methods, the collective shared buffer with dynamic generation method consistently outperforms the other direct methods (even on small problems), and on larger problems achieves over

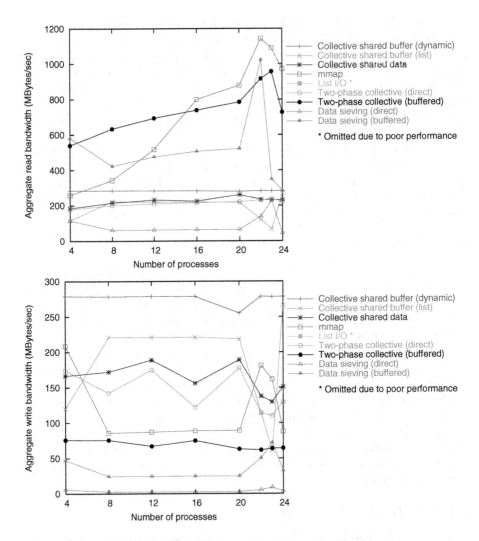

Fig. 1. ROMIO 3D block test performance results

100% of the available bandwidth.[2] Data sieving lags in performance due to repeated reads of the same disk block.

3.3   FLASH I/O Benchmark

The Argonne/Northwestern FLASH I/O benchmark (derived from the FLASH adaptive mesh refinement application[14]) substitutes synthetic data for the original computation, but makes the identical sequence of MPI-IO calls. Each

---

[2] How is this possible? The collective I/O methods read the overlapping data regions from the file only once, yet the benchmark counts the overlapping regions multiple times: $aggregate\_bytes\_read = number\_of\_processes \times data\_per\_process$.

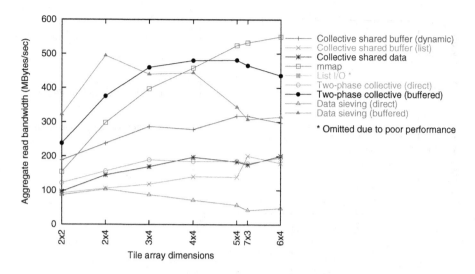

Fig. 2. Tile reader performance results

process contains 80 blocks. Each block is a three-dimensional array of data elements with each surface extended by four ghost cells. Each data element contains 24 variables. Data elements (but not ghost cells) are checkpointed to a file. In the file, data is rearranged so all values of variable 0 are stored first, then variable 1, and so on. Both file and memory datatypes are non-contiguous; each value is 8 bytes and is not contiguous in memory with other values of the same variable. (For our largest problem size the list-based I/O methods use an aggregate $O(10^9)$ list entries requiring twice the memory of the data they describe.)

We explored scalability along two dimensions. The top graph in Figure 3 reports results for a fixed number of processes (22) but a varying block size; file sizes range from 165 MBytes to 15 GBytes. The bottom graph reports results for a fixed block size ($20 \times 20 \times 20$) but a varying number of processes; file sizes range from 469 MBytes to 2.8 GBytes. The graphs show the collective shared buffer with dynamic generation method scaling well on both dimensions and providing the best performance. For the larger process counts it achieves over 90% of the available disk bandwidth, reflecting a 5× improvement over the best existing method (two-phase collective). The buffered methods are limited by cache management overhead and extra copying and have the poorest performance.

4   Conclusion

We explored several new methods to improve MPI-IO in a shared memory computer system. A method that utilizes a shared working buffer, a single aggregator, overlaps I/O and computation via a generalized double-buffering scheme, and reduces startup cost to generate (offset, length) pairs dynamically offered the best aggregate performance for several application I/O patterns.

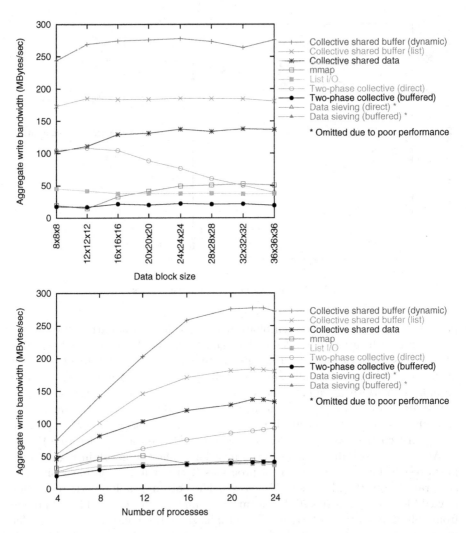

Fig. 3. FLASH I/O performance results: 22 processes (top), 20×20×20 block (bottom)

More generally, we rediscovered two important principles for obtaining good streaming I/O performance: (1) Reduce startup overhead and begin I/O early. (2) Overlap I/O and computation whenever possible. Our new abstraction, the datatype iterator, follows (1): initialization is cheap in contrast to the potentially high cost to generate (offset, length) lists. The sub-buffering mechanism we used in our collective shared buffer I/O methods follows (2).

We utilized the datatype iterator only in a shared memory system. Since the ROMIO two-phase collective uses lists extensively, and our research shows that datatype iterators in conjunction with overlapped I/O and computation can produce better performance with fewer lines of code than lists, an interesting area for future work would be the use of datatype iterators in traditional clusters.

Acknowledgements

Harriet Coverston and Anton Rang also contributed to this work. This material is based on work supported by the US Defense Advanced Research Projects Agency under Contract No. NBCH3039002. Sun, StorageTek, Sun Fire, Sun StorEdge, and Solaris are trademarks or registered trademarks of Sun Microsystems, Inc.

References

1. W. Gropp, S. Huss-Lederman, A. Lumsdaine, E. Lusk, B. Nitzberg, W. Saphir, and M. Snir, *MPI – The Complete Reference*, vol. 2. MIT Press, 1998.
2. R. Thakur, W. Gropp, and E. Lusk, "Data sieving and collective I/O in ROMIO," in *Proceedings of the Seventh Symposium on the Frontiers of Massively Parallel Computation*, pp. 182–189, IEEE Computer Society Press, February 1999.
3. R. Thakur, W. Gropp, and E. Lusk, "On implementing MPI-IO portably and with high performance," in *Proceedings of the Sixth Workshop on Input/Output in Parallel and Distributed Systems*, pp. 23–32, May 1999.
4. A. Ching, A. Choudhary, W.-K. Liao, R. Ross, and W. Gropp, "Efficient structured data access in parallel file systems," in *Proceedings of the IEEE International Conference on Cluster Computing*, pp. 326–335, IEEE Computer Society Press, December 2003.
5. "MPICH2 home page," August 2005. http://www.mcs.anl.gov/mpi/mpich2/.
6. "HPCS – High Productivity Computer Systems," April 2006. http://www.highproductivity.org/.
7. M. Vildibill, "Sun's Hero program: Changing the productivity game," April 2006. http://www.hpcwire.com/hpc/614805.html.
8. R. Ross, N. Miller, and W. Gropp, "Implementing fast and reusable datatype processing," in *Proceedings of the 10th European PVM/MPI Users Group Meeting*, pp. 404–413, Springer-Verlag, October 2003.
9. J. Worringen, J. L. Träff, and H. Ritzdorf, "Fast parallel non-contiguous file access," in *Proceedings of SC2003: High Performance Networking and Computing*, IEEE Computer Society Press, November 2003.
10. J. Worringen, A. Gäer, and F. Reker, "Exploiting transparent remote memory access for non-contiguous- and one-sided-communication," in *Proceedings of the International Parallel and Distributed Processing Symposium*, pp. 163–172, IEEE Computer Society Press, April 2002.
11. "Sun StorageTek QFS software," April 2006. http://www.sun.com/storagetek/management_software/data_management/qfs/.
12. "LAM/MPI parallel computing," April 2005. http://www.lam-mpi.org/.
13. R. Ross, "Parallel I/O benchmarking consortium," August 2005. http://www.mcs.anl.gov/~rross/pio-benchmark/.
14. "ASC center for astrophysical thermonuclear flashes," April 2006. http://flash.uchicago.edu/.

High-Bandwidth Remote Parallel I/O with the Distributed Memory Filesystem MEMFS

Jan Seidel[1], Rudolf Berrendorf[1], Marcel Birkner[1], and Marc-André Hermanns[2]

[1] Department of Computer Science
University of Applied Sciences Bonn-Rhein-Sieg, 53754 St. Augustin, Germany
[2] Central Institute for Applied Mathematics
Research Centre Jülich, 52425 Jülich, Germany

Abstract. The enormous advance in computational power of supercomputers enables scientific applications to process problems of increasing size. This is often correlated with an increasing amount of data stored in (parallel) filesystems. As the increase in bandwith of common disk based I/O devices can not keep up with the evolution of computational power, the access to this data becomes the bottleneck in many applications. MEMFS takes the approach to distribute I/O data among multiple dedicated remote servers on a user-level basis. It stores files in the accumulated main memory of these I/O nodes and is able to deliver this data with high bandwidth.

We describe how MEMFS manages a memory based distributed filesystem, how it stores data among the participating I/O servers and how it assigns servers to application clients. Results are given for a usage in a grid project with high-bandwidth WAN connections.

Keywords: Parallel I/O, Memory filesystem.

1 Introduction

With the increasing problem size of supercomputer applications, the amount of accessed data grows faster than the bandwith of modern disk-based I/O devices. To avoid a bottleneck in data access, there is a great demand on high-performance I/O solutions.

Traditional filesystems that store I/O data on single harddisks like NFS are limiting the I/O performance to the bandwidth of the single I/O bus. Parallel filesystems like GPFS [1] and PVFS2 [2] improve the performance by using multiple nodes for data storage and by allowing concurrent access. These systems have in some situations the drawback that nodes are coupled to form a filesystem statically. This is adequate for static environments but becomes a difficult administrative task for reconfigurable environments like grids, especially if data is distributed over clusters on different locations.

To support the development of portable and performant parallel applications the middleware needs to provide efficient I/O mechanisms. MPI defines a standard for I/O in [3]. ROMIO [4] is an advanced implementation of this standard, which is

B. Mohr et al. (Eds.): PVM/MPI 2006, LNCS 4192, pp. 222–229, 2006.

used in many MPI implementations, e.g. in MPICH [5]. ROMIO uses the "abstract device interface for parallel I/O" ADIO [6] to create an abstraction layer of the underlying filesystem. Filesystems that implement an ADIO device can be used by ROMIO for MPI-IO calls.

In the VIOLA project [7], a german network testbed for grid applications, several clusters are connected by dedicated 10 GBit/s optical WAN connections. For parallel applications, MP-MPICH [8], a special MPI implementation for grid applications, is used to spawn a MPI program over several cluster sites. In VIOLA, dedicated and specialized I/O clusters are used to perform high-bandwidth I/O initiated by I/O clients on different compute clusters.

To exploit the high bandwidth of optical connections disk-based storage is often not adequate as it can not deliver the high bandwidth. With MEMFS we developed a distributed parallel virtual filesystem in main memory that can satisfy very high bandwidth demands. MEMFS can be easily used in reconfigurable grid environments on a demand basis as MEMFS I/O servers are part of the user-level MPI program dedicated to parallel I/O, but working transparently to the user program.

Section 2 introduces the design of MEMFS, describing the filesystem design and multiserver aspects with client-server mapping, data distribution and file locking for consistency reasons. Section 3 shows the current state of our development, presenting performance results. In Section 4 our work is placed in the context with other projects and developments. Section 5 concludes with a summary and future plans.

2 Design

In this section we describe the design of MEMFS for high bandwith remote parallel I/O. The development of a performant interface for parallel remote I/O is divided into two parts and implemented as two separate ADIO devices. TUNNELFS [9] is used to enable transparent access to remote data in a grid. MEMFS is used to store I/O data in the main memory of local or remote nodes. This section presents the design of the MEMFS multiserver environment.

2.1 MEMFS Filesystem

The MEMFS filesystem is designed to store data efficiently in the main memory of one or multiple I/O nodes. It consists of a filetable and the files with the I/O data. The filetable administers the files in the filesystem and contains the information necessary to access the individual files. A file contains besides its I/O data all metadata needed to read and write its content, such as the filename with which it can be accessed and the access mode. The I/O data is managed in blocks that are allocated and deallocated dynamically by using *malloc* and *free*. The metadata is stored on all I/O servers by opening each file on every server. The I/O block size is variable for each file and can be passed to the filesystem as a parameter when opening the file. The MEMFS ADIO device implements all functions as requested by the ADIO interface. MEMFS is explicitly designed for

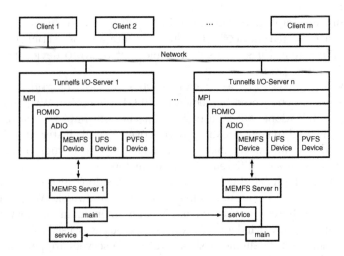

Fig. 1. Layers of the TUNNELFS servers and MEMFS. Multiple MEMFS servers communicate over main and service threads. Clients communicate with servers over the network using the TUNNELFS device.

usage in MPI-IO grid applications and can only be used with MPI-IO calls. Therefore MEMFS is optimized for a small amount of large temporary files accessed by multiple clients in parallel.

2.2   Multiserver Architecture

MEMFS can be started on an arbitrary number of nodes in the grid. With MP-MPICH, the MPI implementation we used, it is possible to start separate programs (user program and I/O server program) and to specify at program startup which grid nodes should be used with each program. Additionally, with a modification to MP-MPICH we are able to hide the I/O servers from the user program, i.e. the user program still works with its MPI_COMM_WORLD on the user nodes.

MEMFS creates a distributed filesystem in the main memory of all MEMFS server nodes. Multiple I/O servers are able to handle requests concurrently. Each server is capable of handling any ADIO request from any client. Currently we use a static client to server mapping, which is described in section 2.3.

The maximum amount of data stored in MEMFS is limited by the accumulated available main memory of the I/O server nodes. The user who starts the application is responsible for limiting data written to MEMFS to this amount. Usually the server processes and the application clients are started on different nodes of the grid, so the memory allocation of the I/O servers does not affect the available memory for the clients.

To support a distributed filesystem each MEMFS server process starts two threads, a *main thread* and a *service thread*. The main thread receives and handles all requests from the MEMFS ADIO device issued by an I/O client. For most requests data has to be transfered to or read from one or more other MEMFS servers as the filesystem is distributed over all servers. In these cases the main

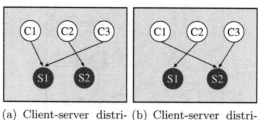

(a) Client-server distri- (b) Client-server distribution for file 1   bution for file 2

Fig. 2. Roundrobin client - server mapping for multiple files

thread sends out requests to other servers. A service thread is responsible for acting upon the requests of MEMFS main threads running on other servers. It continuously listens for these requests and performs the requested operations. When the main thread has received replies for all outstanding requests it returns the results to the client, finishing the operation.

The communication protocol between two servers as illustrated in Fig. 1 is designed in a simple way to avoid deadlocks: The main thread of a server is only allowed to communicate with the service threads of other servers and a service thread is only allowed to handle requests, it cannot issue requests itself. The single MEMFS servers are not currently multithreaded, so there can be only one active ADIO client request per server.

As MEMFS stores data solely in the main memory of the participating nodes it is only available during job execution time. Data that needs to be accessed after job termination must be stored on a persistent filesystem. However, the design of MEMFS is flexible, so it can be used as a fast caching filesystem with appropriate enhancements.

2.3 Client-Server Mapping

When running in conjunction with TUNNELFS we enable different mapping schemes of clients to servers for each file opened with MEMFS. A *mapping* is an assignment such that a client contacts one specific server for all MPI-IO requests and also receives the results of each request from this server. A simple but non-efficient mapping is the *globalmaster assignment*, where all clients are assigned to one MEMFS server. This approach does not parallelize client requests, because the globalmaster server can only handle one request at a time. The standard mapping of MEMFS is the *round robin assignment*, where clients are assigned to servers in a ring manner. This results in an even mapping of clients to servers. Previous assignments are regarded, so the overall number of clients assigned to each server is nearly equal. See Figure 2 for a case where two I/O servers get assigned the same number of clients for both files together. A round robin assignment allows to parallelize the requests from clients, as different clients can contact different servers for their requests concurrently. A sophisticated data distribution scheme between MEMFS servers is necessary to achieve a high

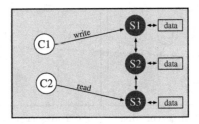

Fig. 3. File consistency in distributed filesystems

parallelization of client requests. Since in our current version each MEMFS server still handles requests serialized, it is important that different clients mostly access data on different servers.

2.4 Data Distribution

Files in MEMFS are distributed to all servers to support parallelism between file accesses of different clients. Currently, we use simple striping between all servers for data distribution. The separate stripes of a file can be accessed by different clients in parallel. The stripe size can be set by each application, since the optimal size is dependent of the file access scheme of the applications' client processes. In an optimal case the I/O data written and read by a client is completely stored on the server it is assigned to. With striping this can only be reached for simple data access schemes of clients. Therefore we will develop more sophisticated data distribution schemes in the future.

2.5 File Locking

The MPI standard defines several conditions under which sequential consistency has to be guaranteed by the filesystem. Sequential consistency for I/O means that for a write operation of client 1 and a read operation of client 2 (see Fig. 3) on the same part of a file the read operation either returns the unchanged data before the write operation or the changed data, but not a mixture of both. In a distributed filesystem write accesses to overlapping regions of a file need to be serialized to guarantee sequential consistency. MEMFS uses a locking mechanism to serialize those requests. The locks are currently managed centrally by one server for each file. In the future, we will work on more sophisticated and scalable locking schemes. The MEMFS locking mechanism distinguishes between write and read accesses. Multiple read accesses to overlapping data regions can be performed concurrently, as no file data is changed. When a write operation is performed neither read nor other write accesses are allowed on the same file region, so they are queued for later execution.

Furthermore locking is only necessary when overlapping file regions are accessed by different clients and at least one of those accesses is a write request. The access scheme of MPI applications however is often restricted to disjunct file regions. For those disjunct accesses no locking mechanism is required to achieve

 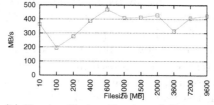

(a) Varying number of clients and servers   (b) Varying filesize. Number of clients and servers fixed to 6

Fig. 4. Achieved I/O bandwidth with TUNNELFS and MEMFS

sequential consistency. The application developer can specify whether the application clients use disjunct or overlapping filesystem accesses by setting an MPI info object. Higher I/O performance can be achieved if the disjunct mode is set because then no overhead through file locking occurs.

3 Results

MEMFS in conjunction with TUNNELFS was evaluated by measuring the I/O performance between different clusters in the VIOLA network. I/O servers and clients where physically separated by an up to 100 kilometres WAN. All server processes were run on one cluster and all clients on one or more other clusters. The server cluster has 6 4-way nodes and supports the use of Myrinet as an internal network device for server-to-server communication, while communication with remote clusters is done through the 10 Gbit/s WAN accessed by 6 GigE network cards, limiting the achievable bandwidth to 750 MB/s. All results were collected with a MPI benchmark program, that measures the performance of standard MPI-IO operations. In this benchmark each client writes data to a shared file with the different MPI-IO operations and then reads the written data back. Currently each client writes to a disjunct file region, so the locking mechanism described in 2.5 was turned off to improve performance. The results shown in Figure 4 (a) were collected by writing and reading 100 MB of data per client. As the figure shows, we nearly reach the maximum of 750 MB/s by placing at least 12 servers on the server cluster (2 server processes per 4-way node) and an equivalent number of clients on another cluster. In Fig. 4 (b) the number of servers and the number of clients is set to a constant value of 6, while the filesize is varying. The stripe size is fixed to 100 MB. The figure shows that the best performance is reached when each server holds exactly one striping block (6 * 100 MB = 600 MB filesize), due to the fact that each client then writes and reads data from exactly one I/O server. Larger files show no significant degregation.

4 Related Work

Creating a filesystem in the main memory is a well known approach for accelerating local I/O. A "Network RamDisk" was developed by Flouris and Markatos [10], creating a distributed filesystem in the main memory of multiple nodes. The Network Ramdisk uses TCP/IP instead of MPI calls for communication, disabling the use of a high-speed special purpose network interconnect like Myrinet. Furthermore the Network Ramdisk does not support distributing the filesystem over several clusters as used in VIOLA. With the increasing use of clusters for high-performance computing many distributed file systems have been developed in the last years. Examples are The General Parallel Filesystem GPFS [1] from IBM and the Lustre file system [11] from Cluster File System, Inc., both designed for very high scalability and performance, or the Google File System [12], explicitly designed for the requirements of Google's applications. Examples for open-source distributed file systems are GFS [13] from Red Hat and PVFS2 [2], which supports an interface for defining file data distribution schemes. All these file systems for clusters have in common that they store data on multiple hard disks accessed in parallel, achieving high I/O performance. But all of them are designed for usage in a static cluster environment, requiring administrative setup of server nodes. Our development in contrast is designed for usage in reconfigurable grid environments, where I/O servers and clients are coupled on a per-job basis.

5   Conclusion and Future Work

We developed a distributed filesystem for main memory that can be accessed by standard MPI-IO calls through the ADIO interface. By introducing the multi-server version of MEMFS the aggregated main memory and network bandwidth of multiple cluster nodes can be used for storage of application data. By using MPI calls for all communication between nodes our approach is independent of the underlying network interconnect. With the implemented ADIO device MEMFS can be used by any MPI-IO application. With TUNNELFS we have designed a flexible infrastructure for high-bandwidth I/O that can be adapted to applications' I/O needs on a per-job level. With MEMFS we added the benefit of flexible user-controlled I/O server configuration based on available ressources rather than a static setup environment, and achieve high performance.

Currently further development of MEMFS focuses on performance optimization, including more efficient and adaptable distribution schemes for the client-server mapping as well as a more sophisticated data distribution than simple striping with a constant stripe size. A possible extension will be the use of MEMFS as a caching system for high-bandwidth I/O during application runtime. Data then can be migrated by MEMFS to a persistent filesystem to become available after job termination.

Acknowledgments

This work was supported within the VIOLA project by the German Ministry of Education and Research under contract number FKZ 01AK605L, and by a research grant of the University of Applied Sciences Bonn-Rhein-Sieg.

References

1. IBM Corporation: IBM General Parallel Filesystem (2006)
   http://www-03.ibm.com/servers/eserver/clusters/software/gpfs.html.
2. PVFS2: Parallel Virtual Filesystem 2 (2006) http://www.pvfs.org/pvfs2.
3. Message Passing Interface Forum (MPIF): MPI-2: Extensions to the Message-
   Passing Interface. University of Tennessee, Knoxville (1996)
   http://www.mpi-forum.org/docs/mpi-20-html/mpi2-report.html.
4. Thakur, R., Ross, R., Lusk, E., Gropp, W.: Users Guide for ROMIO: A High-
   Performance, Portable MPI-IO Implementation. Technical Report ANL/MCS-TM-
   234, Mathematics and Computer Science Division, Argonne National Laboratory
   (2004) http://www-unix.mcs.anl.gov/romio/.
5. Gropp, W., Lusk, E., Doss, N., Skjellum, A.: A High-Performance, Portable Implementation of the MPI Message Passing Interface Standard. Technical report, Mathematics and Computer Science Division - Argonne National Laboratory (1996)
   http://www-unix.mcs.anl.gov/mpi/mpich/.
6. Thakur, R., Gropp, W., Lusk, E.: An Abstract-Device Interface for Implementing
   Portable Parallel-I/O Interfaces. In: Proceedings of the 6th Symposium on the
   Frontiers of Massively Parallel Computation. (1996) 180–187
7. The VIOLA Project Group: Vertically Integrated Optical testbed for Large scale
   Applications (2005) http://www.viola-testbed.de/.
8. Pöppe, M., Schuch, S., Bemmerl, T.: A Message Passing Interface Library for Inhomogeneous Coupled Clusters. In: Proceedings of the IEEE International Parallel
   and Distributed Processing Symposium (IPDPS 2003), Workshop on Communication Architecture for Clusters (CAC 2003), Nice, France (2003)
9. Berrendorf, R., Hermanns, M.A., Seidel, J.: Remote Parallel I/O in Grid Environments. In: Proc. of the Sixth Conference on Parallel Processing and Applied Mathematics (PPAM). Lecture Notes in Computer Science, Poznan, Poland, Springer
   (2005) to appear.
10. Michail D. Flouris and Evangelos P. Markatos: The network ramdisk: Using remote
    memory on heterogeneous nows. In: Cluster Computing: The Journal on Networks,
    Software, and Applications. Volume 2(4)., Baltzer Science Publishers (1999) 281–
    293 http://archvlsi.ics.forth.gr/OS/os.html.
11. Cluster File Systems, Inc.: Lustre file system (2006) http://www.lustre.org/.
12. Ghemawat, S., Gobioff, H., Leung, S.T.: The Google File System. In: Proceedings of
    the nineteenth ACM symposium on Operating systems principles, IEEE Computer
    Society Press (2003) 29 – 43
13. Red Hat, Inc.: Red hat global file system (2006)
    http://www.redhat.com/software/rha/gfs/.

Effective Seamless Remote MPI-I/O Operations with Derived Data Types Using PVFS2

Yuichi Tsujita

Department of Electronic Engineering and Computer Science,
Faculty of Engineering, Kinki University
1 Umenobe, Takaya, Higashi-Hiroshima, Hiroshima 739-2116, Japan
tsujita@hiro.kindai.ac.jp

Abstract. Parallel computation outputs intermediate data periodically, and typically the outputs are accessed for visualization in remote operation. To realize this kind of operations with derived data types among computers which have different MPI libraries, a Stampi library was proposed. For effective data-intensive I/O, a PVFS2 file system has been supported in its remote MPI-I/O operations by introducing MPICH as an underlying MPI library. This mechanism has been evaluated on interconnected PC clusters, and sufficient performance has been achieved with huge amount of data. In this paper, architecture, execution mechanism, and preliminary performance results are reported and discussed.

Keywords: MPI-I/O, Stampi, MPI-I/O process, derived data type, PVFS2.

1 Introduction

Stampi was originally developed to support seamless MPI communications among different MPI libraries by deploying a wrapper interface library on a native MPI library to intermediate MPI communications among different MPI libraries [1]. It also supports MPI-I/O operations both inside a computer and among computers which have different MPI libraries [2].

Among parallel scientific applications, several kinds of I/O interfaces which support a portable data format such as NetCDF [3] were proposed and used. A parallel I/O interface of the NetCDF library named parallel NetCDF (hereafter PnetCDF) [4] was developed with the help of an MPI-I/O interface in vendor's MPI library or MPICH [5]. Although the PnetCDF library succeeded in parallel I/O operations, the same operations among computers which have different MPI libraries each other have not been available. To realize this mechanism, a remote MPI-I/O mechanism of a Stampi library was proposed as an underlying MPI-I/O library. In the PnetCDF library, MPI-I/O functions and MPI functions which create derived data types have been replaced with corresponding Stampi's MPI functions, and derived data types with ROMIO [6] were supported in the remote MPI-I/O operations. For collective I/O, multiple MPI-I/O processes are invoked on PC nodes of a Linux PC cluster. Unfortunately, its performance is poor with such multiple processes on an NFS file system by using ROMIO's MPI-I/O

B. Mohr et al. (Eds.): PVM/MPI 2006, LNCS 4192, pp. 230–237, 2006.

functions. To improve the performance, PVFS2 [7] has been introduced in the remote MPI-I/O mechanism. Remote MPI-I/O operations by using ROMIO with PVFS2 provide sufficient performance even if derived data types are used.

In the rest of this article, architecture and execution mechanism of Stampi, typically details of the remote MPI-I/O mechanism with derived data types, are discussed in Section 2. Preliminary performance results are reported in Section 3. Related work is discussed in Section 4, followed by conclusions in Section 5.

2   Remote MPI-I/O with Derived Data Types

In this section, details of architecture, execution mechanism, and sequence in I/O operations of the library are explained.

2.1   Architecture

Architecture of the I/O mechanism is illustrated in Figure 1. Stampi's intermediate library which has an MPI API is implemented on a native MPI library to relay messages between user processes and an underlying native MPI library, communication library, and I/O systems. High performance MPI communications inside a computer are available by using the native library. While in MPI communications among different MPI libraries, user's counterpart MPI processes are invoked on computation nodes of a remote computer with a remote shell command (rsh or ssh) when a spawn function in the MPI-2 standard is called. Then data communication is carried out by TCP socket connections. If the computation nodes are not able to access outside directly, a router process is invoked on a server node (an IP-reachable node) to relay message data. Then MPI communications among different MPI libraries are available through it.

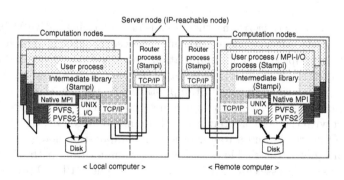

Fig. 1. Architecture of an MPI-I/O mechanism in Stampi

In addition to the communication mechanism, MPI-I/O functions are available both inside a computer (local MPI-I/O) and among computers (remote MPI-I/O). The target remote computer and so on are specified in an `info` object as key values at run time, and the intermediate library identifies which operation is appropriate according to the values. In the local MPI-I/O, I/O operations with a native MPI-I/O library are carried out. If the library is not available, UNIX I/O functions are used instead of it. While remote MPI-I/O operations are carried out by invoking MPI-I/O processes on a remote computer with the similar mechanism used in the MPI spawn function. The invocation is carried out when MPI_File_open() is called inside user processes. The MPI-I/O processes play I/O operations on a target computer by using a native MPI-I/O library or by using UNIX I/O functions if the native one is not available through the intermediate library according to I/O requests from user processes.

2.2 Execution Mechanism

Next, an execution mechanism of the I/O system is explained. Schematic diagram of it is depicted in Figure 2. Firstly, a user issues Stampi's start-up com-

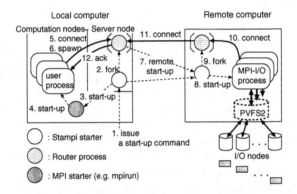

Fig. 2. Execution mechanism of remote I/O operations

mand (Stampi starter; jmpirun), then it calls an underlying native MPI start-up command (MPI starter) such as mpirun. Besides, a router process is also created by the Stampi starter if computation nodes are not IP-reachable nodes. The MPI starter invokes user processes later. Once MPI_File_open() is called in the user processes, either the Stampi starter or the router process invokes another Stampi's starter process on a remote computer which is specified in an info object with the help of a remote shell command. Secondly, the starter kicks off MPI-I/O processes and a router process if computation nodes of the remote computer are not IP-reachable nodes, and a communication path is established among the user processes and MPI-I/O processes. The MPI-I/O processes open a file which is specified in the user processes.

To have higher throughput in collective operations with multiple MPI-I/O processes on different nodes, an attempt to use ROMIO's MPI-I/O functions with PVFS2 support has been done in the remote I/O mechanism. Each MPI-I/O process calls Stampi's MPI-I/O functions, and they call associated ROMIO's MPI-I/O functions. On a PVFS2 file system, the ROMIO's functions with PVFS2 support provide sufficient performance.

In the end of I/O operations, `MPI_File_close()` is called in the user processes. An I/O request and associated parameters of the function are transferred to the MPI-I/O processes, and they close the file at first. Later, they are terminated and the communication path is closed.

As those mechanisms are capsuled in the Stampi's seamless intermediate library, users need not pay attention to the complex mechanisms.

2.3   Execution Sequence of I/O Operations

As an example, time step of execution of `MPI_File_write_all()` with a derived data type is illustrated in Figure 3. In creating a derived data type, firstly

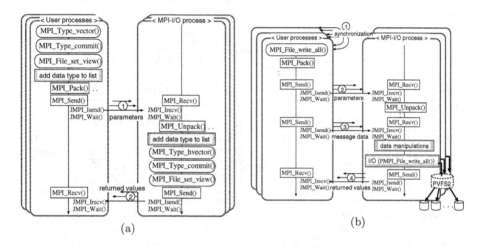

Fig. 3. Execution steps of (a) creation of a derived data type and (b) MPI-I/O operations with a derived data type

`MPI_Type_vector()` and `MPI_Type_commit()` are called in user processes. During these function calls, several parameters such as an old data type, displacement of each data block, and block length are stored in a list-based data type table prepared by a Stampi library. Thus, as much complex the data type is, the more the table is created. Finally, a target derived data type is created, and associated parameters are stored in the table and transferred to corresponding MPI-I/O processes. Once the parameters are stored in the same style table in the MPI-I/O processes, they also create the same data type except that they use `MPI_Type_hvector()`. Later they return a status value whether the operation is completed successfully or not.

In the MPI-I/O operations, user processes issue a collective MPI-I/O function call, then synchronization among them is carried out. If a derived data type is used, user processes extract parameters from their tables. Later, I/O request and associated parameters such as a file name are transferred to corresponding MPI-I/O processes. Later, they call a native MPI-I/O library such as ROMIO, and high performance I/O operation is carried out. Finally, a status value and so on are returned to the corresponding user processes.

3   Performance Evaluation

To evaluate the remote I/O operations with derived data types, performance was measured on interconnected PC clusters. Specifications of the clusters are summarized in Table 1. Each cluster had one server node and four computation

Table 1. Specifications of PC clusters which were used in performance evaluation, where **server** and **comp** in bold font denote server node and computation nodes, respectively

	PC cluster-I	PC cluster-II
server	DELL PowerEdge800 × 1	DELL PowerEdge1600SC × 1
comp	DELL PowerEdge800 × 4	DELL PowerEdge1600SC × 4
CPU	Intel Pentium-4 3.6 GHz × 1	Intel Xeon 2.4 GHz × 2
Chipset	Intel E7221	ServerWorks GC-SL
Memory	1 GByte DDR2 533 SDRAM	2 GByte DDR 266 SDRAM
Disk system	80 GByte (Serial ATA) × 1 (all nodes)	73 GByte (Ultra320 SCSI) × 1 (**server**) 73 GByte (Ultra320 SCSI) × 2 (**comp**)
NIC	Broadcom BCM5721 (on-board)	Intel PRO/1000-XT (PCI-X)
Switch	3Com SuperStack3 Switch 3812	3Com SuperStack3 Switch 4900
OS kernel	Fedora Core 3 2.6.12-1.1381_FC3smp (**server**) 2.6.11 (**comp**)	RedHat Linux 7.3 2.6.12-1.1381_FC3smp (**server**) 2.6.11-1SCOREsmp (**comp**)
Network driver	tg3 version 3.43f (**server**) tg3 version 3.23 (**comp**)	Intel e1000 version 6.0.54 (**server**) Intel e1000 version 5.6.10 (**comp**)
MPI	MPICH version 1.2.7p1	

nodes. Interconnection between the Gigabit Ethernet switches of the clusters was made with 1 Gbps bandwidth via a Gigabit Ethernet switch, Allied Telesis CentreCOM GS908GT.

In a PC cluster-II, PVFS2 (version 1.3.2) was available by dedicating disk spaces (73 GByte each) of four computation nodes. Thus 292 GByte (4 × 73 GByte) was available for the file system. During this test, default stripe size (64 KByte) of it was selected. A router process was not invoked in this test because each computation node was an IP-reachable node.

In this test, performance values of remote I/O operations with derived data types were measured with new and old libraries. In the remote I/O operations, user processes were created on the PC cluster-I, and MPI-I/O processes were invoked on the PC cluster-II to have access to its PVFS2 file system.

MPI-I/O functions were evaluated with derived data types in three block sizes, 16384, 65536, and 262144. The derived data types were constructed from an integer data type (`MPI_INT`) by `MPI_Type_vector()` as shown in Figure 4. Measured

```
comm_size = MPI_Comm_size(MPI_COMM_WORLD);
comm_rank = MPI_Comm_rank(MPI_COMM_WORLD);
nints = datasize/comm_size;
blkcnt = 16384;
MPI_Type_vector(nints/blkcnt, blkcnt, blkcnt*comm_size, MPI_INT, &filetype);
MPI_Type_commit(&filetype);
MPI_File_set_view(fh, blkcnt*sizeof(int)*comm_rank, MPI_INT, filetype,
          ''native'', info);
```

Fig. 4. Pseudo code for creating a derived data type

values are shown in Figure 5. In read operations, I/O throughput values in the old case are saturated around 40 MB/s for a single process, and 32 MB/s for two and four processes. While the values in the new case are around 48 MB/s, 44 MB/s, and 48 MB/s for a single process, two processes, and four processes, respectively. The similar improvement was observed in the new case for write operations. This was due to effective data I/O using PVFS2 functions in ROMIO by introducing an MPICH library in the MPI-I/O processes. Multiple MPI-I/O processes were created on the same number of PC nodes (one MPI-I/O process on each node) in this case, and collective I/O was realized. While the old case provided poor performance because only a single MPI-I/O process was invoked and it used UNIX I/O functions. Besides, serialization of I/O requests from user processes also degraded its performance. From these results, it is confirmed that performance was improved in the new library. It is also noticed that there was not significant difference in performance values for the new library with respect to block size.

4   Related Work

ROMIO provides MPI-I/O operations by using the ADIO interface [8] which provides a common I/O interface to wide variety of I/O systems. Noncontiguous and collective I/O with derived data types is optimized with data sieving and two-phase I/O [9]. MPI-I/O operations to a remote computer where MPICH is available are realized with the help of RFS [10]. An RFS request handler on a remote computer receives I/O requests from client processes and calls an appropriate ADIO library. Another implementation for effective noncontiguous MPI-I/O is listless I/O which is implemented in NEC's MPI/SX library [11]. The listless I/O avoids overheads to represent noncontiguous data types. On the other hand, Stampi itself is not an MPI implementation, and it provides an intermediate library among different MPI libraries by using TCP socket communications for seamless MPI operations on heterogeneous environment. Stampi

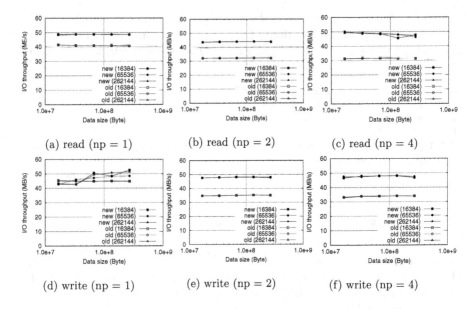

Fig. 5. I/O throughput in remote I/O operations by using new and old MPI-I/O libraries, where **new** and **old** denotes new and old ones, respectively. Besides, **read** and **write** denote MPI_File_read_all() and MPI_File_write_all(), respectively. Numbers in parentheses are block sizes in derived data types.

realizes MPI communications and MPI-I/O operations not only inside the same MPI library but also among different libraries without any attention to heterogeneity in underlying communication and I/O systems.

5   Conclusions

Remote MPI-I/O operations with derived data types on a PVFS2 file system have been realized in Stampi with the help of its remote I/O mechanism. With this library, multiple MPI-I/O processes were created on the same number of nodes each, and effective collective I/O was available. Its I/O performance was improved with the help of PVFS2 functions by using ROMIO in its MPI-I/O processes compared with that of an old library by using UNIX I/O functions.

As a future work, additional support of derived data types such as a distributed array and implementation of the Stampi library in the PnetCDF library as an underlying MPI-I/O library are considered.

Acknowledgments

The author would like to thank Genki Yagawa, director of Center for Computational Science and Engineering (CCSE), Japan Atomic Energy Agency (JAEA), for his continuous encouragement. The author would like to thank the staff at CCSE, JAEA for providing a Stampi library and giving useful information.

This work was partially supported by the Ministry of Education, Culture, Sports, Science and Technology (MEXT), Grant-in-Aid for Young Scientists (B), 18700074 and by the CASIO Science Promotion Foundation.

References

1. Imamura, T., Tsujita, Y., Koide, H., Takemiya, H.: An architecture of Stampi: MPI library on a cluster of parallel computers. In Recent Advances in Parallel Virtual Machine and Message Passing Interface, Volume 1908 of Lecture Notes in Computer Science., Springer (2000) 200–207
2. Tsujita, Y., Imamura, T., Takemiya, H., Yamagishi, N.: Stampi-I/O: A flexible parallel-I/O library for heterogeneous computing environment. In Recent Advances in Parallel Virtual Machine and Message Passing Interface, Volume 2474 of Lecture Notes in Computer Science., Springer (2002) 288–295
3. Rew, R.K., Davis, G.P.: The unidata netCDF: Software for scientific data access. In: Sixth International Conference on Interactive Information and Processing Systems for Meteorology, Oceanography, and Hydrology, American Meteorology Society (1990) 33–40
4. Li, J., Liao, W.K., Choudhary, A., Ross, R., Thakur, R., Gropp, W., Latham, R., Siegel, A., Gallagher, B., Zingale, M.: Parallel netCDF: A high-performance scientific I/O interface. In: SC '03: Proceedings of the 2003 ACM/IEEE Conference on Supercomputing, IEEE Computer Society (2003) 39
5. Gropp, W., Lusk, E., Doss, N., Skjellum, A.: A high-performance, portable implementation of the MPI Message-Passing Interface standard. Parallel Computing **22** (1996) 789–828
6. Thakur, R., Gropp, W., Lusk, E.: On implementing MPI-IO portably and with high performance. In: Proceedings of the Sixth Workshop on Input/Output in Parallel and Distributed Systems. (1999) 23–32
7. PVFS2: http://www.pvfs.org/pvfs2/.
8. Thakur, R., Gropp, W., Lusk, E.: An abstract-device interface for implementing portable parallel-I/O interfaces. In: Proceedings of the Sixth Symposium on the Frontiers of Massively Parallel Computation. (1996) 180–187
9. Thakur, R., Gropp, W., Lusk, E.: Optimizing noncontiguous accesses in MPI-IO. Parallel Computing **28** (2002) 83–105
10. Lee, J., Ma, X., Ross, R., Thakur, R., Winslett, M.: RFS: Efficient and flexible remote file access for MPI-IO. In: Proceedings of the 6th IEEE International Conference on Cluster Computing (CLUSTER 2004), IEEE Computer Society (2004) 71–81
11. Worringen, J., Träff, J. L., Ritzdorf, H.: Fast parallel non-contiguous file access. In: SC '03: Proceedings of the 2003 ACM/IEEE conference on Supercomputing, IEEE Computer Society (2003) 60

Automatic Memory Optimizations for Improving MPI Derived Datatype Performance

Surendra Byna[1], Xian-He Sun[1], Rajeev Thakur[2], and William Gropp[2]

[1] Department of Computer Science, Illinois Institute of Technology, Chicago, IL, USA
{bynasur, sun}@iit.edu
[2] Math. and Computer Science Division, Argonne National Laboratory, Argonne, IL, USA
{thakur, gropp}@mcs.anl.gov

Abstract. MPI derived datatypes allow users to describe noncontiguous memory layout and communicate noncontiguous data with a single communication function. This powerful feature enables an MPI implementation to optimize the transfer of noncontiguous data. In practice, however, many implementations of MPI derived datatypes perform poorly, which makes application developers avoid using this feature. In this paper, we present a technique to automatically select templates that are optimized for memory performance based on the access pattern of derived datatypes. We implement this mechanism in the MPICH2 source code. The performance of our implementation is compared to well-written manual packing/unpacking routines and original MPICH2 implementation. We show that performance for various derived datatypes is significantly improved and comparable to that of optimized manual routines.

Keywords: MPI, derived datatypes, MPI performance optimization.

1 Introduction

MPI derived datatypes [7] enable users to describe noncontiguous memory layouts compactly and to use this compact representation in MPI communication functions. Derived datatypes also enable an MPI implementation to optimize the transfer of noncontiguous data. For example, if the underlying communication mechanism supports noncontiguous data transfers, the MPI implementation can communicate the data directly without packing it into a contiguous buffer. On the other hand, if packing into a contiguous buffer is necessary, the MPI implementation can pack the data and send it contiguously.

In practice, however, many MPI implementations perform poorly with derived datatypes—to the extent that users often resort to packing the data manually into a contiguous buffer and then calling MPI. Such usage clearly defeats the purpose of having derived datatypes in the MPI Standard. Since noncontiguous communication occurs commonly in many applications (for example, Fast Fourier transform, array redistribution, and finite-element codes), improving the performance of derived datatypes has significant value.

The performance of derived datatypes can be improved in several ways. Researchers have used data structures that allow a stack-based approach to parsing a

B. Mohr et al. (Eds.): PVM/MPI 2006, LNCS_4192, pp. 238–246, 2006.

datatype, rather than making recursive function calls, which are expensive [4], [11], [12]. These works improved the performance of derived datatypes to the level of performance with naïve manual implementations for packing noncontiguous data. (We do better than that in this paper.) Wu et al. [13] improved the performance of MPI derived datatypes by taking advantage of the features in InfiniBand to overlap packing and unpacking a message with network communication.

The performance of derived datatypes can be improved further by using optimized algorithms for packing and unpacking of data. Many implementations of derived datatypes use loops in packing/unpacking noncontiguous data. Utilizing data locality in these loops by applying loop optimizations, which a developer cannot easily do without advanced knowledge of memory hierarchy design and optimizations, is beneficial. This area is the focus of our study. These techniques are useful for MPI implementations on various network channels and the performance gain is not limited to fast networks. Our previous work [1] presents the scope of performance improvement by using MPI's profiling interface (PMPI). In this paper, we present automatic selection of optimized packing/unpacking templates within the MPICH2 source code, based on data access patterns, data size, and memory architecture. Ogawa et al. [9] used optimized templates in improving MPI performance for instantiating partial-evaluation code selection in order to reduce software overhead. We, in contrast, use templates to optimize memory performance.

The rest of this paper is organized as follows. In Section 2, we present the design of our optimization mechanism. In Section 3, we describe the implementation details in selecting optimized templates dynamically. In Section 4, we present our experimental results, followed by conclusions in Section 5.

2   Optimization Mechanism

To choose optimized templates automatically, we developed a systematic approach. Our method first retrieves the data access pattern of a derived datatype from user's definition and verifies whether performance improvement is possible with optimizations for a derived datatype before applying them. If improvement is possible, our optimization method uses an analytical model [2] to predict memory access cost and to find optimization parameters with the lowest access cost. These parameters are passed to templates to pack/unpack noncontiguous data.

Overall procedure of optimizing an MPI communication function using derived datatypes has two steps. In the first step, we verify whether a datatype is optimizable or not, and find optimization parameters. In the second step, MPI communication function calls optimized templates automatically.

In MPI programs, after defining a derived datatype, it has to be committed by calling `MPI_Type_commit`. We modified the implementation of the `MPI_Type_commit` function to verify whether optimization is possible. The modified implementation first retrieves the data access pattern, which includes the type of the user-defined datatype, old datatype, strides between consecutive memory accesses, size of the data items, and depth of the derived datatype. If the old datatype is another derived datatype (that is, when a derived datatype is nested), `MPI_Type_commit` retrieves these values for that inner datatype as well. We use the datatype decoder functions of MPI-2, namely `MPI_Type_get_envelope` and `MPI_Type_get_contents` to retrieve the pattern. The overhead of decoding datatypes by using these functions is low.

In order to determine whether a datatype is optimizable or not, the modified `MPI_Type_commit` function verifies a series of heuristics that cause cache misses. It verifies whether the datatype is contiguous or noncontiguous, examines whether the data size is more than cache size, and then calculates the factor of cache and TLB reuse. The optimization method reverts back to the original implementation if it determines that the performance cannot be improved at any of these verifications. We use an optimization flag (`is_optimizable`) to keep track of the results of these verifications. If the performance can be improved, `MPI_Type_commit` determines the optimization parameters and sets the flag `is_optimization` to 1.

We developed optimized templates to pack/unpack noncontiguous data by using various loop optimization methods. In our current implementation, these templates use cache blocking [5], loop unrolling, array-padding optimizations, and software-level prefetching [8].

Various parameters are required in using these optimizations. Examples of optimization parameters are: block size for cache blocking, number of padding elements for array padding, and prefetching distance for software-level prefetching. In our approach, we first select these optimization parameters based on heuristics. To determine if these parameters are optimal, we developed a simple, fast, and accurate memory-access-cost prediction model [2]. This model verifies whether the memory access cost is reduced with the selected parameters. A new set of optimization parameters are selected if the cost is not optimized and the prediction model verifies for lowered cost again.

Examples of optimization parameter selection are as follows. For cache-blocking optimization, the block size is selected in a way that each block fits into the cache memory and virtual-to-physical address mappings of that block fit in the TLB (Translation Look-aside Buffer). For software prefetching, the number of loop iterations needed to overlap a prefetching memory access is called the *prefetching distance* [8]. Assuming memory access latency is l, and the work per loop iteration is w, the prefetch distance is *ceiling (l/w)*. The main loop that packs data is unrolled for all the references that reuse cache lines that are prefetched. An *epilogue loop* is called without prefetching to execute the last few iterations that do not fit in the main loop. We use a special `gcc` function `__builtin_prefetch` to issue these prefetch instructions. A special flag, `-mcpu`, has to be set to compile MPI source code.

In the second step, when the `MPI_Send` function is called to send the data, if the `is_optimization` flag is 1, the `MPI_Send` calls optimized packing templates using the optimization parameters. These templates are also used when the user calls `MPI_Pack` or `MPI_Unpack` to pack or unpack noncontiguous data.

3 Performance Results

We used three sets of benchmarks to evaluate the performance of our optimized implementations.

1. Simple derived datatypes: We chose fixed derived datatypes defined by the SKaMPI benchmark [10]. They describe a memory layout consisting of a number of units of a basic datatype. The number of units depends on the size of data, the size of basic datatype, and strides. We used vector and indexed datatypes.
2. Nested derived datatypes: We use the nested derived datatypes described by Ross et al. in [11]. These datatypes represent a collection of elements from a 3D array. When a 3D array is stored in row-major order, accessing the YZ face and all the YZ faces of the array in X direction is noncontiguous and has poor locality when the size of the YZ face is more than the cache or TLB sizes. We tested a nested datatype describing a 3D cube of YZ planes in the X direction with a vector of vectors (vector of YZ planes in an array).
3. NAS benchmarks: Lu et al. [6] modified four NAS benchmarks to apply MPI derived datatypes for noncontiguous data communication. Among these, LU, BT, and SP have small data transfers and do not benefit from memory optimizations. In the MG benchmark, the data transfers in the comm3 function are noncontiguous and are implemented as packing-then-sent by a sender process and receive-then-unpacking by a receiver. The datatypes described in the modified code are nested datatypes that represent vectors of vectors. We also tested the performance of the matrix transpose operation from the NAS parallel benchmarks' Fourier Transform (FT) program, using MPI derived datatypes. To describe the transpose operation with a derived datatype, we use a datatype that is a vector of vectors (vector of columns in an array).

Except for the NAS MG benchmark, we obtained the performance results of all other benchmarks with an MPI_Send/Recv ping-pong operation. In this operation, a process sends a noncontiguous message that is described by the MPI derived datatypes, and a destination process receives it contiguously. The destination process then sends back the data with the same derived datatype and is received at the first process contiguously. The time is measured at the first process and halved to find the communication cost for one complete data transfer. We ran 20 iterations of each program and calculated the minimum time. We present the performance as transfer rate (MB/s) to normalize the results. The size of the message used in the ping-pong operation is divided by the measured time to find the rate. For the NAS MG benchmark, we compare the execution time of the benchmark.

We compare the performance results for three implementations: manually packing data and sending it (no derived datatypes), MPICH2 version 1.0.3 (unoptimized), and our optimized implementation of the MPICH2 code. The manually implemented pack and unpack codes are written to represent the way a good programmer would write them. Ross et al. [11] showed that the implementation of derived datatypes in MPICH2 outperform those implemented in LAM/MPI. Therefore, we directly compare our results with MPICH2. We compile all manual codes and MPI installations with gcc version 3.2.3 with the flags -O6.

To test the portability of our optimized implementations, we ran these experiments on two different clusters: a 350-node Linux cluster (*jazz*) at Argonne National Laboratory and an 84-node Sun cluster (*sunwulf*) at Illinois Institute of Technology. The nodes of *jazz* have a 2.4 GHz Pentium-4 processor with 1 GB of memory. These processors have 512 KB of built-in L2 cache, with a 64 byte cache line and 8-way associative, a TLB of 128 entries, and a page size of 4 KB. The network interconnect of this cluster is Fast Ethernet. Each node of the *sunwulf* cluster is a Sun Blade-100 workstation with one 500MHz UltraSparc IIe CPU. The L1 cache is 16 KB, with a 16-byte cache line size. The L2 cache has a capacity of 8 MB and its line size is 64 bytes. It has a TLB of 48 entries with 4 KB page size. The network interconnect of *sunwulf* is Gigabit Ethernet.

Figure 1 shows the performance (rate of sending/receiving data in MB/s) of programs using messages formed by vector and indexed datatypes on the *jazz* cluster. Figure 2 shows the performance of the same programs on the *sunwulf* cluster. On both clusters, when the message size is larger than cache size, the performance of the original MPICH2 implementation degrades sharply compared to the manual implementation for both vector and indexed datatypes. With the optimized implementation, this performance is in the same level as that of optimized manual codes. These figures also show that the overhead of optimized implementations is low.

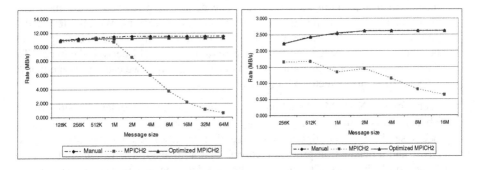

Fig. 1. Bandwidth measurements for vector (left) and indexed (right) datatype on jazz

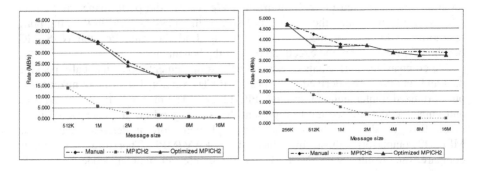

Fig. 2. Bandwidth measurements for vector (left) and indexed (right) on sunwulf

Figure 3 shows the performance of programs communicating messages formed using nested derived datatypes representing a 3D-cube on the *jazz* cluster and Figure 4 shows that on the *sunwulf* cluster. On both clusters, the original MPICH2 performs similar to manual and optimized implementations for smaller data sizes. As the message size (size of 3D cube) becomes larger compared to the L2 cache size, the performance degrades for MPICH2, whereas the optimized implementation maintains superior performance similar to that of the optimized manual program.

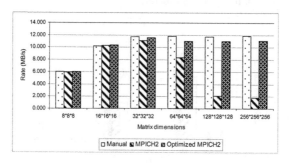

Fig. 3. Bandwidth measurements for the 3D-cube experiment on jazz

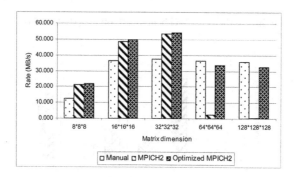

Fig. 4. Bandwidth measurements for the 3D-cube experiment on sunwulf

Figures 5 and 6 show the performance of the NAS MG benchmark on *jazz* and *sunwulf* clusters, respectively. We measured the execution time of the MG benchmark by using 4, 8 and 16 processors with B and C class workloads. The execution time with MPICH2 is higher than that of the original MG benchmark implementation (manual). With optimized MPICH2, the execution time is up to 8% (on average 6%) lower than that of manual implementation, and up to 25% (on average 13%) lower than that of unmodified MPICH2 on the *jazz* cluster. On the *sunwulf* cluster, for 8 and 16 processors, the execution time is up to 12% (on average 7.3%) less than that of the manual implementation. Here, manual implementation is the original NAS MG benchmark, which is not optimized for cache blocking and prefetching. Our optimized MPI derived datatype implementation benefits from using cache blocking in the nested datatypes in the MG benchmark.

Figures 7 and 8 show the performance (rate in MB/s) of the matrix transpose subroutine of NAS FT benchmark on *jazz* and *sunwulf* clusters, respectively. When the message size is larger than the L2 cache size, the rate degrades severely for unmodified MPICH2 because of the large number of cache misses caused by poor data locality. The optimized MPICH2 implementation benefits from using cache blocking in this program. The performance gain is in the range of 50–60% on *jazz* cluster and 50–114% on the *sunwulf* cluster.

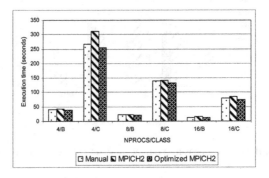

Fig. 5. Execution time of the NAS MG benchmark on jazz (left) and on sunwulf (right)

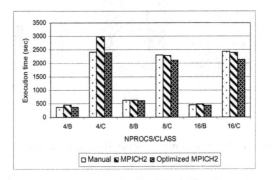

Fig. 6. Execution time of the NAS MG benchmark on sunwulf

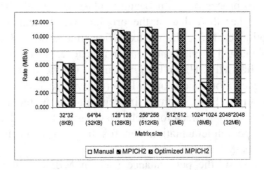

Fig. 7. Bandwidth measurements for matrix transpose experiment on jazz

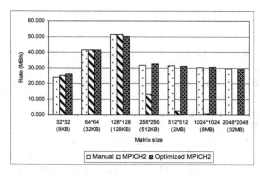

Fig. 8. Bandwidth measurements for matrix transpose experiment on sunwulf

4 Conclusions and Future Work

In this paper, we presented a technique to optimize the performance of MPI derived datatypes. Poor data access performance in dealing with noncontiguous data has been a major performance bottleneck of in packing and unpacking of MPI derived datatypes. Many optimization methods are available in the literature to optimize the data-access performance. However, predicting the optimization parameters with low overhead and automatically applying these optimization strategies is a challenging research issue. We developed models for predicting memory-access cost [2] that can help in dynamically applying optimizations. By combining optimization methods with a memory access model, we have introduced in this paper an approach to optimize memory performance automatically. The optimized implementation of MPI derived datatypes chooses packing templates that are optimized for advanced hierarchical memory systems of modern machines. These templates are parameterized with various architecture-specific parameters (for example, block size and TLB size), which are determined separately for different systems. By using these optimized templates, we obtained significantly higher performance than the existing MPICH2 implementation and manual packing/unpacking by the user. This result is significant because it will improve the performance of `MPI_Pack/Unpack` and MPI communication functions in many applications that use MPI derived datatypes in performing noncontiguous communication. We have shown that our optimized implementations are applicable on multiple architectures (Intel and Sun).

The optimizations described in this paper are not yet incorporated into the MPICH2 release, but we plan to do so. We are also looking at other applications of automatically selecting optimization parameters using the analytical prediction model. For example, in scientific applications, major portion of their run time is spent in executing loops. Using optimized templates can improve the performance of those loops. We are also working on incorporating prefetching strategies within PVFS [3] to improve the performance of data movement.

Acknowledgments. This work was supported in part by the National Science Foundation under NSF grants CNS-0509118, CNS-0406328, EIA-0224377, EIA-0130673, and in part by the Mathematical, Information, and Computational Sciences Division subprogram of the Office of Advanced Scientific Computing Research, Office of Science, U.S. Department of Energy, under Contract W-31-109-ENG-38.

References

1. Surendra Byna, William Gropp, Xian-He Sun, and Rajeev Thakur, "Improving the Performance of MPI Derived Datatypes by Optimizing Memory-Access Cost," In Proceedings of IEEE International Conference on Cluster Computing, December 2003.
2. Surendra Byna, Xian-He Sun, William Gropp and Rajeev Thakur, "Predicting Memory-Access Cost Based on Data-Access Patterns," In Proceedings of IEEE International Conference on Cluster Computing, September 2004.
3. Philip H. Carns, Walter B. Ligon III, Robert B. Ross, and Rajeev Thakur, "PVFS: A Parallel File System for Linux Clusters," In Proceedings of the 4th Annual Linux Showcase and Conference, pages 317--327, Atlanta, GA, 2000, USENIX Association.
4. William Gropp, Ewing Lusk, and Deborah Swider, "Improving the Performance of MPI Derived Datatypes," In Proceedings of the Third MPI Developer's and User's Conference, MPI Software Technology Press, pp. 25–30, March 1999.
5. M. Lam, Edward E. Rothberg, and Michael E. Wolf, "The Cache Performance of Blocked Algorithms," In Proceedings of the Fourth International Conference on Architectural Support for Programming Languages and Operating Systems, pp. 63–74, April 1991.
6. Q. Lu, J. Wu, D. Panda and P. Sadayappan, "Applying MPI Derived Datatypes to the NAS Benchmarks: A Case Study," Technical Report OSU-CISRC-4/04-TR19, Ohio State University.
7. Message Passing Interface Forum, "MPI: A Message-Passing Interface Standard", Version 1.1, June 1995. http://www.mpi-forum.org/docs/docs.html.
8. T. Mowry and A. Gupta, "Tolerating Latency Through Software-controlled Prefetching in Shared-memory Multiprocessors," Journal of Parallel and Distributed Computing, Volume 12, Issue 2, June 1991.
9. H. Ogawa and S. Matsuoka, "OMPI: Optimizing MPI Programs using Partial Evaluation," In Proceedings of IEEE/ACM Supercomputing Conference, Pittsburgh, November 1996.
10. Ralf Reussner, Jesper Larsson Träff, and Gunnar Hunzelmann, "A Benchmark for MPI Derived Datatypes," In Recent Advances in Parallel Virtual Machine and Message Passing Interface, 7th European PVM/MPI Users' Group Meeting, volume 1908 of Lecture Notes in Computer Science, pages 10-17, 2000.
11. R. Ross, N. Miller, and W. Gropp, "Implementing Fast and Reusable Datatype Processing," In Recent Advances in Parallel Virtual Machine and Message Passing Interface, 10th European PVM/MPI Users' Group Meeting, volume 2840 of Lecture Notes in Computer Science, pages 404-413, 2003.
12. Jesper Larsson Träff, Rolf Hempel, Hubert Ritzdorf, and Falk Zimmermann, "Flattening on the Fly: efficient handling of MPI derived datatypes. In Recent Advances in Parallel Virtual Machine and Message Passing Interface, 6th European PVM/MPI Users' Group Meeting, volume 1697 of Lecture Notes in Computer Science, pages 109-116, 1999.
13. Jiesheng Wu, Pete Wyckoff, Dhabaleswar Panda, "High Performance Implementation of MPI Derived Datatype Communication over InfiniBand," In Proceedings of the 18th International Parallel and Distributed Processing Symposium, 2004.

Improving the Dynamic Creation of Processes in MPI-2

Márcia C. Cera, Guilherme P. Pezzi, Elton N. Mathias,
Nicolas Maillard, and Philippe O.A. Navaux

Universidade Federal do Rio Grande do Sul, Instituto de Informática, Porto Alegre, Brazil
{mccera, pezzi, enmathias, nicolas, navaux}@inf.ufrgs.br
http://www.inf.ufrgs.br

Abstract. The MPI-2 standard has been implemented for a few years in most
of the MPI distributions. As MPI-1.2, it leaves it up to the user to decide when
and where the processes must be run. Yet, the dynamic creation of processes, enabled by MPI-2, turns it harder to handle their scheduling manually. This paper
presents a scheduler module, that has been implemented with MPI-2, that determines, on-line (*i.e.* during the execution), on which processor a newly spawned
process should be run. The scheduler can apply a basic Round-Robin mechanism
or use load information to apply a list scheduling policy, for MPI-2 programs with
dynamic creation of processes. A rapid presentation of the scheduler is given,
followed by experimental evaluations on three test programs: the Fibonacci computation, the N-Queens benchmark and a computation of prime numbers. Even
with the basic mechanisms that have been implemented, a clear gain is obtained
regarding the run-time, the load balance, and consequently regarding the number
of processes that can be run by the MPI program.

1 Introduction

In spite of the success of MPI 1.2 [4], one of PVM's features has long been missed in
the norm: the dynamic creation of processes. The success of Grid Computing and the
necessity to adapt the behavior of the parallel program, during its execution, to changing
hardware, encouraged the MPI committee to define the MPI-2 norm. MPI-2 includes
the dynamic management of processes (creation, insertion in a communicator, communication with the newly created processes. . .), Remote Memory Access and parallel
I/O. Although it has been defined in 1998, MPI-2 has lasted to be implemented and
only recently did all major MPI distributions include MPI-2. The notable exception is
LAM-MPI, which has provided an implementation for a few years.

Neither MPI 1.2 nor MPI-2 do define a way to schedule the processes of a MPI
program. The processor on which each process will execute and the order into which
the processes could run are left to the MPI runtime implementation. Yet, in the dynamic case, a scheduling module could help deciding, during the execution, onto which
processor each process should be physically started. This kind of scheduler has been
implemented for PVM [7]. This paper presents such an on-line scheduler for MPI-2.

This contribution is organized as follows: section 2 presents the MPI-2 norm, regarding dynamic process creation, as well as the MPI distributions that implement it,
and how one can schedule such MPI-2 programs. Section 3 presents a simple on-line

B. Mohr et al. (Eds.): PVM/MPI 2006, LNCS 4192, pp. 247–255, 2006.

scheduler, and how it gathers on-line information about the MPI-2 program and the computing resources. Finally, Sec. 4 shows how the scheduler manages to balance the load between the processors for three test applications, and the direct improvement on native LAM-MPI strategies, in terms of number of processes that can be managed, of load balance and of execution time. Section 5 concludes this article.

2 Dynamic Creation of Processes in MPI-2

This section provides a short background on on-line scheduling of dynamically created processes in a MPI program. Section 2.1 introduces some distributions that implement MPI-2 features and details the `MPI_Comm_spawn` primitive. Section 2.2 presents how to schedule such spawned processes.

2.1 MPI-2 Support

Distributions. There are an increasing number of distributions that implement MPI-2 functionalities. LAM-MPI [8] is the first MPI distribution that has implemented MPI-2. Together with it, LAM ships some tools to support the run-time in a dynamic platform: the `lamgrow` and `lamshrink` primitives allow to provide the runtime with information about entering or leaving processors in the MPI virtual parallel machine. MPI-CH is a most classical MPI distribution, yet its implementation of MPI-2 only dates back to January, 2005. Open-MPI is a new MPI-2 implementation based on the experience gained from the development of the LAM-MPI, LA-MPI, PACX-MPI and FT-MPI projects [2]. HP-MPI is a high-performance MPI implementation delivered by Hewlett-Packard, that implements MPI-2 since December, 2005.

Dynamic spawn. MPI-2 provides an interface that allows creating processes during the execution of a MPI program, and letting them communicate by message passing. Although MPI-2 provides more functionality, this article is restricted to the dynamic management of processes and only `MPI_Comm_spawn` will be detailed here (more information may be found in [5] for instance).

`MPI_Comm_spawn` is the newly introduced primitive that creates new processes after a MPI program has been started. It receives as arguments the name of an executable, that must have been compiled as a correct MPI program (thus, with the proper `MPI_Init` and `MPI_Finalize` instructions); the possible parameters that should be passed to the executable; the number of new processes that should be created; a communicator, which is returned by `MPI_Comm_spawn` and contains an inter-communicator so that the newly created processes and the parent may communicate through classical MPI messages. Other parameters are included, but are not relevant to this work.

In the rest of this article, a process (or a group of processes) will be called *spawned* when it is created by a call to `MPI_Comm_spawn`, where the process that calls the primitive is the *parent* and the new processes are the *children*.

2.2 On-Line Scheduling of Parallel Processes

The extensive work on scheduling of parallel programs has yielded relatively few results in the case where the scheduling decisions are taken on-line, *i.e.* during the execution.

Yet, in the case of dynamically evolving programs such as those considered with MPI-2, the schedule must be computed on-line.

The most standard technique is to keep a list of ready tasks, and to allocate them to idle processors. Such an algorithm is called *list scheduling*. The theoretical grounds of list scheduling relies on Graham's analysis [3]. Let T_1 denote the total time of the computation related to a sequential schedule, and T_∞ the critical time on an unbounded number of identical processors. If the overhead O_S induced by the list scheduling (management of the list, process creation, communications) is not considered, then $T_p \leq T_1/p + T_\infty$, which is nearly optimal if $T_\infty \ll T_1$. Note that list scheduling only uses some basic information of "load" about the available processors, in order to allocate tasks to them when they turn idle or underloaded. The scheduler presented in the next section uses such list mechanisms.

3   A Scheduler for MPI-2 Programs

The scheduler is constituted of two main parts: a daemon, that runs during the execution of the application and redefined MPI primitives, to handle the task graph and enable the communication between the MPI processes and the scheduler (Sec. 3.1); and a resource manager (Sec. 3.2). The scheduler must maintain a task graph, in order to compute the best schedule of the processes, a list of ready tasks, and information about the load of the computing resources. The resources manager is responsible for feeding the scheduler with information about the load.

3.1   The Scheduler

The redefined MPI-2 primitives send (MPI) messages to the scheduler process to notify it of each event regarding the program. The scheduler waits for these messages, and when it receives one, it processes the necessary steps: update of the task graph; evolution the state of the process that sent the message; possible scheduling decision.

In this work, scheduling decisions are taken at process creation only: neither preemption neither migration are used by the scheduler, so that there is no relevant event for the scheduler between the creation and the termination of a process. Upon termination of a process, the scheduler only updates its task graph to eliminate the task descriptor associated to the process.

At process creation, the newly created process(es) has(have) to be assigned a processor where it will be physically forked, preferably the less loaded one.

Figure 1 shows the interactions between MPI-2 processes and the scheduler during the dynamic creation of processes. The redefined call to MPI_Comm_spawn by the parent process will establish a communication (arrow (1)) between the parent and the scheduler, to notify the creation of the processes and the number of children that will be created (in the diagram, only one process is created). After receiving the message, the scheduler updates its internal task graph structure, decides on which node the child should physically be created (based, for instance, on information provided by the resource manager), and returns the physical location of the new process to the parent (arrow (2)). The parent process, that had remained blocked in a MPI_Recv, receives

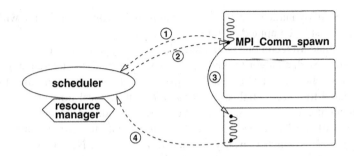

Fig. 1. Interactions between MPI-2 processes and scheduler

the location and physically spawns the child (arrow (3)). The child process will then execute and when it finishes, (*i.e.* calls the redefined MPI_Finalize), it will notify the scheduler (arrow (4)). The scheduler receives the notification and updates the task graph structure.

A very simple scheduling strategy that this scheduler provides a Round Robin allocation of the processes: in the case where it does not have any load information, it is the best choice. Yet, to offer better on-line schedules, a Resource Manager is also provided.

3.2   The Resource Manager

The dynamic resource management includes two main modules: a distributed load monitor that retrieves load metrics from the resources and a manager that coordinates the load monitors and centralizes the information collected by them. The scheduler uses the information to decide how to make effective use of the resources.

Load Monitor. The load monitor consists in daemons that are physically spawned by the resource manager on all the available resources. After being spawned, each monitor cyclically retrieves the usage metrics and gets ready to serve requisitions of the resource manager.

The usage metrics collected by the load monitor are stored in a LRU buffer, so that the oldest values are thrown away when needed. The load, on a given moment, is considered as being the average of the values in the buffer. This mechanism, known as Single Moving Average, is used on Time Series Forecast Models based on the assumption that the oldest values tend to a normal distribution [6]. The average also smoothes the effect of floating variations of the load, that does not characterize the instant resource usage.

Resource Manager. The resource manager module is responsible for the coordination of the load monitors and for keeping a list of the resources updated and ordered by their usages. In order to minimize bottlenecks, communication between the resource manager and the monitor happen on random intervals. The resource manager also offers an interface that makes it possible to use third-party resource managers or monitors. This can be useful for the interaction with other resource managers such as those offered by grid middlewares like Globus [1]. As a simple metrics of the load, the experiments presented hereafter use the CPU usage, multiplied by the average number number of processes run in the last interval.

4   Improving the Creation of Processes with On-Line Scheduling

This section presents and discusses the results obtained with the proposed, simple scheduler, on three test programs: The Fibonacci computation, the N-Queens problem and a simple computation of prime numbers. All experiments have been made with LAM-MPI version 7.1.1, on a Linux-based cluster. The Fibonacci and the N-Queens computations are used to illustrate how one can improve the number of processes that can be spawned with a better schedule (Sec. 4.1). Clearly, this is obtained by a better distribution of the processes, due to a better Round-Robin allocation, when compared with the native LAM-MPI solution. Section 4.2 improves even more for an irregular computation by the use of list-scheduling based on load measurement; the computation of the number of primes in an interval has been used as a synthetic benchmark.

4.1   Round-Robin Algorithm — Spawning More Processes

The Fibonacci Computation. Figure 2 shows how the MPI-2 processes are scheduled on a set of 5 nodes, using a simple Round-Robin strategy implemented in our scheduler, on three different executions of Fibonacci(n) for $n = 7, 10$ and 13. The total number of spawned processes (np) is indicated in the figure. Table 1 gives the numerical values associated to this graphical representation.

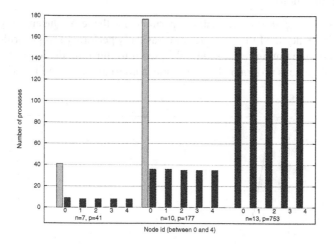

Fig. 2. Load balance of three Fibonacci runs, on 5 nodes. The grey bars show the number of processes on each node, as scheduled by the native LAM-MPI calls. The black bars show the number of processes spawned on each node, when scheduled by a Round-Robin strategy with our solution. The results are given for Fibonacci(7) (41 processes), Fibonacci(10) (177 processes) and Fibonacci(13) (753 processes). Each time when no grey bar appear, no process has been scheduled on the node.

Two main results are illustrated by this experiment: first, LAM's native strategy to allocate the spawned processes is limited: in LAM, there is a native Round Robin mechanism that works well when multiple processes are created by a unique spawn.

Table 1. Number of processes by node, in each one of the three Fibonacci computations presented in Fig. 2, for the native LAM scheduler and for the Round Robin scheduler. Five nodes of the cluster have been used.

	n=7 - np=41					n=10 - np=177					n=13 - np = 753				
Node Number	n0	n1	n2	n3	n4	n0	n1	n2	n3	n4	n0	n1	n2	n3	n4
Native LAM	41	0	0	0	0	177	0	0	0	0	ERROR				
RoundRobin	9	8	8	8	8	36	36	35	35	35	151	151	151	150	150

After having created the first process, the remaining ones will be spawned on successive available nodes. Yet, when the application spawns several processes one by one in successive MPI_Comm_spawn calls, all of them will be allocated to the same node, as shown by the Fibonacci experiment (see Table 1). On the contrary, the simple Round Robin strategy of our scheduler enables a perfect load balance without any intervention of the programmer. Second, due to this native scheme, LAM quickly gets overwhelmed by the number of processes: Fibonacci(13) should create 753 processes. LAM does not support such number of processes in the same node because it hits a limit of file descriptors. According to LAM-MPI user's mailing list, this limit can increase very fast because it is dependent of many factors like system-dependent parameters or number of opened connections.

The N-Queens Computation. The same experiments with the N-Queens program yielded similar results. Since the Fibonacci program basically performs a single arithmetic sum, we did not provide timing results and only used it to illustrate the load balance. In the case of the N-Queens program, the CPU-time is non-negligible and is shown in Fig 3.

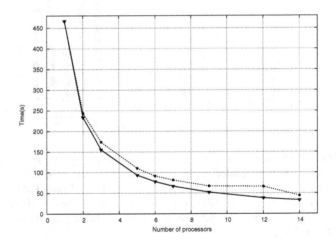

Fig. 3. Timing for the N-Queens application ($N = 18$), with Round-Robin strategy. The solid line shows the ideal run-time (sequential time divided by the number of processors) and the dotted line shows the N-Queens execution time.

The standard backtracking algorithm used to solve the N-Queens problem consists in placing recursively and exhaustively the queens, row by row. With MPI-2, each placement consists in a new spawned task. The algorithm backtracks whenever a developed configuration contains two queens that threaten each other, until all the possibilities have been considered. A maximum depth is defined, in order to bound the depth of the recursive calls. In this test case, the depth has been limited to 1, so that all N tasks are spawned one-by-one since the beginning of the application: thus, a Round-Robin can be performed on the N spawned tasks. As expected, the Round-Robin strategy enables a good speed-up. When LAM is used to run the same application, the native scheduling allocates all spawned processes on the same node, without any parallel gain.

Notice that Fig. 3 does not give any result for 4, 8, 10, 11 and 13 nodes. Actually, the N-Queens test fails for these numbers of nodes, probably due to an internal error in our scheduler. This problem does not invalidate the interest in providing a good Round-Robin mechanism to MPI-2 spawned processes.

4.2 List-Scheduling — Using Dynamic Load Information

Primality Computation is used to test the scheduler with information about the load on each node, in order to decide where to run each process. In this program, the number of prime numbers in a given interval (between 1 and N) is computed by recursive search. As in the Fibonacci program, a new process is spawned for each recursive subdivision of the interval. Due to the irregular distribution of prime numbers and irregular effort to test a single number, the parallel program is natively unbalanced.

Figure 4 shows the run-times *vs.* the workload, as measured by the size N of the interval, with the Round-Robin and List strategies. In the latter case, the resource manager has been used in order to obtain on-line information about the load of the processors. The Round-Robin algorithm maintains a natural load balance between the processors,

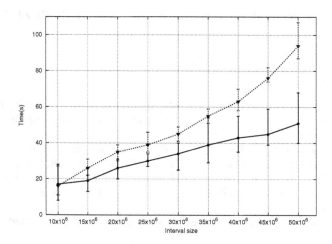

Fig. 4. Timings for the Prime computation, with Round-Robin strategy (dotted line) and List scheduling (solid line)

and the List strategy grants that any under-loaded processor executes more processes. Thus, in both cases, the resources are better employed.

As can be seen, in the case of this irregular computation, the use or our on-line list scheduler with load information enables a consistently better run-time than with Round-Robin, even when the statistical fluctuations are taken into account. Also, it can be seen that the time lasted when used the Round-Robin scheduling increases steady against a smoother increase when using the Resource Manager.

5   Contribution and Future Work

The new functionalities of MPI-2 are highly promising for MPI users who want to benefit from next generation architectures. The possibility of dynamically spawning new processes is specially interesting. This article shows that an extra scheduling daemon, in charge of managing the spawned processes, enables a direct improvement on:

- the load balance of the applications, whether with Round-Robin or by load-balancing schemes based on list scheduling;
- the total number of processes that may be supported by the run-time.

A simple Round-Robin already allows to obtain better results than the native LAM-MPI implementation. A load balancing mechanism is even more powerful.

This promising results show that with very few efforts, a global scheduling tool could be designed and integrated to a MPI distribution, in order to support the execution of dynamic applications, programmed with MPI, in grids. The proposed solution is portable beyond LAM-MPI, since is only uses MPI-2 calls to implement the daemons[1]. The experiments presented here have been made with a simple implementation based on the redefinition of MPI's standard primitives, but its integration inside an open-source distribution is straightforward.

Special thanks: this work has been partially supported by HP Brazil, CAPES and CNPq.

References

1. I. Foster and C. Kesselman. Globus: A metacomputing infrastructure toolkit. *The International Journal of Supercomputer Applications and High Performance Computing*, 11(2):115–128, Summer 1997.
2. E. Gabriel, G. E. Fagg, G. Bosilca, T. Angskun, J. J. Dongarra, J. M. Squyres, V. Sahay, P. Kambadur, B. Barrett, A. Lumsdaine, R. H. Castain, D. J. Daniel, R. L. Graham, and T. S. Woodall. Open MPI: Goals, concept, and design of a next generation MPI implementation. In *Proceedings, 11th European PVM/MPI Users' Group Meeting*, pages 97–104, Budapest, Hungary, September 2004.
3. R. Graham. Bounds on multiprocessing timing anomalies. *SIAM J. Appl. Math.*, 17(2):416–426, 1969.

---

[1] Actually, the only specific part is the determination of the PID of the processes, which is easily done in other MPI distributions.

4. W. Gropp, E. Lusk, and A. Skjellum. *Using MPI: Portable Parallel Programming with the Message Passing Interface*. MIT Press, Cambridge, Massachusetts, USA, oct 1994.
5. W. Gropp, E. Lusk, and R. Thakur. *Using MPI-2 Advanced Features of the Message-Passing Interface*. The MIT Press, Cambridge, Massachusetts, USA, 1999.
6. J. D. Hamilton. *Time Series Analysis*. Princeton University Press, 1994.
7. J. Ju and Y. Wang. Scheduling pvm tasks. *SIGOPS Oper. Syst. Rev.*, 30(3):22–31, 1996.
8. J. M. Squyres and A. Lumsdaine. A Component Architecture for LAM/MPI. In *Proceedings, 10th European PVM/MPI Users' Group Meeting*, number 2840 in Lecture Notes in Computer Science, pages 379–387, Venice, Italy, September / October 2003. Springer-Verlag.

Non-blocking Java Communications Support on Clusters

Guillermo L. Taboada, Juan Touriño, and Ramón Doallo

Department of Electronics and Systems
University of A Coruña, Spain
{taboada, juan, doallo}@udc.es

Abstract. This paper presents communication strategies for supporting efficient non-blocking Java communication on clusters. The communication performance is critical for the overall cluster performance. It is possible to use non-blocking communications to reduce the communication overhead. Previous efforts to efficiently support non-blocking communication in Java have led to the introduction of the Java NIO API. Although the Java NIO package addresses scalability issues by providing select() like functionality, it lacks support for high speed interconnects. To solve this issue, this paper introduces a non-blocking communication library to efficiently support specialized communication hardware. This library focuses on reducing the startup communication time, avoiding unnecessary copying, and overlapping computation with communication. This project provides the basis for a Java Message-passing library to be implemented on top of it. Towards the end, this paper evaluates the proposed approach on a Scalable Coherent Interface (SCI) and Gigabit Ethernet (GbE) testbed cluster. Experimental results show that the proposed library reduces the communication overhead and increases computation and communication overlapping.

1 Introduction

There is a growing interest shown by scientific and enterprise community in commodity clusters. The reason is that they deliver outstanding parallel performance at a competitive cost. A cluster consists of computing nodes connected together by a network fabric—usually a high-performance interconnect like SCI, Myrinet, or GbE. Scalability is a key factor to confront new challenges in cluster computing—it depends heavily not only on the network fabric, but also on the communication middleware.

This growing need of efficient communication middleware has led the community to devote significant efforts on this subject, although almost exclusively on native protocols. A thorough work focused on native protocols is that of Verstoep et al. [1], where several implementation issues are studied in order to obtain an efficient use of Myrinet. In this study, a non standard user level communication interface is implemented varying reliability protocols, maximum transfer unit, multicast protocols and studying Serial Direct Memory Access (SDMA)-based

B. Mohr et al. (Eds.): PVM/MPI 2006, LNCS 4192, pp. 256–265, 2006.

versus Processor Input/Output (PIO)-based message passing and remote-memory copy. The proposed approach inherits some optimizations from [1].

Despite the dominance of native protocol optimizations, the increasing interest in Java for high performance computing has recently increased the need for efficient Java communication middleware. This efficiency is of critical importance on clusters, especially on System Area Networks (SANs). In such environments, the overall performance is quite sensitive to the communication overhead [2]. As Java does not provide direct SAN protocols support, socket libraries and IP emulation layers have to be implemented on top of the high performance low-level SAN protocols. Moreover, communication is a major bottleneck in parallel Java applications. Thus, supporting efficient non-blocking communication on clusters, especially on SANs, appears to be a key objective to improve Java communication efficiency. As High Performance Cluster support has been traditionally focused on the blocking Java Remote Method Invocation (RMI) a follow-up aimed at supporting efficient non-blocking Java communications on clusters appears to be a promising research topic. This paper reports on the results obtained from the implementation of a non-blocking Java communication library with High Performance Cluster support.

1.1   Related Work

Previous efforts at obtaining non-blocking Java communications, NBIO (`http://www.eecs.harvard.edu/~mdw/proj/java-nbio/`) and Jaguar [3] have led to the introduction of some facilities in Java NIO to address scalability issues in server applications. Current efforts in non-blocking Java communications are more oriented to support communication for higher level libraries rather than constitute a messaging system *per se*. Therefore, their importance is centred around their projects. This is the case for `mpjdev` [4] used in HPJava [5] and of `xdev` used in a Java messaging-passing system, MPJ Express [6]. The `xdev` library is highly scalable due to the use of Java NIO and an efficient buffering scheme [7], supporting also Myrinet communications. mpiJava [8] is an object-oriented Java wrapper library to MPI implementations providing similar performance to native MPI implementations. Thus, non-blocking primitives present in native MPI implementations can be used efficiently in Java. Another Java Message-passing library that support non-blocking communication and Myrinet clusters is MPJ/Ibis [9].

2   Efficient Communication Libraries on Clusters

In the context of High Performance Cluster Computing the use of Network Interface Cards (NICs) is an attractive option as they offload communication processing from the host CPU. This helps in freeing up valuable CPU cycles for application processing. Moreover, higher performance in terms of both latency and bandwidths can be reached with these network fabrics, although this performance is usually only obtained by using their own efficient protocols. Figure 1 shows an overview of some protocols on SCI and GbE. Given components are colored in dark grey, whereas contributions presented in this paper are depicted in light grey.

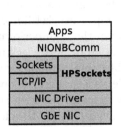 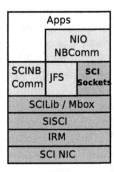

Fig. 1. Overview of communication libraries on popular cluster interconnects

Regarding SCI, the `IRM` driver interacts directly with the hardware, whereas `SISCI` provides resource management and a higher level API. This library implements basic mechanisms to share memory segments between nodes and to transfer data between them. `SCILib` is a communication protocol that offers unidirectional message queues. Depending on the message size `SCILib` presents three communication protocols: *inline, short* (both one-copy protocols) and *long* (zero-copy protocol). `Mbox` is a library that provides with remote interrupt mechanisms, so the target side can wait explicitly for an event or register a callback routine, whereas the initiator side triggers the event. SCI SOCKET [10] is a High Performance Socket implementation on SCI obtaining startup times as low as $4\mu s$ on commodity clusters.

Regarding GbE, its socket implementations are usually not very efficient. Various projects tried to reduce the overhead of these protocols by means of High Performance Sockets implementations—much like SCI SOCKETS on SCI. These High Performance Sockets projects are usually lightweight communication protocols focused on reducing latency by removing buffering overheads and protocol processing. In this context, some efforts include FastSockets [11], SOVIA [12], Sockets over GbE [13], and GAMMAsockets [14].

3   Designing Java Communication Libraries on Clusters

A non-blocking Java communication library, named `NBComm`, has been designed for efficient use of Java on clusters. This library abstracts the lower network layer and supports higher middleware libraries or runtime systems. As a result, such systems and libraries can be easily ported to different interconnects. This implementation constitutes the basis for a Java Message-passing library, as it provides a communication library with efficient non-blocking primitives along with good performance on different cluster interconnects.

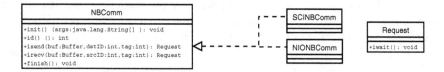

Fig. 2. `NBComm` API

This library is focused on reducing latency, avoiding unnecessary copying, and computation/communication overlapping. Figure 2 shows its object diagram, which consists of `NBComm`, the abstract communicator class that defines the general behaviour of the communication methods, and two implementation classes, `SCINBComm` and `NIONBComm`, for supporting different communication libraries. In this case, `SCINBComm` follows a native approach, implementing communications in native code over `SCILib` with a lightweight Java layer on top of it, whereas `NIONBComm` is a pure Java NIO-based solution. These classes implement the general behaviour in function of the underlying communication libraries: `init()` initialises the communicator object and `finish()` finalizes the communicator object; `id()` gets the identification for each process; `iwait()` waits for the completion of a communication; and `isend()` and `irecv()` perform communication using a *direct* `ByteBuffer` (a Java NIO buffer) which belongs to the class `Buffer`. These buffers can be accessed directly, and more efficiently, from native applications as they may reside outside of the normal garbage-collected heap. The `Buffer` class is similar to the Java NIO Buffer.

Listing 1.1. Non-blocking communications code example

```java
public static void main( String args []) throws Exception{
    int tag=10, size=10, capacity=40;
    int [] data = new int [ size ];
    NBComm nbComm = NBCommFactory.getNBComm("sci");
    nbComm.init ( args );
    int myId = nbComm.id ();
    int peer = 1−myId;
    Buffer buf = new Buffer (BufferFactory.getBuffer (capacity ));
    if (myId==0){
        buf.write (data ,0 ,data.length );
        Request req = nbComm.isend (buf ,peer ,tag );
        req.iwait ();
    } else if (myId==1) {
        Request req = nbComm.irecv (buf ,peer ,tag );
        req.iwait ();
        buf.read (data ,0 ,data.length );
    }
    nbComm.finish ();
}
```

Listing 1.1 shows a code example of a parallel application that uses SCINBComm (getNBComm("sci")). This application performs a non-blocking point-to-point communication. The init() and finish() functions serve as barrier because these methods do not return the control to the application until all processes involved in the parallel application have reached those points.

4   Implementing Efficient Non-blocking Communication

NBComm uses a dedicated thread for communication (receptor_thread) which is responsible for receiving messages. This thread is implemented in SCINBComm in native code whereas in pure Java for NIONBComm. Listing 1.2 shows its operation pseudocode.

There are two possible ways to implement message arrival notification depending on the implementation. The first, the native solution, is through a callback() function or through an event that is registered for being triggered every time a message arrives. The second, the pure Java solution, is checking arrival notification using Java NIO Selector. Each message is uniquely identified by <srcid,tag>, and irecv() requests posted and not actually received are in the posted_messages linked list.

Listing 1.2. Pseudocode of the receptor_thread operation

```
WHILE NBComm.finish() is not called
  IF pending_messages = 0 THEN
      wait until message arrival notification
  END IF
  receive message header
  check if this message irecv has been posted
  IF posted THEN
      receive message data in the irecv Buffer buf
      delete irecv post from posted_messages
  ELSE
      receive message data in temporal buffer
      add received post to posted_messages
  END IF
  notify the message reception to the waiting requests
END WHILE
```

A problem in this implementation is that NBComm subclasses replicate some code as it appears as Java code in NIONBComm and as native code in SCINBComm. Thus, this code can not be factorized in the superclass NBComm making it harder to maintain the source code. A proposed solution consists of moving the interconnection hardware support to a lower API level (Java sockets) and using NIONBComm over these low level libraries.

4.1   Java Sockets with SAN Support

In order to support high performance interconnection technologies on Java sockets, a High Performance Java socket implementation, called *Java Fast Sockets (JFS)*, has been developed. JFS aims to be efficient and portable by providing two alternative solutions using pure Java and JNI wrappers to low-level SAN protocols. In the presence of these SAN protocols, JFS uses the JNI approach. Otherwise, it uses the pure Java solution. Moreover, the use of the new Java NIO capabilities, such as new data containers (*direct* `ByteBuffer`), new I/O channels, selectors and selection keys, can optimize performance in JFS. Finally, by setting the default `SocketImplFactory` to a factory that returns JFS sockets, every socket operation in an application can transparently use JFS.

4.2   Native Java Communication Support

In the design of native support of `SCINBComm` and in JFS, both libraries use communication mechanisms implemented by the underlying libraries. High Performance Clusters usually provide several protocols depending on the message size, as communication performance depends on the trade-off between latency and protocol processing overhead. Thus, one-copy protocol trades off high CPU load for low latency, whereas zero-copy protocol cuts down system load (high bandwidth rates with low CPU loads). A sensible choice between protocols involves using one-copy protocol for latency sensitive applications, and zero-copy protocol for applications with high bandwidth requirements. On SCI, native libraries resort to `SCILib`, implementing the non-blocking semantic on top of this blocking layer by means of threads. The protocol choice can be configured by the user.

5   Performance Evaluation

In this section an evaluation of `NBComm` implementations is presented. Additionally, mpiJava non-blocking communication over MPICH on GbE has also been tested for comparison purposes. SCI-MPICH [15] is not supported by mpiJava in our testbed. In order to evaluate the performance, half of the round trip time of a ping-pong test (hereafter called latency) is measured. Moreover, two specific non-blocking communication benchmarks including a communication/computation overlapping test and an overlapping communications test are used.

5.1   Experiment Configuration

Our testbed consist of two dual-processor nodes (PIV Xeon at 2.8 GHz with hyper-threading disabled and 2GB of memory) interconnected via SCI and GbE. The SCI NIC is a D334 card plugged into a 64bits/66MHz PCI, whereas the GbE is a Marvell 88E8050 with an MTU of 1500 bytes. The OS is linux CentOS 4.2 with kernel 2.6.9 and compilers gcc 3.4.4 and Sun JDK 1.5.0_05. The SCI libraries are SCI SOCKETS/DIS 3.0.3. mpiJava version 1.2.5 runs on top of MPICH 1.2.5.

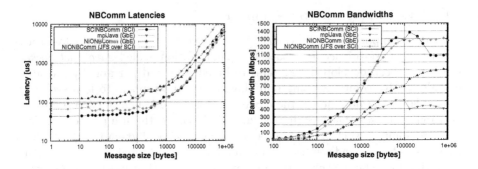

Fig. 3. Measured latencies and bandwidths of NBComm implementations

5.2   Performance Results

Figure 3 shows experimentally measured latencies and bandwidths of NBComm implementations on SCI and GbE as a function of the message length. The bandwidth graph (right side) is useful to compare long-message performance, whereas latency graph (left side) serves to compare short-message performance (note that their scale is logarithmic). In order to analyse the overhead imposed by NBComm, experimental results from SCILib (library used by SCINBComm), JFS, and Java sockets (libraries used by NIONBComm) are shown in Figure 4.

The two lower graphs of Figure 4 show the latency and bandwith of NIONBComm using GbE. In addition, the graphs also show the latency and bandwith of the raw Java sockets. The difference between the performance of NIONBComm and Java sockets shows the imposed overhead. This overhead is aproximately $60\mu s$ in terms of latency. As can be seen from the two upper graphs in Figure 4, SCINBComm obtains lower startup time than NIONBComm over JFS on SCI. The overheads in latency imposed by the NBComm layer are around $40\mu s$ and $58\mu s$ over SCILib and JFS respectively. Bandwidth performance is quite similar except for messages larger than 256KB where the pure Java implementation outperforms the native implementation. As expected, SCINBComm obtains better results in general than NIONBComm using JFS on SCI. However, the performance gain is due to the use of JNI. JFS has an asymptotic bandwidth similar to native sockets and startup times as low as $8\mu s$. Some experimentally measured examples of latency reduction have been observed: a 64Kb message in the SCI testbed where the reception is posted after receiving the message has $t_{isend} = 156\mu s$, $t_{send} = 308\mu s$, $t_{irecv} = 3\mu s$, $t_{recv} = 308\mu s$ and $t_{iwait} = 2\mu s$. The sender process obtains a time gain of $152\mu s$ (49%), apart from not having to wait to send, and the receiver process obtains a time gain of $303\mu s$.

The CPU overlap test determines the amount of software overhead involved in sending and receiving messages. The benchmark code consists of inserting gradually increasing computation between the calls that initiate and complete a non-blocking send or receive operation. By determining the maximum amount of computation that can be overlaped with communication the computation/-communication overlapping parameter can be obtained. This assumes that the

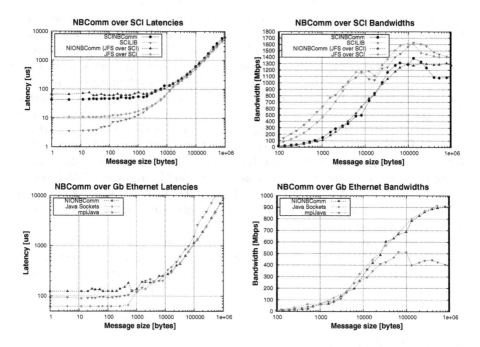

Fig. 4. Measured latencies and bandwidths of NBComm vs. underlying libraries

computation cost does not affect the measured communication time. The results obtained by benchmarking 1Kb messages show that a 37% of the communication time can be overlapped with computation in SCINBComm. A 6% and a 40% performance improvement is obtained for NIONBComm and mpiJava respectively. Native-based solutions provides a higher degree of computation/communication overlapping.

The overlapping communications test benchmarks the overlap of communication with additional communication. Rather than filling idle CPU time with computation, as in the previous test, it can be used to send additional messages. It has been experimentally observed that sending 8 simultaneous 1Kb messages helps achieve a latency reduction of 44% in SCINBComm, a 33% in NIONBComm, and a 49% in mpiJava.

6   Conclusions

Communication performance is critical for the overall system cluster performance. In this scenario non-blocking communications can significantly reduce the communication overhead. Nevertheless, the definition of an efficient non-blocking Java communication library with cluster support poses an important number of implementation issues. These can be summarized in designing the solution for receiving messages, notify the arrival of messages, the study of the efficiency of data movements and the API definition. This Java communication library can use Java sockets implementations or native communication libraries specialized for SAN systems. This paper has presented a non-blocking Java communication library (NBComm) that resolves efficiently numerous design issues aforementioned and provides cluster support. This library aims at reducing the startup time of communications, avoiding unnecessary copying and overlapping computation and communication. In the design of the library a thread is devoted to receive messages (receptor_thread). The approach followed also ensures that unnecessary copying is avoided writing directly to a buffer of type direct ByteBuffer and DMA is used for messages longer than 8KB. This library implements different solutions depending on the underlying communication libraries—SCINBComm is implemented for using SCI native communication libraries and NIONBComm for using Java sockets. A High Performance Java socket implementation JFS can also be used as communication layer for NIONBComm, providing additionally access to SCI for this solution.

The use of non-blocking communication can gain significant improvements with respect to the use of blocking communication in parallel applications. It has been experimentally assessed that non-blocking communication is specially advantageous, obtaining latency reductions and overlapping computation with communication, yielding communication overhead reductions up to 50%.

Acknowledgments

This work was funded by the Ministry of Education and Science of Spain under Project TIN2004-07797-C02 and under a FPU grant AP2004-5984.

References

1. K. Verstoep, R. Bhoedjang, T. Rühl, H. Bal, and R. Hofman. Cluster Communication Protocols for Parallel-programming Systems. *ACM Transactions on Computer Systems*, 22(3):281–325, 2004.
2. G. L. Taboada, J. Touriño, and R. Doallo. Performance Analysis of Java Message-Passing Libraries on Fast Ethernet, Myrinet and SCI Clusters. In *Proc. 5th IEEE International Conference on Cluster Computing (CLUSTER'03)*, pages 118–126, Hong Kong, China, 2003.
3. M. Welsh and D. E. Culler. Jaguar: Enabling Efficient Communication and I/O in Java. *Concurrency: Practice and Experience*, 12(7):519–538, 2000.
4. S. B. Lim, B. Carpenter, B. Fox, and H.-K. Lee. A Low-Level Communication Library for Java HPC. In *Proc. 6th International Conference on Algorithms and Architectures for Parallel Processing (ICA3PP'05)*, *LNCS 3719, Springer-Verlag*, pages 429–434, Melbourne, Australia, 2005.
5. H.-K. Lee, B. Carpenter, G. Fox, and S. B. Lim. HPJava: Programming Support for High-Performance Grid-Enabled Applications. *International Journal of Parallel Algorithms and Applications*, 19(2–3):175–193, 2004.
6. M. Baker, B. Carpenter, and A. Shafi. MPJ Express: Towards Thread Safe Java HPC. In *Proc. 8th IEEE International Conference on Cluster Computing (CLUSTER'06)*, Barcelona, Spain, 2006.

7. M. Baker, B. Carpenter, and A. Shafi. An Approach to Buffer Management in Java HPC Messaging. In *Proc. 6th International Conference on Computational Science (ICCS'06), LNCS 3992, Springer-Verlag*, pages 953–960, Reading, UK, 2006.

8. M. Baker, B. Carpenter, G. Fox, S. Ko, and S. Lim. mpiJava: an Object-Oriented Java Interface to MPI. In *Proc. 1st International Workshop on Java for Parallel and Distributed Computing (IPPS/SPDP'99), LNCS 1586, Springer-Verlag*, pages 748–762, San Juan, Puerto Rico, 1999.

9. M. Bornemann, R. V. van Nieuwpoort, and T. Kielmann. MPJ/Ibis: A Flexible and Efficient Message Passing Platform for Java. In *Proc. 12th European PVM/MPI Users' Group Meeting, (PVM/MPI'05), LNCS 3666, Springer-Verlag*, pages 217–224, Sorrento, Italy, 2005.

10. F. Seifert and H Kohmann. SCI SOCKETS - A Fast Socket Implementation over SCI. http://www.dolphinics.com/pdf/whitepapers/sci-socket.pdf. [Last visited: July 2006].

11. S. H. Rodrigues, T. E. Anderson, and D. E. Culler. High-Performance Local-Area Communication With Fast Sockets. In *Proc. Winter 1997 USENIX Symposium*, pages 257–274, Anaheim, CA, 1997.

12. J.-S. Kim, K. Kim, and S.-I. Jung. SOVIA: A User-level Sockets Layer Over Virtual Interface Architecture. In *Proc. 3rd IEEE International Conference on Cluster Computing (CLUSTER'01)*, pages 399–408, New Port Beach, CA, 2001.

13. P. Balaji, P. Shivan, P. Wyckoff, and D. K. Panda. High Performance User Level Sockets over Gigabit Ethernet. In *Proc. 4th IEEE International Conference on Cluster Computing (CLUSTER'02)*, pages 179–186, Chicago, IL, 2002.

14. S. Petri, L. Schneidenbach, and B. Schnor. Architecture and Implementation of a Socket Interface on top of GAMMA. In *Proc. 28th IEEE Conference on Local Computer Networks (LCN'03)*, pages 528–536, Bonn, Germany, 2003.

15. J. Worringen and T. Bemmerl. MPICH for SCI-connected Clusters. In *SCI Europe'99*, pages 3–11, Toulouse, France, 1999.

Modernizing the C++ Interface to MPI

Prabhanjan Kambadur, Douglas Gregor,
Andrew Lumsdaine, and Amey Dharurkar

Open Systems Laboratory, Indiana University
{pkambadu, dgregor, lums, adharurk}@osl.iu.edu

Abstract. The Message Passing Interface (MPI) is the *de facto* standard for writing message passing applications. Much of MPI's power stems from its ability to provide a high-performance, consistent interface across C, Fortran, and C++. Unfortunately, with cross-language consistency at the forefront, MPI tends to support only the lowest common denominator of the three languages, providing a level of abstraction far lower than typical C++ libraries. For instance, MPI does not inherently support standard C++ constructs such as *containers* and *iterators*, nor does it provide seamless support for user-defined classes. To map these common C++ constructs into MPI, programmers must often write non-trivial boiler-plate code and weaken the type-safety guarantees provided by C++. This paper describes several ideas for modernizing the C++ interface to MPI, providing a more natural syntax along with seamless support for user-defined types and C++ Standard Library constructs. We also sketch the C++ techniques required to implement this interface and provide a preliminary performance evaluation illustrating that our modern interface does not imply unnecessary overhead.

1 Introduction

Ever since its standardization as a message passing API, the Message Passing Interface (MPI) [1] has rapidly become the most popular API for writing message passing applications. MPI's popularity is partly due to its ability to support C, Fortran and C++ [7]. To provide such language interoperability, MPI presents a language-independent communications interface that can be realized with concrete bindings in all three languages. While C and Fortran provide a similar level of abstraction and support for user-defined types, C++ supports a much higher level of abstraction, due to language features such as object-oriented programming through classes, generic containers through templates, and operator overloading. These features lend themselves to a more expressive interface than that provided by MPI. The following code illustrates how a modern C++ interface can simplify the use of MPI, by allowing the user to transmit even complicated data types—here, a list of lists—with the same syntax and semantics as primitive data types:

```
std::list<std::list<int> > lst;
mpi::Send(lst, dest, tag, comm);
```

B. Mohr et al. (Eds.): PVM/MPI 2006, LNCS 4192, pp. 266–274, 2006.

```
std::list<std::list<int> > ls; // initialize ls
int num_lists = ls.size();
MPI_Send(&num_lists, 1, MPI_INT, dest, tag, comm);
std::list<std::list<int> >::iterator out_iter = ls.begin();
while (out_iter != ls.end()) {
    std::vector<int> buffer(out_iter->begin(), out_iter->end());
    MPI_Send((void*)&buffer.front(),out_iter->size(), MPI_INT, dest, tag, comm);
    ++out_iter;
}
```

Fig. 1. Interfacing C++ with MPI

The equivalent code written for the existing MPI interface is shown in Figure 1. Serialization of std::list<std::list<int> > is performed manually, because MPI is unable to cope with linked lists.[1]

We present several ideas for modernizing the MPI interface in C++ without incurring unnecessary penalties due to abstraction. To achieve our goals, we apply Generic Programming techniques. By contrast, previous efforts such as MPI++ [5] and OOMPI [2] relied on object-oriented techniques to achieve the same goals. Generic Programming is an emerging software development paradigm that equally emphasizes efficiency and re-usability. While early uses of the Generic Programming paradigm were restricted to sequential libraries such as the Standard Template Library (STL), the success of these libraries has fueled an interest in developing generic high performance scientific computing libraries [3].

2   A Modern C++ Interface

A modern C++ interface to MPI should retain the flavor of MPI, but provide support for the idioms common to modern C++ programs, including seamless support for user-defined types and the containers and iterators of the STL. We describe how a modern C++ interface to MPI would express point-to-point and collective operations and how it might interact with user-defined types.

2.1   Point-to-Point Interface

A C++ point-to-point interface can improve over the existing MPI point-to-point interface in several ways. First, since type information can be deduced using C++ template mechanisms, the MPI_Datatype argument is no longer required. Second, certain function parameters (such as the number of elements being transmitted) can be provided with default values, so they may be omitted by the user. Third,

---

[1] The method described in Figure 1 is one amongst many ways to serialize containers. The other methods of serializing containers are at least as arcane as the one shown.

in true C++ fashion, we can permit communication of STL *containers, iterator ranges*, and user-defined data types through the same mechanism. Finally, the **void*** values used to pass references to data into the existing MPI interface can be replaced with C++ references, improving type safety. The following code transfers a std::list from rank 1 to rank 0. In this example, the interface abstracts away the need to (de-)serialize the list and communicate the size, type, and dimensionality of the data.

```
// Sender: Rank 0                    // Receiver: Rank 1
std:: list <int> data;              std:: list <int> data;
/* fill    data list */             mpi:: Recv(data, 0, msg_tag);
mpi:: Send(data, 1, msg_tag);
```

In C++, *iterators* are generalizations of pointers that are used to interface between containers and algorithms. The following code sample transfers contents of an iterator range from rank 0 to rank 1. Here, the sender parses integers from standard input (cin) using an istream_iterator, sending the results to rank 1, which receives the data into an std::vector. This example also demonstrates that non-blocking communication can be carried out in much the same way as in the existing MPI interface.

```
// Sender: Rank 0                           // Receiver: Rank 1
mpi:: Send(istream_iterator <int>(cin),    std:: vector <int> data;
           istream_iterator <int>(),        mpi:: Request req =
           1, mst_tag);                          mpi:: Irecv (std:: back_inserter (data),
                                                             0, msg_tag);
                                            // Do some other work
                                            req. Wait();
```

2.2   Collectives

In addition to point-to-point operations, MPI provides a rich set of collective operations. A modern C++ interface to MPI can improve these collectives using the same techniques applied to point-to-point operations. For instance, the following code reads a line of input from the user on the rank 0 process and broadcasts the result to all processes:

```
string input;
if (comm.rank() == 0) std::cin >> input;
mpi::Broadcast(input, 0, comm);
```

Several MPI collectives, such as MPI_Reduce and MPI_Scan, have an operation parameter that specifies how values from different processes will be combined. Varying the operation parameter can produce different results, such as computing the product or global minimum of the values stored on each process. Following current C++ practice as established by the STL, collectives should accept a

```
struct employee_record {
    std::string employee_name;
    int employee_id;
    std::list<std::string> address;
};
template<class Archiver>
    void serialize(Archiver & ar, employee_record & rec, const unsigned int version) {
        ar & rec.employee_name & rec.employee_id & rec.address;
    }
```

Fig. 2. Serializing User-defined Datatypes

function object argument that performs the requested operation. The STL provides function objects for many common reduction operations, including sums (std::plus), products (std::multiplies), and logical combinators (std::logical_and, std::logical_or); a modern C++ interface to MPI would provide additional function objects that match the remaining reduction operations provided by MPI. More importantly, users must be free to define their own functions and function objects for reduction, either for user-defined or built-in C++ types. The following example computes the longest string prefix common to my_string on every processor:

```
string common_prefix(const string& s1, const string& s2) {
    if (s1.size() <= s2.size())
        return string(s1.begin(), mismatch(s1.begin(), s1.end(), s2.begin()).first);
    else return common_prefix(s2, s1);
}
string global_common_prefix = mpi::Allreduce(my_string, &common_prefix);
```

2.3   User-Defined Types

Section 2.1 illustrates how a modern interface can support containers and iterators with a simple abstract interface. However, this interface needs to be extended to support all data types, including user-defined types. To support the most complicated data types requires serialization. We adopt the interface provided by the Boost Serialization Library (BSL) [4], both for its simplicity and its built-in support for the containers and iterators of the STL.

Figure 2 illustrates how one would provide serialization functions for a user-defined type. Serializing employee_record by hand is non-trivial, requiring code to serialize strings of unknown lengths and linked lists. However, defining the serialization behavior for the BSL is rather simple, requiring only the definition of a serialize() function template. serialize() accepts an Archiver argument and a reference to the object that will be (de-)serialized. The function body itself describes which data members are to be serialized and de-serialized. The operator & is bi-directional and acts like the output streaming **operator<<** during serialization and input **operator>>** during de-serialization.

```
struct primitive_tag {};
struct serialized_tag {};
struct container_tag : serialized_tag {};
template <typename T> struct object_traits {typedef serialized_tag object_category;};
template <>
  struct object_traits<int> {
    typedef primitive_tag object_category;
    static inline MPI_Datatype get_mpi_type() { return MPI_INT; }
  };
template <typename T, typename Alloc>
  struct object_traits<std::vector<T, Alloc>>{typedef container_tag object_category;};
```

Fig. 3. Traits Classes and Mapping to MPI Datatypes

The Archiver is the class that determines the exact means by which data is serialized and de-serialized. Archiver is a template parameter because there are many ways to archive data. For instance, to serialize data for transmission via MPI, one could either use the binary archiver provided by the BSL or implement a new archiver based on MPI_Pack/MPI_Unpack. An auxiliary benefit of defining serialize() for user-defined types is the ability to (de-)serialize via other BSL archivers, for storing objects into files in a variety of formats.

3   Implementation Strategies

A modern C++ interface to MPI can seamlessly support primitive, library-defined, and user-defined data with a simple, concise syntax familiar to C++ programmers. However, the benefits of such an interface are lost if abstraction penalties affect performance. In particular, communicating primitive datatypes through this mechanism should be as fast as raw MPI. In this section, we elaborate on the C++ mechanisms that can be used to realize our interface.

3.1   Function Specialization

A modern C++ MPI interface should provide a uniform interface to all datatypes regardless of whether they need to be serialized or not. To deliver the best performance possible for each datatype, we must pick the most *specialized* function for each datatype at compile time. For example, if the datatype being sent is an **int**, then we send it as a MPI_INT, whereas a list of integers would need to be serialized and transmitted as MPI_BYTE.

We can apply certain C++ template idioms common to Generic Programming to ensure that the most specialized functions are chosen at compile time, in particular, traits classes and tag dispatching. Traits classes are a means of extracting information from types at compile time. Information is encoded in traits classes in the form of nested types, which allows them to be queried at compile time. Serialization information can be extracted from the object_traits class, illustrated in Figure 3. For primitive data types the nested get_mpi_type() function provides the MPI_Datatype that corresponds to each primitive C++ type.

```
template <typename T>
inline void Send(const T& var, int dest, tag tag, MPI_Comm comm) {
  typedef typename object_traits<T>::object_category object_category;
  send_impl(var, dest, tag, object_category());
}
template <typename T>
void send_impl(const T& var, int dest, tag tag,
               MPI_Comm comm, primitive_tag) {
  MPI_Send ((void*)var, 1, object_traits<T>::get_mpi_type(), dest, tag, comm);
}
template <typename T>
void send_impl(const T& var, int dest, tag tag,
               MPI_Comm comm, serialized_tag) {
  // Serialize var and then MPI_Send
}
```

Fig. 4. Tag Dispatching Using Type Traits

Tag dispatching queries traits at compile time to choose the most specialized implementation for a call to a given user-level function. Figure 4 demonstrates tag dispatching for the Send() operation. There are two underlying implementations for this version of Send(). The first send_impl() is used for primitive datatypes, which directly invokes MPI_Send with the corresponding MPI datatype. The second send_impl() is used for objects that need to be serialized. The top-level Send() function selects among the two send_impl() implementations by passing the object_category for type T as the final parameter, in effect using the tag types primitive_tag and serialized_tag as a compile-time switch.

3.2 Collectives

The collectives described in Section 2.2 provide functionality beyond what is available with existing MPI collectives. For instance, the common string prefix example is an Allreduce() over strings, but each string must be serialized for transmission via MPI. Thus, our C++ interface to MPI collectives requires re-implementation of MPI's collective operations for serialized data types.

To ensure that the new implementations of these collective operations are as efficient as the existing MPI implementation's collectives, function specialization can be used. For instance, consider how the following two calls to Allreduce() would be implemented most efficiently with the existing MPI interface:

```
int local_int = ... , result;
mpi::Allreduce(local_int, result, std::plus<int>());
struct hash_int { int operator()(int x, int y); };
mpi::Allreduce(local_int, result, hash_int());
```

In this first case, we are computing the sum of the integers stored in each process. This operation can be implemented most efficiently with a single call

MPI_Allreduce(..., MPI_INT, MPI_SUM, ...), which is likely optimized within the existing MPI implementation. In the second case, we are still performing reduction on integers, but we have provided a user-defined function object. Here, we can still take advantage of any communication optimizations provided by MPI_Allreduce(), but instead of passing a built-in MPI operation such as MPI_SUM, we will create our own with MPI_Op_create().

4   Performance

Throughout the paper, we have claimed that our modernized C++ interface will not incur any performance penalties for primitive datatypes. To test the validity of our claims, we implemented a subset of the point-to-point interface described and compared the NetPIPE latency and bandwidth numbers with those of raw MPI. We added two new tests to NetPIPE that use our interface with primitive and serialized types. The serialized sends involved replacing the **char** type used for communication buffers in NetPIPE with a simple Char class, which is a lightweight wrapper over a **char** value that requires serialization.

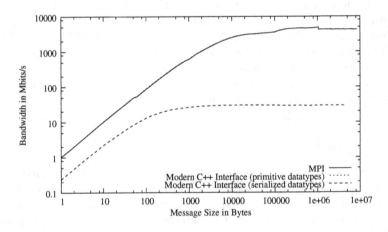

Fig. 5. NetPIPE numbers for our modernized C++ interface to MPI and MPI

Tests were run on the Odin cluster at Indiana University, in which each compute node contained 8 GB of RAM and two AMD Opteron processors running at 2 Ghz, each with a 1 MB cache. The compute nodes are connected by Mellanox Infiniband cards. The Odin cluster is running Red Hat Enterprise Linux with kernel version 2.6.9-22.0.1.ELsmp, GCC version 3.4.4, and Open MPI 1.0.1.

Performance results are shown in Figure 5. The NetPIPE latency and bandwidth numbers with our modern MPI interface exactly match those of raw MPI when transmitting primitive types. When data requires serialization, both latency and bandwidth are affected. This is because serialization incurs additional cost of calls to the serialize function, two memory copies and one extra (MPI_Send, MPI_Recv) pair to communicate data size. However, the performance results for serialized communication shown in Figure 5 are preliminary and can be vastly improved. We used a vanilla Binary Archive class in our experiments. Instead, a customized MPI Archive class that can pack data into a MPI implementation's underlying data format would greatly reduce serialization costs. Also, the Char example demonstrates the worst case scenario for serialization as there are millions of tiny objects that need to be serialized.

5   Related Work

Many attempts at providing a higher level of abstraction for MPI within C++, such as MPI++ [5] and OOMPI [2], are based on object-oriented design and implementation techniques. These libraries also support user-defined data types and provide a more clean and concise interface to MPI from C++. Skjellum et al. provide a comprehensive object-oriented analysis of MPI [6].

Object-oriented techniques have some limitations that can impact both the usability and the performance of libraries. Object inheritance is used heavily to provide an uniform programming interface. Thus, to send or receive a user-defined type such as employee_record (Figure 2), one must derive it from a given abstract base class, overriding one or more of its virtual methods. However, this presupposes that the user-defined type in question can be modified: if the type comes from another library (e.g., the std::vector type from the C++ Standard Library), or for some reason cannot derive from the abstract base class (e.g., because it is not allowed to have a virtual function table), that type cannot be used with the library. In our approach, we apply Generic Programming techniques to transmit any user-defined type without altering the type itself. Moreover, virtual functions incur run-time penalties, both due to indirection required and due to missed opportunities for compiler optimization. By contrast, we avoid these additional levels of indirection through the use of traits and tag dispatching.

6   Conclusions

We have shown that the C++ interface to MPI can be modernized in several ways. This interface can provide a more natural C++ syntax for message passing, with seamless support for user-defined data types and the programming style encouraged by the STL. Moreover, this interface can be provided without sacrificing performance when transmitting primitive types, so that the modern C++ interface only incurs a performance penalty for serialization only when serialization is necessary. In the future, we hope to implement the remaining ideas in this paper, especially the interface to collective operations using function objects. We will also investigate ways to reduce the overhead associated with serialization and improve the performance of non-blocking operations.

Acknowledgements

The authors thank Jeremiah Willcock and Brian Barrett for their insightful comments throughout this project. This work was supported by NSF grants EIA-0202048 and EIA-0131354 and a grant by the Lilly Endowment.

References

1. *MPI-1.1 Specification*, http://www.mpi-forum.org/docs/docs.html.
2. J. M. Squyres B. C. McCandless and Andrew Lumsdaine. Object Oriented MPI (OOMPI): a class library for the Message Passing Interface. In *Second MPI Developer's Conference: Notre Dame, IN, USA*. IEEE, 1996.
3. Lie-Quan Lee. *Generic Programming for High-Performance Scientific Computing*. PhD thesis, University of Notre Dame, 2002.
4. Robert Ramsey. *Boost Serialization Library*. http://www.mpi-forum.org/docs/docs. html, 2003.
5. Anthony Skjellum, Ziyang Lu, Purushotham V. Bangalore, and Nathan E. Doss. Explicit parallel programming in C++ based on the message-passing interface (MPI). In Gregory V. Wilson, editor, *Parallel Programming Using C++*. MIT Press, 1996.
6. Anthony Skjellum, Diane G. Wooley, Ziyang Lu, Michael Wolf, Purushotham V. Bangalore, Andrew Lumsdaine, Jeffrey M. Squyres, and Brian McCandless. Object-oriented analysis and design of the Message Passing Interface. In *j-CCPE, 13(4): 245–292*, 2001.
7. Jeffrey M. Squyres, Bill Saphir, and Andrew Lumsdaine. The Design and Evolution of the MPI-2 C++ Interface. In *Proceedings, 1997 International Conference on Scientific Computing in Object-Oriented Parallel Computing*, Lecture Notes in Computer Science. Springer-Verlag, 1997.

Can MPI Be Used for Persistent Parallel Services?

Robert Latham, Robert Ross, and Rajeev Thakur

Mathematics and Computer Science Division
Argonne National Laboratory
Argonne, IL 60439, USA
{robl, rross, thakur}@mcs.anl.gov

Abstract. MPI is routinely used for writing parallel applications, but it is not commonly used for writing long-running parallel services, such as parallel file systems or job schedulers. Nonetheless, MPI does have many features that are potentially useful for writing such software. Using the PVFS2 parallel file system as a motivating example, we studied the needs of software that provide persistent parallel services and evaluated whether MPI is a good match for those needs. We also ran experiments to determine the gaps between what the MPI Standard enables and what MPI implementations currently support. The results of our study indicate that MPI can enable persistent parallel systems to be developed with less effort and can provide high performance, but MPI implementations will need to provide better support for certain features. We also describe an area where additions to the MPI Standard would be useful.

1 Introduction

Achieving good performance on today's high-end computers involves effectively utilizing a variety of network interconnects, a large number of compute resources, and high-quality algorithms. Application developers make heavy use of libraries and tools to manage this complexity while still delivering high performance. For their work, Parallel application writers commonly choose the message-passing model, embodied by the MPI Standard [10]. MPI defines a rich API that can be used across many disparate hardware platforms and provides many useful features such as datatype packing, collective communication, nonblocking communication, and dynamic process management. High-quality MPI implementations further provide heterogeneous communication and deliver high performance.

Parallel system services, as opposed to applications, are usually not written in MPI. One would imagine, however, that MPI's portability, performance, and features should make it an attractive candidate for implementing parallel system services as well. Why, then, don't services use MPI? Could they? We investigate these issues in detail in this paper. For concreteness, we use the parallel file system PVFS2 [12] as an example for studying the needs of such software. We have been heavily involved in the development of PVFS2 and are familiar with its requirements. PVFS2 and its predecessor, PVFS [2], represent a decade of

B. Mohr et al. (Eds.): PVM/MPI 2006, LNCS 4192, pp. 275–284, 2006.

parallel file system research and engineering. PVFS2 was written to deliver high performance at scales of hundreds of servers and tens of thousands of clients and has done so on some of the world's fastest and largest classes of supercomputers, such as IBM BG/L, Cray XT-3, and large Linux clusters.

We first give a brief overview of PVFS2 and its architecture. Then, using PVFS2 as an example, we study the needs of software for persistent parallel services and examine how well MPI is equipped to meet those needs. We find in most cases that the MPI Standard supports the features we need. Some helpful features, however, are not available in some commonly deployed MPI implementations. We also describe an area that would benefit from additions to the MPI Standard.

2   PVFS2: A Persistent Parallel Service

A *persistent parallel service* is system software that manages multiple hardware components to provide a single logical resource for use by parallel applications. It is persistent in the sense that it exists beyond the life of a single application, typically running for weeks or months at a time. A parallel file system is an example of a persistent parallel service.

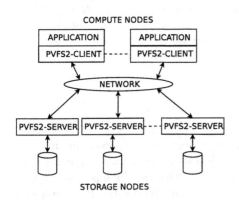

PVFS2 [12] is a high-performance parallel file system being developed as a joint project by Argonne National Laboratory, Clemson University, and the Ohio Supercomputer Center. PVFS2 comprises multiple persistent servers. File striping across these servers enables multiple clients to access different parts of a file in parallel, resulting in high performance. PVFS2 software on the client side hides all these details from the client and instead presents a single logical view of a file.

Fig. 1. PVFS2 architecture: `pvfs2-client` forwards kernel-level requests to `pvfs2-server` processes running on the servers. In turn, `pvfs2-server` deals with managing data on storage devices.

PVFS2 provides many features such as native support for popular networking technologies (e.g. Myrinet, InfiniBand, and TCP/IP), multiple APIs (POSIX, MPI-IO), user-controlled striping of files across nodes, a well-defined interface for describing new data distribution schemes, support for heterogeneous clusters, and distributed metadata. It uses commodity network and storage hardware and is easy to install (no kernel patch). The familiar UNIX file tools (such as `ls`, `cp`, and `rm`) can be used on PVFS2 files and directories.

In the following sections, we use PVFS2 as an example to study the common needs of persistent parallel services and then investigate how well MPI supports those features.

3   Service Identification

Any persistent service needs to handle the important issue of locating the servers. For traditional network services, the IP address and port number are often listed in a configuration file. PVFS2 follows a similar approach. The configuration files for PVFS2 servers list all the servers that form the parallel file system. Each server reads this list at startup. A PVFS2 client uses its own configuration file to locate PVFS2 servers (see Figure 2). This file resembles a Unix /etc/fstab file and provides the network address of any one of the PVFS2 servers, a mount point on the client system, and a few other parameters. The client inquires with the listed server about the file system, obtains a complete listing of all the servers, and then begins interacting with the file system.

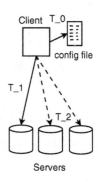

Fig. 2. Client establishing connections to PVFS2 servers. The client has to consult a configuration file and connect to one of the PVFS2 servers before discovering where the other servers are.

If PVFS2 used MPI, it could use MPI's features that enable service identification. The MPI name publishing interface (MPI_PUBLISH_NAME, MPI_LOOKUP_NAME) provides a method for clients and servers to exchange information. Clients could use a well-known key to discover an initial contact point. This key would provide service discovery that is independent of the underlying network interconnect or the MPI implementation. Clients would be insulated from server changes, be it a different port, host, or even interconnect, without system administrators needing to update client-side configuration files. MPI might still need some sort of configuration information, but at least we would be able to concentrate that information into a single source, instead of one source for MPI and another for PVFS2.

In practice, however, MPI implementations currently do not support this functionality as well as needed. For this functionality to be usable, MPI implementations must support name publishing and resolution across independently started MPI processes – PVFS2 servers are not restarted with every new client application. We ran tests with several commonly deployed MPI implementations and found that they support this mode of operation, but only under certain conditions (summarized in Table 1). For example, the processes must be part of the same MPD ring in MPICH2 [11], and Open MPI [7] programs require special measures when launching the orted daemons. This additional component (MPD or orted) must also be persistent and able to tolerate node failure.

Table 1. Capabilities of MPI implementations. An ideal implementation would have a Y in all columns.

Feature	MPICH2 1.0.3	Open MPI 1.0.1	BGL-MPI V1R2M1
Published name appears to other singleton processes	N	N	N
Connect/Accept work under singleton MPI_INIT	N	N	N
MPI_COMM_JOIN works under singleton MPI_INIT	Y	N	N
Does not require a previously established MPI environment (e.g. lamboot, MPD, others)	N	N	N
MPI datatype processing supports heterogeneous architectures	N	N	N
Support for external32	N	N	N

4   Establishing Connection

After clients have discovered what services are running, they need to connect to those services. The traditional Unix socket model has the familiar TCP accept/connect handshake. Other protocols have analogous mechanisms. PVFS2 uses an abstraction that is layered on top of the connection mechanisms of multiple networks, providing portability.

The use of MPI could simplify this process greatly. MPI's dynamic process functionality supports two different ways for clients to establish communication with servers. One approach has the server process call MPI_COMM_ACCEPT, waiting for a corresponding client-side call to MPI_COMM_CONNECT. MPI_COMM_JOIN provides another approach for two processes that already share a UNIX network socket to establish MPI communication. In both cases, the functions returns an MPI intercommunicator, over which the clients and servers can communicate. Furthermore, the accept/connect functions in MPI are *collective*. A group of clients can connect to a group of servers at the same time, and the resulting intercommunicator can be used for communication between any client and any server.

These MPI functions provide a simpler interface than do the corresponding Unix socket ones, abstracting away details such as allocating a socket and setting protocol-specific values in data structures. In addition, they are portable: the MPI implementation takes care of implementing the connection mechanism over the underlying network protocol, freeing the system software developer from the effort.

Taking an MPI approach to client connections introduces a few challenges, however. The accept/connect method needs the name of an open MPI port. If the name-publishing interface in an MPI implementation works across independently launched MPI programs (as described in Section 3), MPI_PUBLISH_NAME and MPI_LOOKUP_NAME can be used to obtain the MPI port name. Otherwise, unwieldy implementation-specific strings would have to be passed around by hand. MPI_COMM_JOIN does not have a dependency on the name-publishing interface. For situations where the name publishing approach is not feasible, this allows a UNIX socket and familiar IP and port locations to be used for service identification. The socket is used only for the initial handshake; all other communication goes over the native transport used by the MPI implementation.

5   Fast Data Transfer

A persistent parallel service needs fast data transfer between clients and servers. PVFS2 has a few specific needs in this area.

- It needs fast communication of data between clients and servers over a number of different networking technologies, using the fastest protocol for each network, for example TCP over Ethernet, GM or MX over Myrinet, the native InfiniBand protocol over InfiniBand.
- For control messages between client and server (not for data), it needs support for heterogeneity, because clients and servers could run on different architectures. For example, Argonne's IBM BG/L system has a mix of PPC64, PPC32, and IA32 nodes.
- It needs support for communicating noncontiguous data efficiently.
- It needs support for asynchronous communication.

A substantial amount of code has been written in PVFS2 to support these needs. PVFS2 uses an abstraction called the Buffered Message Interface (BMI) [3] for portable high-performance communication over multiple networks. For control messages, PVFS2 defines an encoding scheme that converts all commands to a fixed-length, little-endian format, which allows PVFS2 clients and servers to have any mix of byte endianess or word size. (Defining this encoding correctly took many iterations.) PVFS2 implements its own way of communicating noncontiguous data, which required several thousand lines of code.

MPI is a perfect fit for all these requirements. MPI provides a portable interface for communication, and MPI implementations do the job of implementing that interface efficiently on the underlying network. The MPI Standard supports heterogeneous communication through the use of MPI datatypes. MPI implementations, however, vary in their support for heterogeneity. For example, MPICH-1 does support heterogeneous mode architectures, whereas MPICH-2 and Open MPI at present do not. The MPI Standard is limited in that there is no universal way to express certain sized types, such as 64-bit integers, and PVFS2 file handles are 64-bit values. Nonetheless, we could use MPI_LONG_LONG, which is often 64 bit; if not, we could use two MPI_INT types. MPI also supports communication of noncontiguous data through derived datatypes. Additionally, the MPI_TYPE_CREATE_STRUCT routine provides a way to create a user-defined MPI datatype out of arbitrary application data types. MPI implementations, however, have historically not performed well on derived datatypes. Nonetheless, various research efforts have demonstrated that derived datatypes can be implemented in a way that delivers good performance [13,15]. We hope MPI implementations will devote effort to optimizing derived datatypes. MPI also supports nonblocking communication, which allows us to overlap communication with disk I/O.

These features of MPI make it ideally suited for use in data communication, although better support is needed from implementations in the areas of communication between heterogeneous nodes.

6   Fault Tolerance

Any persistent parallel software needs to be resilient against faults as far as possible. The robustness depends on how well the software itself is designed and implemented and on the robustness of the external components that the software uses.

In a cluster environment, each PVFS2 server represents a potential point of failure, and error recovery becomes an important consideration. To that end, the PVFS2 system operate in a stateless manner: there are no locks to revoke or leases to offer, and client tracking is not necessary. This stateless nature makes recovering from server failure much easier. PVFS2 can retry operations in order to hide transient problems. If a server failure occurs, PVFS2 operations will time out and return an error to the caller. If a server has been restarted (by hand or perhaps by a failover script), the newly restarted server will be able to service the client request.

If PVFS2 were implemented by using MPI, it would require the MPI implementation to be resilient against failure. The MPI Standard itself does not say much about fault tolerance; it is left as a quality of the implementation. But MPI does have some features that can help in writing resilient programs. For example, MPI has a well-defined mechanism for error returns from functions, and users can specify their own error handlers. The default error handler is that the entire job aborts on error, but users can change that to "errors return" or define their own error handler. MPI also has the notion of intercommunicators for two groups of processes (for example, clients and servers) to communicate. When two independently started processes connect to each other and communicate over the intercommunicator, the failure of one process need not cause the other process to die.

Most MPI implementations, unfortunately, are not robust against errors. For example, if the connection between two processes is lost, the entire MPI job may abort; or if a single process is killed, the entire MPI job may get killed. This kind of failure will not be good for a parallel file system that uses MPI. Although there are some efforts at building fault-tolerant MPI implementations [1,6], more work is needed in this area.

Another area where MPI can help is in the parity calculation for a software-RAID like approach providing fault-tolerance for data stored on the parallel file system. Gropp et al. [8] proposed a *lazy redundancy* scheme that makes use of both MPI-IO consistency semantics and the MPI collective functions MPI_REDUCE_SCATTER and MPI_REDUCE. Implementing this scheme becomes much easier when PVFS2 servers are based on MPI, because the servers could simply use these collective calls (more on this in Section 7).

The processes providing the parallel service can only communicate with each other once they have established an MPI communicator. At one extreme we could establish many two-process communicators. Having all these communicators makes the system resilient to failure. If a process dies, communication can be carried out over a different communicator. On the other hand, so many communicators greatly complicates any all-to-all or one-to-many messaging algorithms. At the other extreme we could establish an all-encompassing communicator spanning all processes. In exchange for simplified communication, such a system would be more fragile. If any process died, the surviving process would need to detect that failure and coordinate the creation of a new all-encompassing communicator Further, we would need this reconstruction process to maintain the properties of MPI communicators (context, fixed identifiers) that make them so useful.

7   Collective and Aggregate Operations

In PVFS2, many operations require multiple steps performed across many servers. Creating a new file requires instantiating a single metadata entry and a data file entry on each server. Removing a file requires removal of the corresponding metadata and directory entries, followed by removal of the data file from each server. A stat system call needs to collect partial file size information from each server before returning the total size of a file.While the client code makes just one function call for these operations, the underlying library carries out a one-to-

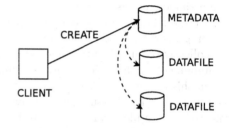

Fig. 3. An aggregate operation lets a single create request initiate creation of the metadata entry and datafile entries on each server. The servers could potentially be better connected to each other than clients (as in a WAN), yielding fewer messages, better performance, and lower latency.

many operation. The client library posts these messages as nonblocking sends to the servers and waits for their response.

An alternative approach would have clients send a single "create file" message to one of the servers and have servers then orchestrate actions on the client's behalf, as described in [4]. This approach simplifies the synchronization of operations and leads to the natural use of structured communication patterns such as broadcasting an operation request by using a tree-based algorithm as shown in Figure 4(b). We call these higher-level messages "aggregate operations" because they result in a collection of operations across multiple servers.

Aggregate operations also make deployment over the wide-area more efficient. We can easily imagine a topology where the servers are located near to each other while the clients may be quite far away, network-wise. These aggregate messages mean fewer network round trips between clients and servers and lower latency.

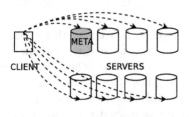
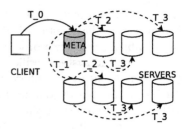

(a) Independent file removal. 8 timesteps

(b) Collective file removal. 3 timesteps

Fig. 4. File removal requires deletion of the data file on each server. The independent approach has little room for optimization, requires careful coordination to keep metadata consistent, and needs $O(N)$ timesteps to complete. The collective approach simplifies metadata updates and requires only $O(log(N))$ timesteps.

The servers can exchange messages with each other over their local network and send a single response over the long-haul, high-latency link.

MPI is well known for its collective operations, such as broadcast, allreduce, and scatter/gather. Many implementations have optimized collective operations [14]. The collective communication operations in MPI are defined to be collective over a communicator; all processes in the communicator must call them. In an application, this requirement is easy to meet. In PVFS2, however, the servers do not know which client will issue the collective operation, for example, which client will want to delete a file. PVFS2 needs to be able to respond to unpredictable client requests. In an MPI environment, servers would naturally post nonblocking collective calls or a broadcast with a "wildcard" (ANY_SRC) root that would be specified later. These calls, however, do not exist in MPI; MPI collectives are blocking calls. While there was a proposal in the MPI-2 Forum for nonblocking collectives, these did not make it into the final standard. The MPI forum decided those who needed nonblocking collectives could implement them with a thread which in turn called the blocking collective equivalent. In a server environment, however, spawning a thread for each potential client becomes untenable as the number of clients scales to the thousands and beyond. Some implementations have extensions that support these features, for example, in IBM's MPI [9] (although it has been deprecated). We are investigating the issue of how nonblocking (or wildcard) collectives could be supported as an extension to MPI, what their semantics would be, and how they could be implemented efficiently. Further, we will have to address how to provide an efficient collective implementation while also solving the fault tolerance issues brought up in Section 6. We plan to develop a prototype implementation to explore these issues.

8   Conclusions

Writing parallel system software can be a significant undertaking. A production parallel file system such as GPFS, GFS, Lustre, or PVFS2 takes many years to develop and stabilize. Much of this effort goes into implementing many of the features that MPI already supports, and this duplicate effort could be avoided. While there are some challenges in implementing system software using MPI today, they are due mainly to the limitations of MPI implementations rather than deficiencies in the MPI Standard itself. At the same time, the addition of nonblocking collectives to MPI would make it an even more natural basis for building parallel system software.

The requirements we have discussed apply to more than just PVFS2 or other parallel file systems. For example, resource managers could use MPI dynamic process functions to launch parallel jobs (via MPI_COMM_SPAWN), and system monitoring daemons could use MPI datatypes and support for heterogeneous communication to monitor disparate resources. Desai et al. [5] used MPI to implement a variety of system-level application utilities, such as file staging, file synchronization, and a parallel shell.

We note that using MPI for implementing persistent system services does not restrict user applications to being MPI applications. The PVFS2 client could determine whether MPI has been initialized (by calling MPI_INITIALIZED) and then call MPI_INIT if it hasn't been. Clients and servers can then communicate using MPI even if they were not started as MPI programs. (Again, all implementations need to support this feature of MPI, called "singleton init.") We would expect that MPI-using applications would call MPI_INIT before making any system service calls. It would of course be an error for an application to call MPI_INIT twice.

In summary, we would like to implement PVFS2 using MPI. We hope MPI implementers will take up the challenge and develop high-quality implementations that can be used to develop system software such as a parallel file system.

Acknowledgments

This work was supported by the Mathematical, Information, and Computational Sciences Division subprogram of the Office of Advanced Scientific Computing Research, Office of Science, U.S. Department of Energy, under Contract W-31-109-Eng-38.

References

1. George Bosilca, Aurelien Bouteiller, Franck Cappello, Samir Djilali, Gilles Fedak, Cecile Germain, Thomas Herault, Pierre Lemarinier, Oleg Lodygensky, Frederic Magniette, Vincent Neri, and Anton Selikhov. MPICH-V: Toward a scalable fault tolerant MPI for volatile nodes. In *Supercomputing '02: Proceedings of the 2002 ACM/IEEE Conference on Supercomputing*, pages 1–18, Los Alamitos, CA, 2002. IEEE Computer Society Press.
2. Philip H. Carns, Walter B. Ligon III, Robert B. Ross, and Rajeev Thakur. PVFS: A parallel file system for Linux clusters. In *Proceedings of the 4th Annual Linux Showcase and Conference*, pages 317–327, Atlanta, GA, October 2000. USENIX Association.

3. Phillip Carns. Design and analysis of a network transfer layer for parallel file systems. Master's thesis, Clemson University, Clemson, S.C., July 2001.
4. Phillip H. Carns. *Achieving Scalability in Parallel File Systems*. PhD thesis, Dept. of Electrical and Computer Engineering, Clemson University, Clemson, SC, May 2004.
5. Narayan Desai, Rick Bradshaw, Andrew Lusk, and Ewing Lusk. MPI cluster system software. *Lecture Notes in Computer Science*, (3241):277–286, September 2004. 11th European PVM/MPI Users' Group Meeting.
6. Grahm Fagg and Jack Dongarra. FT-MPI: Fault tolerant MPI, supporting dynamic applications in a dynamic world. *Lecture Notes in Computer Science*, pages 346–353, 2000. 7th European PVM/MPI Users' Group Meeting.
7. Edgar Gabriel, Graham E. Fagg, George Bosilca, Thara Angskun, Jack J. Dongarra, Jeffrey M. Squyres, Vishal Sahay, Prabhanjan Kambadur, Brian Barrett, Andrew Lumsdaine, Ralph H. Castain, David J. Daniel, Richard L. Graham, and Timothy S. Woodall. Open MPI: Goals, concept, and design of a next generation MPI implementation. In *Proceedings, 11th European PVM/MPI Users' Group Meeting*, pages 97–104, Budapest, Hungary, September 2004.
8. William D. Gropp, Robert Ross, and Neill Miller. Providing efficient I/O redundancy in MPI environments. *Lecture Notes in Computer Science*, 3241:77–86, September 2004. 11th European PVM/MPI Users' Group Meeting.
9. International Business Machines Corporation. *IBM Parallel Environment for AIX 5L: MPI Subroutine Reference*, third edition, April 2005.
10. Message Passing Interface Forum. MPI-2: Extensions to the message-passing interface, July 1997. http://www.mpi-forum.org/docs/docs.html.
11. MPICH2. http://www.mcs.anl.gov/mpi/mpich2.
12. The PVFS2 parallel file system. http://www.pvfs.org/pvfs2.
13. Robert Ross, Neill Miller, and William Gropp. Implementing fast and reusable datatype processing. *Lecture Notes in Computer Science*, 2840, September 2003. 11th European PVM/MPI Users' Group Meeting.
14. Rajeev Thakur, Rolf Rabenseifner, and William Gropp. Optimization of collective communication operations in MPICH. *International Journal of High-Performance Computing Applications*, 19(1):49–66, Spring 2005.
15. Jesper Larsson Traff, Rolf Hempel, Hubert Ritzdorf, and Falk Zimmermann. Flattening on the fly: Efficient handling of MPI derived datatypes. In *PVM/MPI 1999*, pages 109–116, 1999.

Observations on MPI-2 Support for Hybrid Master/Slave Applications in Dynamic and Heterogeneous Environments

Claudia Leopold and Michael Süß

University of Kassel, Research Group Programming Languages/Methodologies
Wilhelmshöher Allee 73, D-34121 Kassel, Germany
{leopold, msuess}@uni-kassel.de

Abstract. Large-scale MPI programs must work with dynamic and heterogeneous resources. While many of the involved issues can be handled by the MPI implementation, some must be dealt with by the application program. This paper considers a master/slave application, in which MPI processes internally use a different number of threads created by OpenMP. We modify the standard master/slave pattern to allow for dynamic addition and withdrawal of slaves. Moreover, the application dynamically adapts to use processors for either processes or threads. The paper evaluates the support that MPI-2 provides for implementing the scheme, partly referring to experiments with the MPICH2 implementation. We found that most requirements can be met if optional parts of the standard are used, but slave crashes require additional functionality.

Keywords: dynamic process management, malleability, adaptivity, hybrid MPI/OpenMP, master/slave pattern.

1 Introduction

Traditionally, MPI programs have used a fixed number of homogeneous processes. Modern architectures and especially grids, in contrast, are characterized by dynamic and heterogeneous resources: Nodes can crash, be withdrawn by the scheduler in favor of higher-priority jobs, or join a running computation after having finished a previous task. Moreover, different nodes may comprise a different number of processors.

The ability of applications to dynamically adapt to a changing number of processors is often denoted as *malleability*. This term goes back to Feitelson and Rudolph [1], who classify jobs as rigid, moldable, evolving, or malleable. Both evolving and malleable jobs change the number of processors during execution, evolving jobs for internal reasons such as requesting additional processors for a complicated subcomputation, and malleable jobs in reaction to changes caused by the environment.

Many architectures combine shared-memory within the nodes and distributed-memory in-between the nodes. They can be programmed in a hybrid style, using

B. Mohr et al. (Eds.): PVM/MPI 2006, LNCS 4192, pp. 285–292, 2006.

MPI processes that are composed of threads. Whether or not the processors of a node are more profitably used for processes or threads, depends on the application. It may be useful to change this assignment dynamically. We call this feature *process-thread adaptivity*.

This paper evaluates the support for malleability and process-thread adaptivity that is provided in MPI-2, mainly through the dynamic process management functions. We base our discussion on a hybrid MPI/OpenMP application from the simulation domain, which is described in Sect. 2. The application uses a master/slave scheme, in which slaves correspond to MPI processes that internally deploy a different number of OpenMP-threads.

Previous work by the same authors has shown that additional processes can be dynamically incorporated into this application [2]. The present paper adds the aspect of process-thread adaptivity, and discusses the case of slaves leaving the computation prematurely. We show that MPI-2 provides sufficient support for process-thread adaptivity if the implementation covers some optional parts of the standard. Evolving programs are supported as well, but the case of a slave leaving the computation abruptly can not be handled appropriately, and we discuss possible workarounds.

Sect. 2 of the paper starts with an outline of the application, including parallelization and deployment of hybrid processes. Then, Sect. 3 explains at an algorithmic level our modifications of the master/slave scheme to handle dynamic and heterogeneous resources. The realization of this scheme in hybrid MPI-2/OpenMP is the topic of Sects. 4–6: Sect. 4 recalls the program structure for incorporating additional processes, Sect. 5 discusses process-thread adaptivity, and Sect. 6 is devoted to node withdrawals. Related work is reviewed in Sect. 7, and the paper finishes with conclusions in Sect. 8.

2   Application and Experimental Setting

The example program, called WaterGAP, computes current and future water availability worldwide [2]. WaterGAP partitions the surface area of continents into equally-sized grid cells. Based on input data for climate, vegetation etc., it simulates the flow of water, both vertically (precipitation, transpiration) and horizontally (routing through river networks), over a period of several years. The program has been written in C++.

WaterGAP uses two levels of parallelism: a master/slave scheme implemented with MPI at the outer level, and data parallelism implemented with OpenMP-threads at the inner level [2]. The master/slave scheme relies on the observation that the set of grid cells is naturally partitioned into basins that do not exchange water with other basins. Thus, the overall computation is divided into independent tasks that correspond to one basin each. Task sizes are known in advance, but range from a few very large tasks to many small ones.

Scalability of the master/slave scheme is limited, since the program can not run faster than the time needed to compute the largest basin. Therefore, data parallelism is used to speed up the computation of large basins internally. Data parallelism yields lower speedups than master/slave parallelism [2], i.e., if a multi-processor node is assigned one large basin, it finishes earliest when using a multi-threaded process. If the same node is assigned several small basins, it finishes earlier when using several single-threaded processes. Therefore, we use a different number of threads for different processes.

Experiments were carried out on the compute cluster of the University of Kassel, a Linux cluster that comprises a large number of double-processor nodes, and one eight-processor node. On this architecture, a maximum speedup of 22 was achieved with 32 processors [2]. Here, the largest basin was computed by a multi-threaded process, other large basins were computed by double-threaded processes, and the small basins were computed by single-threaded processes.

In all experiments, we used the Portland Compiler, and the MPICH2 [3] implementation of MPI-2 (release 1.0.3, process manager mpd, compiled with Portland compiler). Experiments were carried out both interactively and through the batch system. We experimented with both C and C++ bindings of the MPI functions.

3   Dynamic and Heterogeneous Master/Slave Scheme

The standard master/slave scheme uses one master and several slaves. The master starts computation by sending a task to each slave. Whenever a slave has finished its task, it reports the result back to the master and gets the next task, until all tasks have been processed. We modify the scheme to incorporate:

- dynamic arrival of slaves,
- arrival of more powerful slaves that can take over expensive tasks, and
- sudden or announced withdrawal of slaves.

The first case is easy to handle at an algorithmic level: the master adds the slave to its pool of communication partners, and sends a task. The other two cases require task reassignment. While one can think of very sophisticated and efficient schemes, we restrict our considerations to a simple scheme here that is sufficient to identify and study essential requirements for MPI support:

After creation, a new process connects to the master and requests work. The master assigns the tasks by size, starting with the largest task. To keep track of the state of computation, it stores for each process: task currently assigned to, size of this task (in grid cells), and number of processors. The latter is sent to the master with the slave's work request.

Although tasks are assigned strictly in order of decreasing size, an assignment may be a better or worse fit. A good fit maps a large basin to a process with many processors, or a small basin to a process with a single processor. Architecture-specific thresholds specify the meaning of terms large etc. In our setting, basins are classified as large, medium, or small, depending on the number of grid cells; slaves are classified as powerful (8 processors), normal (2 processors), or weak (1 processor). We speak of a good fit for combinations large-powerful, medium-normal, and small-weak.

One case of a bad fit assigns a large basin to a weak slave (combinations large-weak, large-normal, and medium-weak). Here, the slave starts computing, but when a more powerful slave arrives later on, the master reassigns the basin. As MPI-2 provides no means for the master to signal this event to the first slave, the slave occasionally asks whether there was a reassignment. If so, it abandons its work and requests a next task from the master.

The reverse case that a small or medium basin is assigned to a powerful slave (combinations small-powerful, small-normal, and medium-powerful), occurs only when all larger basins have already been assigned before. Hence, after receiving the basin, the slave splits itself up into multiple processes. One process computes the basin, and the others request more work from the master, i.e., become separate slaves. The splitting generates weak processes when the assigned basin is small, and normal processes when it is medium.

Moreover, tasks are reassigned when the master learns that a slave has died, and will therefore not finish its task, or when the task pool is empty, but some results have not been received yet. When several slaves are computing the same basin and one has found the result, the others are abandoned as soon as they report back.

One case needs particular consideration: reassignment of a large basin (from a dead slave) after powerful slaves have been split up into groups of weak ones. The master stores the grouping of processes, keeping the original process as a leader. To assign a task to the group, it requests all processes except the leader to exit (when they report having finished their present task). Then, it assigns the task to the leader, who spawns new threads.

4   Dynamic Integration of Processes

A program version that allows for dynamic integration of slaves has been described by us [2]. It uses an additional process, called server, that invokes the accept function and helps in communicator construction. All communication is accomplished through intracommunicators that connect two processes each: master and slave, or master and server. The construction of a single communicator for all processes proved difficult, since communicator constructor functions are blocking and collective. Busy slaves can not call these functions, except in a separate thread, which would interfere, however, with the internal OpenMP structure for data parallelism.

All occurrences of `MPI::COMM_WORLD` had to be replaced by pairwise communicators. Since there is no `MPI::ANY_COMM`, the master waits for a message from any communicator with loop

```
while (!isMessage) {
    rank = (rank + 1) % total;
    isMessage = comms[rank].Iprobe(...);
}
```

where `comms` is an array of all intracommunicators.

5   Process-Thread Adaptivity

The modified master/slave scheme poses two requirements:

- A slave that runs on a multi-processor node must be able to dynamically spawn either processes or threads on the same node.
- A slave must be able to exit computation after receiving a termination request from the master.

For the first requirement, a slave must know how many processors it owns. Although the OpenMP function `omp_get_num_procs` yields the number of physical processors, it is possible that only part of them are available to the application. Thus, information must be passed from the resource manager (e.g. batch system) to the MPI application. MPI-2 defines a constant `MPI::UNIVERSE_SIZE` for that purpose, but leaves it to the implementation to set its value or not. The MPICH2 implementation sets the value to parameter `usize` of `mpiexec`. We use this parameter to provide to each slave the number of processors it owns.

Spawning the corresponding number of threads is a simple call to the OpenMP function `omp_set_num_threads`. Processes are spawned with `MPI::Comm::Spawn`. The new processes can not rely on `MPI::UNIVERSE_SIZE`, but get the number of processors from their parent, through an argument of the spawn function. These processes also differ from the processes started with `mpiexec` in that they are connected to their parent with a communicator. We close this communicator immediately, and then handle all processes the same way.

Threads are always spawned on the same node. Processes, in contrast, may be spawned on any node that is available to the MPI system. This placement may be inappropriate as the system can not take the existence of threads into account (especially if they have not been created yet). To keep track of the available resources, we always spawn processes on the same node as their parent. MPI-2 supports that with the reserved `info` key `host`, which is an optional part of the standard again.

The second issue (slaves exit computation) is easy to resolve. As will be further discussed in the next section, the slave first disconnects from the rest of the program, by closing the master-slave intracommunicator, and then calls Finalize. Since this function is collective over the set of connected processes only, the slave returns immediately.

6   Termination of Processes

For evolving processes, i.e., program-initiated termination, the exit of slaves is easy. The principle has already been explained in the last paragraph. It relies on the fact that a slave is connected to the rest of the program through a single communicator between master and slave only. All other communicators are closed immediately after their creation. Thus, the slave can disconnect, without enforcing any other process to participate in this blocking and collective operation. Note that a process is connected to all processes in `MPI::COMM_WORLD`, but we start each process with a separate call to `mpiexec`, and so `MPI::COMM_WORLD` is a singleton.

After termination, the master must exclude the slave from its pool of communication partners, since Iprobe does not work with a null communicator. Also, the basin must be reassigned to another slave. With this scheme, a slave may leave computation at any time, either in reaction to a termination request by the master, or on request of the resource manager (provided that the resource manager can pass the request to the slave).

Implementation of malleability, in contrast, is problematic. When a slave suddenly dies, it is not able to call Disconnect nor Finalize. The MPI standard states that "'if a process terminates without calling Finalize, ... the effect on connected processes is not defined"'. Thus, it may happen that a single faulty process brings the whole application down.

According to our experiments, the MPICH2 implementation is more robust. When a slave dies, the master's message-waiting loop (see Sect. 4) continues without any problem, just not receiving messages from the dead slave anymore. We tested this feature by running each process in a separate window, killing one with `Ctrl-C` (during a computation phase), and observing the output. The behavior was the same in the batch system, with an exit call in one slave's code.

Using the reassignment scheme described in Sect. 3, the program manages to compute all tasks and generate the complete output. Nevertheless, we did not find a correct way to finish the program. The MPI standard requires that each process calls Finalize, which is a collective and blocking operation over connected processes. While a slave can disconnect from the rest of the program and terminate as described above, the master can not disconnect from a dead slave. Consequently, its call to Finalize does not return. The standard defines the function `MPI::Abort` to kill processes, but the behavior of this function is not specified in detail. In our experiments, this function did not return either. The only way we found to let the program return, was to omit the Finalize call from the master. Then termination works fine, except for an error message, but this workaround of course conflicts with the standard.

The termination problem can probably be solved by clarifying the behavior of `MPI::Abort`. An alternative solution relies on a communicator clean function that eliminates all dead processes from the communicator, i.e., live processes are disconnected from dead ones, and dead processes do not need to take part in any future collective operation. Such a function may either be provided by the MPI API, or be invoked implicitly by the MPI implementation. The implicit variant is already provided by Fault Tolerant MPI or FT-MPI [4]. It comfortably solves our termination problem since after cleaning, the master can call Finalize. In FT-MPI, communication functions return an error code after a communication partner has crashed. This mechanism solves a second problem: notification of the master in the event of slave death. As the master regularly contacts all slaves in the message-waiting loop, it learns about the crash soon and can reassign the basin immediately. Unfortunately, FT-MPI supports only part of MPI-2.

7   Related Work

The process termination aspect of malleability has been discussed under the heading of fault tolerance, e.g. in a survey paper by Gropp and Lusk [5], and in FT-MPI [4].

Much work on malleability was carried out in the scheduling community, where it was shown that malleability significantly improves system throughput in both supercomputers and grids [6,7]. Two approaches for making MPI programs malleable have been followed: 1) checkpointing, i.e., interrupting the program, saving its state, and later restarting it with a different number of processes [7], and 2) folding, i.e., using a fixed number of processes, and coping with changes in the number of processors by varying the number of processes per processor [8]. We are not aware of other experience reports on making an application malleable with the MPI-2 dynamic process management routines.

Outside MPI, research on handling node crashes with the master/slave scheme has been done with PVM [9] and Java [10]. The more general divide-and-conquer pattern is considered by Wrzesińska et al., in a Java-based framework [11]. They suggest a scheme to avoid redoing work that another process already did before crashing. None of this work considers multi-threaded processes or process-thread adaptivity.

Except for malleability, hybrid MPI/OpenMP programming is well understood [12,13], including dynamic variations in the number of threads per process for better load balancing [14].

8   Conclusions

This paper has discussed MPI-2 support for dynamic and heterogeneous processes, on the basis of a hybrid master/slave application. The master/slave scheme was modified to dynamically add processes, and reassign tasks when powerful slaves arrive or slaves exit. We observed that MPI-2 supports integration of slaves and process-thread adaptivity, provided that the implementation covers optional parts of the standard: the constant `MPI::UNIVERSE_SIZE` and the `info` key `host`. Termination, in contrast, requires active participation of a slave, or functionality beyond the MPI standard to eliminate dead slaves from a communicator, and to notify the master after slave crashes.

In experiments, the malleable program performed better than the original one, mainly because it started before all desired resources were available. Malleability and process-thread adaptivity come at the price of higher programming overhead and a performance penalty. For the master/slave example, the programming overhead was reasonably low, but this may be different for applications that require algorithmic changes such as data redistribution. The performance penalty is due to the overhead for additions and withdrawals of nodes, the need to use pairwise communicators instead of `MPI_COMM_WORLD`, the bookkeeping overhead at the master, and task reassignment costs. Our application has a high computation-to-communication ratio, and thus the overhead was not an issue.

Future research may address improvements of the simple master/slave scheme referred to in this paper. For instance, the master may restrict use of multi-threaded slaves to basins that would otherwise delay the overall computation. It may also cooperate with the resource manager to get a forecast of resources. Notification of slaves after reassignment may use one-sided communication instead of pairwise communication-based polling. Furthermore, checkpointing may be integrated. Finally, the scheme may be refined to handle the case that the master dies. Sophisticated master/slave patterns may be implemented in a skeleton library, which may extend to other malleable patterns.

References

1. Feitelson, D.G., Rudolph, L.: Toward convergence in job schedulers for parallel supercomputers. In: Job Scheduling Strategies for Parallel Processing, Springer LNCS 1162 (1996) 1–26
2. Leopold, C., Süß, M., Breitbart, J.: Programming for malleability with hybrid MPI-2 and OpenMP: Experiences with a simulation program for global water prognosis. In: High Performance Computing & Simulation Conference. (2006) 665–670.
3. Gropp, W., et al.: MPICH2 User's Guide, Version 1.0.3. (November 2005) Available at http://www-unix.mcs.anl.gov/mpi/mpich2.
4. Fagg, G.E., et al.: Process fault-tolerance: Semantics, design and applications for high performance computing. Int. Journal of High Performance Computing Applications 19(4) (2005) 465–478
5. Gropp, W., Lusk, E.: Fault tolerance in message passing interface programs. Int. Journal of High Performance Computing Applications 18(3) (2004) 363–372
6. Kalé, L.V., Kumar, S., DeSouza, J.: A malleable-job system for timeshared parallel machines. In: IEEE/ACM Int. Symp. on Cluster Computing and the Grid. (2002) 230–237
7. Vadhiyar, S.S., Dongarra, J.J.: SRS: A framework for developing malleable and migratable parallel applications for distributed systems. Parallel Processing Letters 13(2) (2003) 291–312
8. Utrera, G., Corbalán, J., Labarta, J.: Implementing malleability on MPI jobs. In: Proc. Parallel Architectures and Compilation Techniques. (2004) 215–224
9. Goux, J.P., et al.: An enabling framework for master-worker applications on the computational grid. In: IEEE Int. Symp. on High Performance Distributed Computing. (2000) 43–50
10. Baratloo, A., et al.: Charlotte: Metacomputing on the web. In: Int. Conf. on Parallel and Distributed Computing Systems. (1996) 181–188
11. Wrzesińska, G., et al.: Fault-tolerance, malleability and migration for divide-and-conquer applications on the grid. In: IEEE Int. Parallel and Distributed Processing Symposium. (2005)
12. Smith, L., Bull, M.: Development of mixed mode MPI/OpenMP applications. Scientific Programming 9(2–3) (2001) 83–98
13. Rabenseifner, R.: Hybrid parallel programming on HPC platforms. In: European Workshop on OpenMP. (2003) 185–194
14. Spiegel, A., an Mey, D.: Hybrid parallelization with dynamic thread balancing on a ccNUMA system. In: European Workshop on OpenMP. (2004) 77–82

What MPI Could (and Cannot) Do for Mesh-Partitioning on Non-homogeneous Networks

Guntram Berti and Jesper Larsson Träff

C&C Research Laboratories, NEC Europe Ltd.
Rathausallee 10, D-53757 Sankt Augustin, Germany
{berti, traff}@ccrl-nece.de

Abstract. We discuss the *mesh-partitioning* load-balancing problem for non-homogeneous communication systems, and investigate whether the MPI *process topology functionality* can aid in solving the problem. An example kernel shows that specific communication patterns can benefit substantially from a non-trivial MPI topology implementation, achieving improvements beyond a factor of five for certain system configurations. Still, the topology functionality lacks expressivity to deal effectively with the mesh-partitioning problem. A mild extension to MPI is suggested, which, however, still cannot exclude possibly sub-optimal partitioning results. Solving instead the mesh-partitioning problem outside of MPI requires knowledge of the communication system. We discuss ways in which such could be provided by MPI in a portable way. Finally, we formulate and discuss a more general *affinity scheduling problem*.

1 Introduction

Applications involving large datasets are often parallelized using a data partitioning approach, as in mesh-based solution of partial differential equations. This leads to the following *mesh-partitioning problem*: A large mesh, represented as an undirected, weighted *problem graph* $G = (V, E, w)$ with edge (and possibly vertex) weights w is to be mapped onto a smaller set of processors P, such as to minimize application run time. This is approximated by minimizing communication costs, which are assumed to be a function of the value of the *edge cut* (sum of weights of edges in G crossing processor boundaries), while keeping computational load evenly distributed. Although this commonly used model is at best an approximation to the communication cost optimization problem (e.g. network contention is very hard to capture, communication volume may easily be overestimated, etc., see [4]), we will stick to it here.

Assuming a homogeneous, fully connected system, a common approach to solving the mesh-partitioning problem is to partition G into $|P|$ approximately equal-sized subsets, minimizing the value of the edge cut, i.e. finding a mapping $\pi : V \mapsto P$ such that

$$\sum_{\pi(u) \neq \pi(v)} w(u, v) \quad \text{is minimal} \tag{1}$$

B. Mohr et al. (Eds.): PVM/MPI 2006, LNCS 4192, pp. 293–302, 2006.

under the *balancing condition* (which can be relaxed) that

$$|\pi^{-1}(p)| \leq \lceil |V|/|P| \rceil \tag{2}$$

This *graph partitioning problem* is NP-complete [3], but many good heuristics exist [2,5,8,11], and are implemented in a number of libraries [7,10,16,17].

Most of these algorithms can be extended to handle non-homogeneous *processor computing powers*, but it is more difficult to handle systems with non-homogeneous *communication systems*, for instance with a mesh or torus topology, or with a hierarchical structure like clusters of SMP nodes. The simple graph partitioning approach is not adequate here, since vertices of G with "heavy" edges might end up on processors connected by "weak" communication links. There is obviously no way a graph partitioner can exclude this possibility without additional knowledge of the underlying system. Different approaches to tackling this problem have been proposed and discussed.

In [16] the authors model the communication system as a complete *host graph* $H = (P, C, c)$ with a cost function c on edges $(p_0, p_1) \in C$ reflecting the "cost" of communication between processors p_0 and p_1. Deriving the costs $c(p_i, p_j)$ is not straightforward and to some extent even application-dependent. The authors favor a *quadratic path length* (QPL) metric, leading to a *network cost matrix* (NCM) penalizing connections going over many hops of the *physical* network. The *mapping problem* is then defined as a generalization of the partitioning problem (1): Find $\pi : V \mapsto P$ such that

$$\sum_{\pi(u) \neq \pi(v)} w(u, v) c(\pi(u), \pi(v)) \quad \text{is minimal} \tag{3}$$

subject to the balancing condition (2).

To solve this problem, they extend their homogeneous multi-level heuristic by mapping the coarsest problem graph to the host graph in an approximately optimal way (the exact solution is equivalent to the *quadratic assignment problem* and again NP-complete). The Kernighan-Lin heuristic used to derive partitions of the finer problem graph levels is modified to take the modified cost function and the resulting larger set of potential moves into account.

The same model of the communication system is used in [9]. However, they first start with a complete conventional partitioning, and then use the host graph to guide an incremental improvement of the partitioning. Still other heuristics for solving the mapping problem were given in [5,6,10]. In contrast, the *Dynamic Resource Utilization Model* (DRUM) [1] uses measurements to derive a hierarchical scalar characterization of compute nodes, merging both computing power and network bandwidth into a single "power" value per node. As the hierarchy is explicit in the abstract model, general-purpose partitioners such as Zoltan [17] can be instrumented to use specific partitioning strategies at each level [1].

In all cases cited above, the description of the hardware architecture and the related network performance parameters have to be set up manually. The network models discussed so far represent compromises, aiming to be simple enough for the underlying optimization approach. The NCM ignores hierarchical structures, thus excluding level-specific partitioning. On the other hand, the averaging DRUM model loses some fine-grained local structure and seems less suited for e.g. mesh architectures.

Applications requiring graph partitioning are frequently using the *Message-Passing Interface* (MPI) for process communication. MPI, being strictly a communication interface, has no functionality for solving the mesh-partitioning problem. Since the MPI interface has no notion of "cost" of communication, MPI also cannot supply the knowledge of the underlying system required to construct the weighted host graph needed by a partitioning/mapping package. However, the internal assumptions about the underlying system present in any MPI implementation could potentially be made useful to solve the mesh-partitioning problem. This could be done either *implicitly* via the *graph topology functionality* of MPI, using the two-stage approach to the mesh-partitioning problem discussed and evaluated in sections 2 and 3. An orthogonal solution, discussed in Section 4, is to make the assumptions of the MPI implementation *explicitly* accessible in an abstract, portable and non-constraining fashion, to be used to construct the desired host graph for a mapping package. Finally, in Section 5, we take a broader view and ask if graph partitioning does not solve a too narrow problem altogether.

2   Mesh-Partitioning with MPI Process Topologies

Although not capable of solving the mesh-partitioning problem, MPI defines functionality to solve a *process re-mapping problem* that could be used as the second stage in a *two-stage approach*: first partition the mesh into $|P|$ subsets V_i, $i = 0, \ldots |P| - 1$ assuming a homogeneous communication system, second find an optimal mapping of the $|P|$ subsets onto the set of processors.

The *graph topology functionality* of MPI [12, Chapter 6] makes it possible to specify a *non-weighted communication graph*, abstracting the communication pattern of the $|P|$ processes. The MPI implementation in turn can use this information to create a new communicator representing a process remapping which is best suited for the given communication graph on the given system. It is up to the MPI implementation to provide a suitable remapping (which could be just the identity mapping). While the two-stage approach has often been discussed, e.g. in [16], using the MPI topology functionality for the second process remapping step has apparently not been considered previously.

Assuming that the MPI implementation at hand has a non-trivial implementation of the topology functionality, the problem arises how to specify the communication graph of the $|P|$ processes. Putting an edge between two processes whenever there is an edge in G between two partitions is likely to lead to a communication graph overstating weak connections, possibly to the point of being a complete graph without information. Using edge weights corresponding to communication load between partitions would be an informative alternative, but is unfortunately not permitted by the MPI functionality. Instead, an edge could be put if the total weight of edges between two partitions exceeds a certain *threshold*.

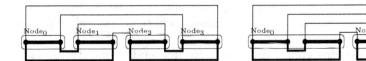

Fig. 1. Algorithm with hybercubic communication pattern. Communication intensity and/or volume decrease with increasing hypercube dimension. Heavier edges denote heavier communication. Left: optimal mapping of the algorithm onto a $2 + 2 + 2 + 2$ processor SMP cluster. Right: optimal mapping onto a $4 + 4$ processor cluster.

As the examples below will show both the threshold solution and the two-stage approach itself have limitations. We assume an SMP system with a marked difference in communication performance between processes on the same vs. on different SMP nodes. Similar examples can be constructed for systems with other, non-homogeneous interconnects.

The first example shows that unweighted graphs and thresholds are too weak to enforce an optimal mapping, unless complete knowledge of the underlying system is available, thus defying the idea of a portable, system-independent solution to the mesh-partitioning problem: Selecting the correct threshold *a priori* without knowledge of the target system configuration is not possible.

Example 1. Consider a hypercube algorithm with strong communication along dimension 0, less strong along dimension 1, etc. that we want to map onto an SMP system. Clearly, the processors should be mapped such that as many of the lower-dimensional, heavily communicating edges are inside SMP nodes, with higher-dimensional, weaklier communicating edges between nodes, cf. Figure 1.

Consider first the two-dimensional case of 4 processes to be mapped onto a $2 + 2$ processor cluster. Selecting a threshold resulting in edges along dimension 0 only, would make it possible for the MPI implementation to place pairs of connected processors on the same node, such that the heaviest communication takes place inside SMP nodes. On the other hand, selecting a lower threshold and having edges both along dimension 0 and dimension 1 would make it impossible for the MPI implementation to make the right decision since each process would be marked as communicating with two other processes.

Moving to three dimensions, for a $4 + 4$ processor cluster the best threshold would put edges along dimension 0 and 1. For a $2 + 2 + 2 + 2$ processor cluster the best threshold would put edges only along dimension 0.   □

The problem is aggravated for systems with more than two layers of communication. In such cases even *with* knowledge of the underlying system, it is in general not possible with an unweighted graph to provide enough information to the MPI implementation to permit an optimal solution.

Hierarchical structures are found e.g. in multi-physics codes, where coupling within the "single-physics" cores occurs much more often than across sub-problem boundaries. If such problems are part of a larger application, we get yet another weaker level of coupling. Concrete examples are found in climate research, where

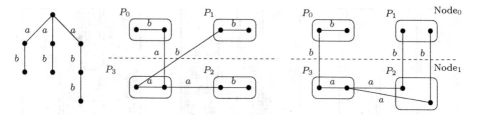

Fig. 2. Weighted 8 node tree (left). The optimal partition (middle) has cut weight $2a + b$. No matter how the 4 sets of this partition are mapped onto the $2 + 2$ processors, at least one a-edge crosses SMP nodes. A worse partition with cut weight $2a+3b$ (right) can be mapped such that the weight of edges crossing SMP nodes is only $3b$.

different models are coupled to achieve a more comprehensive global model [15]. For instance, we may have 3D/3D coupling of flow and chemistry components for both air and ocean, each coupling to their respective spatial neighbor partitions and a coupling occurring with lower frequency via the ocean/air interface.

A natural modification of the MPI graph topology mechanism would be to allow weighted graphs to model the intensity of communication along the edges. As the next examples show, the two-stage approach is strictly weaker than a direct solution of the processor mapping problem: A graph partitioner without knowledge of the target system (as modeled by the host graph) cannot compute the most suitable partition for the system.

Example 2. We consider a weighted tree of 8 nodes as shown in Figure 2. There are two different edge weights a and b with $a > b$. The minimum cut partition has cut weight $2a + b$ but is not optimal for mapping onto a $2 + 2$ processor SMP cluster, since at best an a and a b edge cross between nodes. Instead, the suboptimal partition with cut weight $2a + 3b$ is better suited, since the weight of the edges between processes on different SMP nodes can be arranged to be only $3b$. For appropriate values of bandwidth and edge weights, the ratio in communication load between the two partitionings can become arbitrarily large.   □

Example 3. Example 2 may seem artificial. Figure 3 shows that the two-stage approach can give arbitrarily bad results even for mesh-based graphs.   □

From the examples two conclusions can be drawn:

1. The non-weighted MPI topology functionality does not provide enough information for optimal process remapping in case of different communication requirements between different processes. This could easily be remedied by allowing *weighted communication graphs* in the MPI functionality. This and other problems (lack of scalability, lack of control of optimization criterion, etc.) was discussed in [14].
2. Even with weighted graphs the two-stage approach to mesh-partitioning may deliver arbitrarily bad solutions. In [16], the two-stage approach is shown to

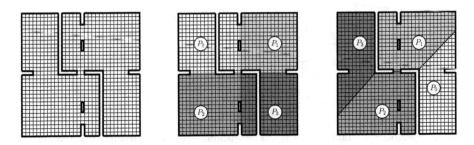

Fig. 3. Left: "Jagged" mesh. Middle: Homogeneous optimal 4-way partitioning. Right: Optimal partitioning for a 2+2 SMP cluster $\{P_0, P_1\}, \{P_2, P_3\}$, having minimal coupling between both SMP nodes.

be worse by a factor of about 2-3 on average for benchmark meshes on cluster architectures, using weighted edge cut as measurement (no actual timings are given).

Even though the two-stage approach is inferior to a direct solution, it can be a viable and user-friendly option in cases where the partitioning of the problem is fixed, as the next section will show.

3 An Application Kernel

To illustrate the possible performance benefits achievable by mesh partitioning using the (theoretically sub-optimal) two-stage approach with the final process remapping done by the MPI topology functionality, we consider a *communication kernel* with the hypercube communication pattern described in Example 1. This pattern is assumed to be the outcome of the first stage mesh partitioning, and the second stage consists in a process remapping to fit the target system. This is carried out by defining a communication graph which can be input to MPI to perform the appropriate process remapping. The kernel is written such that the process to processor mapping implied by MPI_COMM_WORLD (where MPI processes are distributed consecutively in increasing MPI rank order over the SMP nodes) is unsuited for SMP systems: the most frequent communication will be between processes on different nodes. In the kernel the communication frequency along hypercube edges increases exponentially with decreasing dimension of the edge. Increase factor as well as size of the data sent along the dimensions can be varied, but will not be of concern here.

In order to gain any effect a non-trivial implementation of the MPI graph topology functionality is required. This is fulfilled by MPI/SX [13], and the measurements shown in Table 1 have been conducted on a four node, 32 processor NEC SX-8 system. The difference in communication bandwidth between processes on the same SMP node and processes on two different nodes is about a factor of two, but more importantly, if several processes on a node attempt to communicate with processes on other nodes at the same time, the communication is serialized. Thus,

Table 1. Running time (in micro-seconds) for the hypercube kernel on an NEC SX-8. The first two columns describe the SMP configuration, and the remainder give the running time on various MPI process distributions: MPI_COMM_WORLD communicator, a random communicator, and communicators created by the topology functionality. Here topo[i] denotes a communication graph with edges along hypercube dimensions $0, \ldots, i-1$. The largest improvements over MPI_COMM_WORLD the distribution is shown in bold, and ranges from a factor of two to a factor of more than 5.

Processes	Distribution	WORLD	random	topo[1]	topo[2]	topo[3]	topo[4]	topo[5]
8	8	5632	5620	5627	5958	5639		
	4 + 4	21021	22399	6653	**6517**	20987		
	2 + 2 + 2 + 2	18402	20379	6223	**6221**	18970		
16	8 + 8	321742	223179	66703	**55329**	68053	321769	
	4 + 4 + 4 + 4	221754	185305	62101	**51588**	65404	222387	
	1 + 7 + 1 + 7	291283	212762	166729	**160646**	166331	291008	
	2 + 6 + 2 + 6	265642	225786	**65491**	159097	65844	295374	
	3 + 5 + 3 + 5	239661	221891	170463	**157388**	164895	271440	
	8 + 4 + 4	320393	218326	68943	**53494**	66942	320363	
32	8 + 8 + 8 + 8	1090388	1197901	460068	**228929**	251979	532812	1093170

mapping heavily communicating processes to the same node is doubly beneficial. As detailed in Example 1, in the absence of edge weights in the MPI topology functionality, graph edges must be chosen to reflect the SMP system. Too few edges (e.g. only along hypercube dimension 0) can give sub-optimal improvement, and too many edges makes too many processes indistinguishable such that a good remapping cannot be guaranteed.

Table 1 gives some results of running the kernel on various number of processes and distributions over the SMP nodes. In each case good results are achieved when each subcube of the communication graph fits onto one SMP node. Bad results are generally achieved when the subcubes are too large for the SMP nodes (e.g. column topo[4], corresponding to communication graphs with 16 process subcubes to be mapped onto 8 process SMP nodes), and good or even best results are achieved with smaller sized subcubes than the size of the SMP nodes (most of the best results are in column topo[2]). The best overall improvements exceed a factor 5 for distributions with 16 and 32 processes.

4   Portable MPI Topology Introspection

As shown, using the MPI topology functionality to solve the mesh-partitioning problem has inherent limitations, even if weighted graphs would be allowed. The alternative is to do the mesh-partitioning completely outside of MPI. This requires information on the communication system, either *a priori*, by measuring, or both. Measuring alone is of limited value on a loaded system, and in general has difficulties capturing effects of contention (cf. end of Section 5 for possible solutions).

Instead, we propose to leverage the implicit knowledge on the system which is present in any MPI implementation (no matter how rudimentary). For an

MPI *communicator* we model the part of the communication system used by the processes in the communicator as a complete graph with multiple weight functions. Nodes correspond to processes, with edge weights modeling either the number of abstract *hops* between two processes, or the *relative bandwidth*, or *relative latency*, ... A hop measures the number of communication layers between two processes. Processes on the same node of an SMP cluster would be one hop distant (the number of hops from a process to itself being 0), and processes on different node would be two hops distant. In a 1D linear array, each process (in MPI_COMM_WORLD) has two neighbors which are one hop away, two neighbors that are two hops away and so on.

We believe that it is possible to make this abstract representation available to an MPI application (e.g. mesh-partitioner) in a meaningful and portable way, regardless of the actual system. We suggest the following functionality.

- Functions returning the number of neighbors of the calling process that are exactly n hops away, the list of such neighbors, and the maximal hop distance to any other process in the communicator.
- Functions returning the hop distance, relative bandwidth and relative latency between the calling process and any other process in the communicator. Relative bandwidth, e.g., could be expressed as the ratio to the bandwidth for the process communicating with itself. This issue is bound to be contentious.
- A function returning the maximum number of simultaneous communication operations to processes at a given hop distance.
- Possibly more involved functions for estimating the effects of contention, e.g. returning the load of the communication path between the calling process and any other process in the communicator, given that (a) the two processes are the only processes communicating, (b) the load under the worst bisection with the two processes belonging to different parts.

These proposals are portable in the sense that each processor is only required to be able to return information about its own neighborhood (relative to the given communicator). A trivial implementation is possible, and would map all processes as being one hop away.

This functionality would clearly make it possible to build the concrete graphs as used in the NCM approach [16], or to construct hierarchical models like DRUM, as well as other imaginable representations. For instance, a hierarchical graph can be built by an application as follows:

1. Get all neighbors with hop distance 1,
2. Compute local graph components
3. While the graph is not connected:
    (a) introduce a new hierarchical node for the current component
    (b) Get all neighbor graph components for the next larger hop distance n
    (c) Compute the resulting larger components

5   Beyond Partitioning: Affinity Scheduling

The discussion so far assumed that a host graph $H = (P, C)$ is given. In general, however, H is only a subgraph of the (available part of the) global machine graph H^G, and is selected by a system scheduler, typically based on the number of compute nodes specified by the user. This places the burden of specifying an adequate subset of the machine on both the user (who may not know about it), and the scheduler (who does not know about the application). Giving the scheduler more knowledge about the resource requirements of an application, it could choose an optimal subset of the machine matching high-level user preferences:

- The user could demand just enough processors to finish within one hour
- The application is partitioned into largely independent tasks and can therefore be distributed to weakly connected nodes
- It is found that the application will not achieve good parallel performance on the currently available set of nodes, and it is scheduled for a later time

These tasks cannot by solved in the narrow frame of graph mapping. Instead, we propose to consider the following *optimal subgraph scheduling problem*: Given the time dependent *global machine graph* $H^G(t) = (P^G(t), C^G(t))$, $t > 0$ ("free processors at time t"), a *utilization cost function* $K = K(P, \tau, t)$ (cost for using processor set $P \subset P^G(t)$ for duration interval $[t, t + \tau]$), and a *user preference function* $\Phi = \Phi(K, t)$ (preference of finishing the task until time t with total cost K), find a *starting time* t_0 and a mapping $\pi : V \mapsto P^G(t_0)$ such that

$$\Phi\left(K(\pi(V), t_0, T_{app}), t_0 + T_{app}\right) \quad \text{is minimized} \quad (T_{app} = T_{app}(\pi(V))) \qquad (4)$$

Here, the total (expected) time T_{app} is an application-specific performance estimation based on the partitioning and the available network (sub)topology. In (4), a hidden constraint is that the subgraph $H^G(t)$ must be available for all times $t = t_0 + \tau, 0 \leq \tau \leq T_{app}(\pi(V))$.

For homogeneous architectures, $K(P, \tau, t) = \alpha P \tau$ and $\Phi(K, T) = KT$ would be reasonable choices. Changing Φ, a user could slant the result in favor of cheaper or faster computation. Information about $H^G(t)$ is generally available only in the system scheduler, thus, a solution to problem (4) would have to access this information. Using scheduler information together with actual network measurements might also permit to estimate the bandwidth available to new applications, thus combining the advantages of static and dynamic network information.

6   Summary

We investigated two orthogonal paths to solving the mesh-partitioning problem for systems with a non-homogeneous communication system. A two-stage approach, consisting of ordinary graph partitioning followed by a remapping relying on the MPI topology functionality, and probably requiring the least change on behalf of the application, is limited by the restriction to non-weighted communication graphs of the MPI standard. As an orthogonal approach, we discussed additional, portable, system-independent MPI functionality, which could aid the application programmer in constructing the desired graph model of the system to be used as input to sophisticated mesh-partitioners. An artificial, but not unrealistic kernel showed the large potential gains by performing an appropriate process mapping.

References

1. K. D. Devine, E. G. Boman, R. T. Heaphy, B. A. Hendrickson, J. D. Teresco, J. Faik, J. E. Flaherty, and L. G. Gervasio. New challenges in dynamic load balancing. *Appl. Numer. Math.*, 52(2–3):133–152, 2005.
2. C. M. Fiduccia and R. M. Mattheyses. A linear-time heuristic for improving network partitions. In *19th ACM/IEEE Design Automation Conference (DAC)*, pages 175–181, 1982.
3. M. R. Garey, D. S. Johnson, and L. J. Stockmeyer. Some simplified NP-complete graph problems. *Theoretical Computer Science*, 1(3):237–267, 1976.
4. B. Hendrickson and T. G. Kolda. Graph partitioning models for parallel computing. *Parallel Computing*, 26:1519–1534, 2000.
5. B. Hendrickson and R. Leland. A multi-level algorithm for partitioning graphs. In *Proceedings of the 1995 ACM/IEEE Supercomputing Conference*, 1995.
6. B. Hendrickson, R. Leland, and R. V. Driessche. Skewed graph partitioning. In *Proc. 8th SIAM Conference on Parallel Processing for Scientific Computing*, 1997.
7. G. Karypis. METIS. http://www-users.cs.umn.edu/~karypis/metis/.
8. B. W. Kernighan and S. Lin. An efficient heuristic procedure for partitioning graphs. *Bell System Technical Journal*, 49:291–307, 1970.
9. I. Moulitsas and G. Karypis. Architecture aware partitioning algorithms. Technical Report DCT Research Report 2006/02, Digital Technology Center, University of Minnesota, Jan. 2006.
10. F. Pellegrini and J. Roman. SCOTCH: A software package for static mapping by dual recursive bipartitioning of process and architecture graphs. In *High-Performance Computing and Networking (HPCN), Europe*, volume 1067 of *Lecture Notes in Computer Science*, pages 493–498. Springer-Verlag, 1996.
11. J. E. Savage and M. G. Wloka. Parallelism in graph-partitioning. *Journal of Parallel and Distributed Computing*, 13:257–272, 1991.
12. M. Snir, S. Otto, S. Huss-Lederman, D. Walker, and J. Dongarra. *MPI – The Complete Reference*, volume 1, The MPI Core. MIT Press, second edition, 1998.
13. J. L. Träff. Implementing the MPI process topology mechanism. In *Supercomputing*, 2002. http://www.sc-2002.org/paperpdfs/pap.pap122.pdf.
14. J. L. Träff. SMP-aware message passing programming. In *Eigth International Workshop on High-level Parallel Programming Models and Supportive Environments (HIPS03)*, pages 56–65, 2003.
15. S. Valcke, D. Declat, R. Redler, H. Ritzdorf, R. Vogelsang, and P. Bourcier. The PRISM coupling and I/O system. In *Proceedings of VECPAR 2004*, 2004.
16. C. Walshaw and M. Cross. Multilevel mesh partitioning for heterogeneous communication networks. *Future Generation Comput. Syst.*, 17(5):601–623, 2001.
17. Zoltan: Data-management services for parallel applications. http://www.cs.sandia.gov/Zoltan/.

Scalable Parallel Trace-Based Performance Analysis

Markus Geimer, Felix Wolf, Brian J. N. Wylie, and Bernd Mohr

John von Neumann Institute for Computing (NIC)
Forschungszentrum Jülich, 52425 Jülich, Germany
{m.geimer, f.wolf, b.wylie, b.mohr}@fz-juelich.de

Abstract. Automatic trace analysis is an effective method for identifying complex performance phenomena in parallel applications. However, as the size of parallel systems and the number of processors used by individual applications is continuously raised, the traditional approach of analyzing a single global trace file, as done by KOJAK's EXPERT trace analyzer, becomes increasingly constrained by the large number of events. In this article, we present a scalable version of the EXPERT analysis based on analyzing separate local trace files with a parallel tool which 'replays' the target application's communication behavior. We describe the new parallel analyzer architecture and discuss first empirical results.

1 Introduction

Event tracing is a well-accepted technique for post-mortem performance analysis of parallel applications. Time-stamped events, such as entering a function or sending a message, are recorded at runtime and analyzed afterwards with the help of software tools. For example, graphical trace browsers like VAMPIR [1] and PARAVER [2], allow fine-grained investigation of execution behavior using a zoomable time-line display.

However, in view of the large amounts of data usually generated, automatic off-line trace analyzers, such as the EXPERT tool from the KOJAK toolset [3,4], can provide relevant information more quickly by automatically searching traces for complex patterns of inefficient behavior and quantifying their significance. In addition to usually being faster than a manual analysis performed using trace browsers, this approach is also guaranteed to cover the entire event trace and not to miss any pattern instances.

Unfortunately, sequentially analyzing a single trace file does not scale to applications running on thousands of processors. Even if access locality is exploited, the amount of main memory might not be sufficient to store the current working set of events. Moreover, the amount of trace data might not even fit into a single file, which already suggests to perform the analysis in a more distributed fashion.

In this paper, we describe how the pattern search can be done in a more scalable way by exploiting both distributed memory and parallel processing capabilities available on modern large-scale systems. Instead of sequentially analyzing a single global trace file, we analyze separate local trace files in parallel by *replaying* the original communication on as many CPUs as have been used to execute the target application itself.

We start our discussion with a review of related work in Section 2, followed by an overview of our trace analyzer's new parallel design in Section 3, where it is also compared to the previous sequential design. Then, in Section 4, we discuss the parallel

B. Mohr et al. (Eds.): PVM/MPI 2006, LNCS 4192, pp. 303–312, 2006.

pattern-analysis mechanism in more detail, before we show preliminary experimental results that already demonstrate the improvement over the sequential analysis in Section 5. Finally, in Section 6 we conclude the paper and outline further improvements.

2 Related Work

Wolf et al. [5] review a number of approaches addressing scalable trace analysis. Dynamic periodicity detection in OpenMP applications [6] avoids recording redundant performance behavior, while the frame-based SLOG trace-data format [7] supports scalable visualization. Important to our particular approach has been the distributed trace analysis and visualization tool VAMPIR Server[8], which provides parallel trace access mechanisms, albeit targeting a 'serial' human client in front of a graphical trace browser as opposed to fully automatic and parallel trace analysis. A tree-based main memory data structure for event traces called cCCG [9] allows potentially lossy compression of trace data while observing specified deviation bounds.

Non-trace-based on-line performance tools, such as Paradyn [10] or Periscope [11], that analyze performance data in real-time address scalability by employing hierarchical networks for efficient reduction and broadcast operations between back-end processes and the tool front-end. The particular way patterns are specified and implemented in EXPERT was stimulated by the APART Specification Language (ASL) [12], which provides a formal notation to describe performance properties of parallel applications. Other ASL-inspired work includes JavaPSL [13], a Java version of ASL, and the aforementioned Periscope tool. KappaPI 2 [14] sequentially searches trace files of message-passing applications for patterns very similar to those used in our approach, but in KappaPI 2 emphasis is put on generating recommendations on how to improve the performance using knowledge of bottleneck use cases.

3 Overview of Parallel Trace Analysis

Instead of sequentially analyzing a single and potentially large global trace file, we analyze multiple local trace files in parallel based on the same parallel programming paradigm as the one used by the target application. For the sake of simplicity, we currently have restricted ourselves to handle only single-threaded MPI-1 applications, which implies that our parallel analyzer is an MPI-1-based program as well. The analyzer is executed on as many CPUs as have been allocated for the target application, allowing to run it within the same batch job as the application itself. Using an allocation with a different (smaller) number of CPUs for the analysis would require a separate batch job introducing typically significant additional waiting time in the performance analysis workflow. Figure 1 depicts the analysis workflow along with responsible components in comparison to the sequential analysis implemented by EXPERT.

The parallel analyzer itself uses a distributed memory approach, where each process reads only the trace data that was recorded for the corresponding process of the target application. This specifically addresses scalability with respect to wider traces, this is, those from larger numbers of processes. Since longer traces can be handled by selective tracing — i.e., by recording events only for code regions of particular interest — we

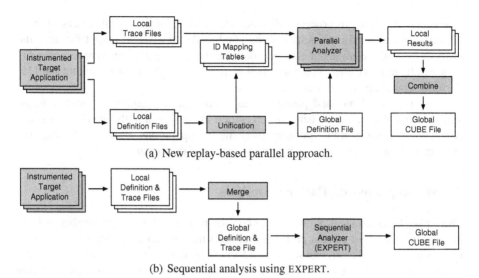

(a) New replay-based parallel approach.

(b) Sequential analysis using EXPERT.

Fig. 1. Schematic overview of the new parallel analysis work flow (a) in comparison to the previous sequential analysis (b). Stacked rectangles denote multiple instances of files or applications executed in parallel.

assume that the local trace data can be completely held in the main memory of the compute nodes. This has the advantage of having efficient random-access to individual events, whereas this is often not the case when dealing with a global trace file.

The actual analysis can then be accomplished by performing a *parallel replay* of the application's communication behavior. The central idea behind this replay-based analysis approach is to analyze a communication operation using an operation of the same type. For example, to analyze a point-to-point message, the event data necessary to analyze this communication is also exchanged in point-to-point mode between the corresponding analysis processes. To do this, the new analysis traverses local traces in parallel and meets at the synchronization points of the target application by replaying the original communication. How this idea can be used to search for complex patterns of inefficient behavior will be described in more detail in Section 4.

The event records stored in the individual per-process trace files use local identifiers to refer to static program entities, such as source-code regions or MPI communicators. Therefore, these local identifiers are mapped onto unique, global identifiers for the exchange of trace data between analysis processes. In the sequential analysis this mapping is part of the *Merge* step. In the parallel approach, this is similarly accomplished by performing a preprocessing step using a separate program that sequentially unifies the definitions of the per-process traces and generates a global definitions file that is shared between all analysis processes. To avoid reading the entire local trace files to extract definition records, we have modified the KOJAK measurement system to write definition and event records into separate files. The *Unification* step also creates a set of mapping tables that the analysis processes use to convert local into global identifiers while reading their local event data.

Each parallel analysis process only calculates a subset of the overall analysis report. Therefore, these local reports have to be combined into a single output file after the analysis has completed. In our current prototype, the individual analysis processes write their results to local files, which are then merged into a global CUBE output file [15] during a separate postprocessing *Combine* step.

These sequential pre- and postprocessing steps can be optimized in several ways, among which the most promising option is their integration into the analyzer and concomitant parallelization to minimize costly file I/O operations. However, detailed discussion of these optimizations is beyond the scope of this paper.

4   Message Passing Pattern Analysis

The replay-based analysis approach can be used to search for a large number of inefficiency patterns. Our current prototype supports the full range of MPI-1 performance metrics offered by the original sequential EXPERT tool, with the exception of *Late Receiver, Messages in Wrong Order* that is rarely significant in practice. A representative subset of these patterns is diagrammed in Figure 2. Their detection algorithms will be used to illustrate the parallel analysis mechanism below.

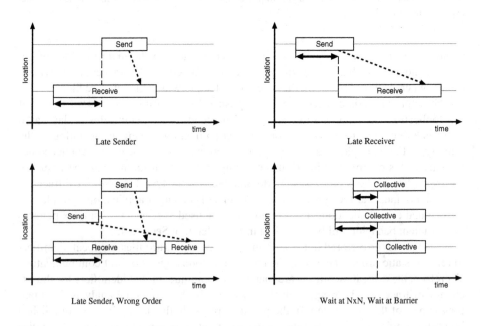

Fig. 2. Patterns of inefficient behavior

4.1   Point-to-Point Communication

As an example for inefficient point-to-point communication, we consider the so-called *Late Sender* pattern. Here, a receive operation is entered by one process before the corresponding send operation has been started by the other. The time lost due to this pattern is therefore the difference between the timestamps of the enter events of the MPI function instances which contain the corresponding message send and receive events. The complete Late Sender pattern consists of four events, specifically the two enter events and the respective message send and receive events.

During the parallel replay, the detection of this performance problem is triggered by the point-to-point communication events involved (i.e., send and receive). That is, when a send event is found by one of the processes, a message containing this event as well as the associated enter event is created. This message is then sent to the process representing the receiver using a point-to-point operation. To ensure the correct matching of send and receive events, we use equivalent tag and communicator information to perform the communication.

When the receiver reaches the receive event, the aforementioned message containing the *remote constituents* of the pattern is received. Together with the locally available constituents (i.e., the receive and the enter events), a Late Sender situation can be detected by comparing the timestamps of the two enter events and calculating the time spent waiting for the sender. This approach relies on the availability of a synchronized clock: otherwise linear interpolation of timestamps [16] is used, but alternative methods of time correction are being considered.

The detection of the *Late Receiver* pattern is very similar and straightforward to implement. However, to avoid sending redundant messages while executing the detection algorithms for the different performance problems related to point-to-point communication, we exploit specialization relationships between patterns and reuse results obtained on higher levels of the hierarchy. This is implemented using a sophisticated event notification and call-back mechanism similar to the publish-and-subscribe approach presented in [4]. For this pattern the severity is calculated by the receiver but attributed to the sender's location. To avoid the additional overhead of transferring the calculated waiting time back to the sender, it is stored as a *remote result* at the receiving process.

By contrast, detecting the *Late Sender, Messages in Wrong Order* pattern is more difficult. This pattern describes the situation that during a Late Sender pattern, another message is waiting to be received by the same destination but which was sent earlier. To detect it, we would need a global view of the messages currently in transit while assessing the Late Sender situation, which is not available in a parallel implementation. Therefore, each analysis process keeps track of the last occurrences of the Late Sender pattern found in its local trace using a ring buffer. If a receive event is encountered during the replay, we compare the timestamps of the corresponding send event and those of the buffered Late Sender occurrences. If the Late Sender's send operation starts after the send event associated with the current receive, the Late Sender instance is classified as a Wrong Order situation and removed from the buffer. Note that this approach does not guarantee to find all occurrences of this pattern, although empirical results suggest that the coverage of our method is sufficient in practice.

4.2 Collective Communication and Synchronization Operations

The second important type of communication operations are MPI collective operations. As an example of a related performance problem, we discuss the detection of the *Wait at* $N \times N$ pattern, which quantifies the waiting time due to the inherent synchronization in N-to-N operations, such as MPI_Allreduce.

While traversing the local trace data, all processes involved in a collective operation will eventually reach their corresponding collective exit events. After verifying that it relates to an N-to-N operation, accomplished by examining the associated region identifier, the analyzer invokes the detection algorithm, which determines the latest of the corresponding enter events using an MPI_Allreduce operation. After that, each process calculates the local waiting time by subtracting the timestamp of the local enter event from the timestamp of the enter event obtained through the reduction operation. The group of ranks involved in the analysis of the collective operation is easily determined by re-using the communicator of the original collective operation.

Very similar algorithms can be used to implement patterns related to 1-to-N, N-to-1 and barrier operations. As with point-to-point operations, a single MPI call is used to calculate the asscociated waiting times. Only barrier operations, for which the analyzer also calculates asymmetries that occur when leaving the operation, require two calls.

5   Results

To evaluate the effectiveness of parallel analysis based on a replay of the target application's communication behavior, a number of experiments with our current prototype implementation have been performed at a range of scales and compared with the sequential EXPERT tool. To facilitate a fair comparison, a restricted version of EXPERT was used that provides only the functionality of our parallel prototype, i.e., support for MPI-2, OpenMP, and SHMEM pattern analysis was disabled.

Measurements were taken on the IBM BlueGene/L system at Forschungszentrum Jülich (JUBL), which consists of 8,192 dual-core 700 MHz PowerPC 440 compute nodes (each with 512 MBytes of memory), 288 I/O nodes, and p720 service and login nodes each with eight 1.6 GHz Power5 processors [17]. The system was running the V1R2 software release with GPFS parallel filesystem configured with 4 servers. A dedicated partition consisting of all of the compute nodes was used for the parallel analyses, whereas the sequential programs (pre- and postprocessing, and EXPERT) ran on the lightly-loaded login node. Two applications with quite different execution and performance characteristics have been selected for detailed comparison.

The ASC benchmark SMG2000 [18] is a parallel semi-coarsening multigrid solver, which uses a complex communication pattern. The MPI version performs a lot of non-nearest-neighbor point-to-point communication operations (and only a negligible number of collective communication operations) and can be considered to be a stress-test for the memory and network subsystems of a machine. To investigate *weak scaling* behavior, a fixed $64 \times 64 \times 32$ problem size per process with five solver iterations was configured, resulting in a nearly constant application run-time as additional CPUs were used. Because the number of events traced for each process increases with the total number of processes, the aggregate trace volume increases faster than linearly.

The second case, PEPC-B [19], uses a locally-developed parallel tree code for computing long-range forces in N-body particle systems applied in this case to beam-plasma interactions. With a fixed problem size consisting of one million charged particles updated for 10 steps, increasing the number of CPUs reduces overall run-time as a demonstration of *strong scaling* behavior. By contrast to the SMG2000 benchmark, it uses a significant proportion of collective communication and synchronization operations.

Figure 3 charts wall-clock execution times for the uninstrumented applications and their analysis with a range of process numbers on JUBL. The 8-fold doubling of process numbers necessitates a log–log scale to show the corresponding range of times, particularly for the old sequential analysis (which furthermore becomes impractical for the largest traces). The figure shows the total time needed for the parallel analysis including the aforementioned sequential steps, the time taken by the parallel analysis without sequential steps, and the time taken by the parallel replay itself without file I/O. Due to the often considerable variation in the time for file I/O (e.g., depending on overall filesystem load) the times reported are the best of several measurements.

While the set of execution traces from 1,024 PEPC-B processes only reached 400 MBytes aggregate size (56 million events in total), the corresponding execution traces from 1,024 SMG2000 processes were 10 GBytes (a total of 1,886 million events). The largest set of execution traces from 16,384 SMG2000 processes amounted to 230 GBytes (over 40,000 million events in total). Both applications have communication characteristics that result in individual process traces being considerably smaller or larger than the average.

File I/O can be seen to command increasing proportions of the analysis time, however, future versions of the parallel analysis will reduce this overhead by parallelizing

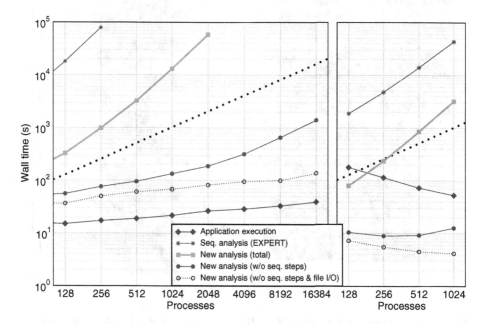

Fig. 3. Execution times for SMG2000 (left) and PEPC-B (right) and their analysis using the sequential EXPERT and new prototype at a range of scales. Linear scaling is the bold dotted line.

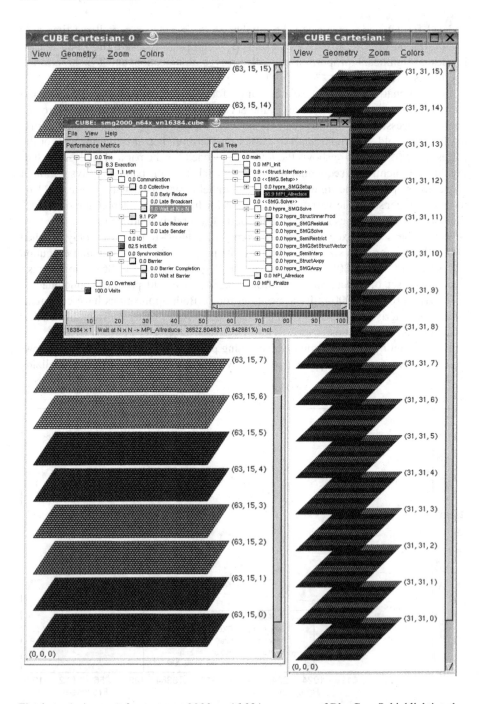

Fig. 4. Analysis report for ASC SMG2000 on 16,384 processors of BlueGene/L highlighting the distribution of the *Wait at N x N* performance metric in the SMG.Setup MPI_Allreduce on the physical machine topology distribution (left) and MPI process topological distribution (right).

the currently sequential pre- and postprocessing steps and thereby eliminating the need to read and write intermediate data files. By contrast, the actual procedure of replaying and analyzing the event traces, the focus of this paper, exhibits a satisfactory scaling behavior up to very large configurations. On account of its replay-based nature, the time needed for this part of the analysis procedure depends on the communication behavior of the target application. Since communication is a key factor in the scaling behavior of the target application as well, similarities can be seen in the way both curves evolve as the number of processes increases.

Notably, the total time for the new analysis approach is already more than one order of magnitude faster than the sequential analysis based on EXPERT, which makes it possible to examine wider (and longer) parallel traces in a reasonable time.

While SMG2000 is a reasonable test case for examining the scaling behavior of performance analysis to large scales, as a well-optimized benchmark application, the analysis results are of little interest (see Figure 4). On the other hand, PEPC-B is a relatively new application which has recently been scaled in size and the performance report shows that communication and load imbalance have become increasingly important issues.

6 Conclusion and Future Work

We have presented a novel approach for automatically analyzing event traces of large-scale applications based on exploiting the distributed memory capacity and the parallel processing capabilities of modern supercomputing systems. Instead of sequentially analyzing a single and potentially large global event trace file, we analyze separate local trace files with an analyzer, that is a parallel application in its own right, replaying the target application's communication behavior. This approach has been elaborated to implement the detection algorithms for a variety of performance problems related to the use of the MPI-1 parallel programming interface. In the future, we plan to add support for additional APIs, such as OpenMP and MPI-2, and will investigate using a smaller number of processes for the replay analysis than were used for the measurement, to provide greater analysis flexibility.

To evaluate the scalability of our approach, we have performed experiments with different applications using our prototype implementation on up to 16,384 CPUs. Although the overall analysis time is currently dominated by the sequential parts of the procedure and associated file I/O, the new approach is already more than one order of magnitude faster than the sequential analysis carried out by the EXPERT tool, thereby enabling analyses at scales that have been previously inaccessible.

Since the remaining sequential overhead can be reduced by integrating and parallelizing the pre- and postprocessing parts to eliminate the need to read and write intermediate data files, these early results point to further improvements that can be realized based on the new approach, as we focus on these parts of the analysis work flow. The all-in-memory analysis (perhaps using cCCGs) will also be explored for opportunities to facilitate the detection of new and more complex performance problems.

References

1. Nagel, W., Weber, M., Hoppe, H.C., Solchenbach, K.: VAMPIR: Visualization and Analysis of MPI Resources. Supercomputer **63, XII**(1) (1996) 69–80
2. Labarta, J., Girona, S., Pillet, V., Cortes, T., Gregoris, L.: DiP : A Parallel Program Development Environment. In: Proc. 2nd Int'l Euro-Par Conf. (Lyon, France), Springer (1996)
3. Wolf, F., Mohr, B.: Automatic performance analysis of hybrid MPI/OpenMP applications. Journal of Systems Architecture **49**(10-11) (2003) 421–439
4. Wolf, F., Mohr, B., Dongarra, J., Moore, S.: Efficient Pattern Search in Large Traces through Successive Refinement. In: Proc. European Conf. on Parallel Computing (Euro-Par, Pisa, Italy), Springer (2004)
5. Wolf, F., Freitag, F., Mohr, B., Moore, S., Wylie, B.: Large Event Traces in Parallel Performance Analysis. In: Proc. 8th Workshop on Parallel Systems and Algorithms (PASA, Frankfurt/Main, Germany). Lecture Notes in Informatics, Gesellschaft für Informatik (2006)
6. Freitag, F., Caubet, J., Labarta, J.: On the Scalability of Tracing Mechanisms. In: Proc. European Conference on Parallel Computing (Euro-Par, Paderborn, Germany). Lecture Notes in Computer Science 2400, Springer (2002)
7. Wu, C.E., Bolmarcich, A., Snir, M., Wootton, D., Parpia, F., Chan, A., Lusk, E., Gropp, W.: From Trace Generation to Visualization: A Performance Framework for Distributed Parallel Systems. In: Proc. SC2000 (Dallas, TX, USA). (2000)
8. Brunst, H., Nagel, W.E.: Scalable Performance Analysis of Parallel Systems: Concepts and Experiences. In: Parallel Computing: Software Technology, Algorithms, Architectures and Applications, Elsevier (2004) 737–744
9. Knüpfer, A., Nagel, W.E.: Construction and Compression of Complete Call Graphs for Post-Mortem Program Trace Analysis. In: Proc. of the International Conference on Parallel Processing (ICPP, Oslo, Norway), IEEE Computer Society (2005) 165–172
10. Roth, P.C., Miller, B.P.: On-line automated performance diagnosis on thousands of processes. In: ACM SIGPLAN Symposium on Principles and Practice of Parallel Programming (PPoPP'06, New York City, NY, USA). (2006)
11. Fürlinger, K., Gerndt, M.: Distributed Application Monitoring for Clustered SMP Architectures. In: Proc. 9th Int'l Euro-Par Conf. (Klagenfurt, Austria), Springer (2003)
12. Fahringer, T., Gerndt, M., Mohr, B., Wolf, F., Riley, G., Träff, J.L.: Knowledge Specification for Automatic Performance Analysis. Technical Report FZJ-ZAM-IB-2001-08, ESPRIT IV Working Group APART, Forschungszentrum Jülich (2001) Revised version.
13. Fahringer, T., Seragiotto, Jr., C.: Modelling and Detecting Performance Problems for Distributed and Parallel Programs with JavaPSL. In: Proc. SC2001 (Denver, CO, USA). (2001)
14. Jorba, J., Margalef, T., Luque, E.: Performance Analysis of Parallel Applications with KappaPI 2. In: Proc. Parallel Computing 2005 (ParCo, Málaga, Spain). (2006)
15. Song, F., Wolf, F., Bhatia, N., Dongarra, J., Moore, S.: An Algebra for Cross-Experiment Performance Analysis. In: Proc. Int'l Conf. on Parallel Processing (ICPP, Montreal, Canada), IEEE Computer Society (2004)
16. Wolf, F.: Automatic Performance Analysis on Parallel Computers with SMP Nodes. PhD thesis, RWTH Aachen, Forschungszentrum Jülich (2003) ISBN 3-00-010003-2.
17. The BlueGene/L Team at IBM and LLNL: An overview of the BlueGene/L supercomputer. In: Proc. SC2002 (Baltimore, MD, USA), IEEE Computer Society (2002)
18. Advanced Simulation and Computing Program: The ASC SMG2000 Benchmark Code. http://www.llnl.gov/asc/purple/benchmarks/limited/smg/ (2001)
19. Gibbon, P.: PEPC: A Multi-Purpose Parallel Tree-Code. http://www.fz-juelich.de/zam/pepc/ (2005)

TAUg: Runtime Global Performance Data Access Using MPI

Kevin A. Huck, Allen D. Malony, Sameer Shende, and Alan Morris

Performance Research Laboratory
Department of Computer and Information Science
University of Oregon, Eugene, OR, USA
{khuck, malony, sameer, amorris}@cs.uoregon.edu
http://www.cs.uoregon.edu/research/tau

Abstract. To enable a scalable parallel application to view its global performance state, we designed and developed *TAUg*, a portable runtime framework layered on the TAU parallel performance system. TAUg leverages the MPI library to communicate between application processes, creating an abstraction of a global performance space from which profile views can be retrieved. We describe the TAUg design and implementation and show its use on two test benchmarks up to 512 processors. Overhead evaluation for the use of TAUg is included in our analysis. Future directions for improvement are discussed.

Keywords: parallel, performance, runtime, MPI, measurement.

1 Introduction

Performance measurement of parallel applications balances the need for fine-grained performance data (to understand relevant factors important for improvement) against the cost of observation (measurement overhead and its impact on performance behavior). This balance becomes more delicate as parallel systems increase in scale, especially if the scalability of the performance measurement system is poor. In practice, measurements are typically made for post-mortem analysis [1,2], although some tools provide online monitoring[3] and analysis for purposes of performance diagnosis [4,5] and steering [6,7,8,9]. For any performance experiment, the performance measurement system is an intimate part of the application's execution and need/cost tradeoffs must take this into account.

Scalable efficiency necessitates that performance measurements be made concurrently (in parallel threads of execution) without centralized control. The runtime *parallel performance state* can be considered to be logically a part of the application's global data space, but it must be stored distributively, local to where the measurements took place, to avoid unnecessary overhead and contention. Measurement tools for post-mortem analysis typically output the final performance state at the end of program execution. However, online tools require access to the distributed performance state during execution.

B. Mohr et al. (Eds.): PVM/MPI 2006, LNCS 4192, pp. 313–321, 2006.

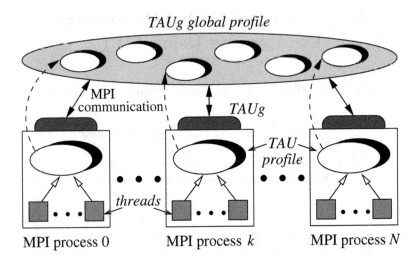

Fig. 1. TAUg System Design

In this paper, we consider the problem of runtime support for application-level access to global parallel performance data. Our working assumption is that the importance of online performance data access is decided by the application, but will depend directly on the efficiency of the solution. It is equally important that the solution be as portable as possible, flexible enough to accommodate its use with different parallel computation scenarios, and scalable to large numbers of processes. The main challenges are in defining useful programming abstractions for coordinated performance data access, and in creating necessary infrastructure that meets portability and efficiency objectives.

We describe a solution for use with the TAU parallel performance system called *TAUg*. The TAUg design targets MPI-based applications (see §2) and utilizes MPI in its default implementation (see §3) for portability and scalability. The initial version of TAUg was tested with ASCI benchmarks sPPM and Sweep3D and a synthetic load balancing simulation. The results are reported in §4. Discussion of the TAUg approach and our future goals are discussed in §5. Related work is in §6, and §7 gives concluding remarks.

2   Design

In our approach to the TAUg system design, we first identified the desired operational abstraction, and second, considered how best to implement it with MPI. Figure 1 shows these two perspectives. The bottom part of the figure represents what TAU produces as a profile for each process. The TAU profile is an aggregation of individual thread profiles. TAUg provides the abstraction of a globally-shared performance space, the TAUg global profile. The dashed lines represent the promotion of each process profile into this space. TAUg uses MPI to create this global abstraction on behalf of the application.

2.1 Views and Communicators

In TAU, events are defined for measurement of intervals (e.g., entry and exit events for subroutines) or atomic operations (e.g., memory allocation events). In TAUg, the *global performance space*, representing all events (interval and atomic events) profiled on all processes and threads, is indexed along two dimensions. The first dimension is called the *TAUg (global) performance view*, and represents a subset of the performance profile data being collected (i.e., a subset of the TAU profiled events). In our initial implementation, a view can specify only one event, whose profile gives the performance for that event measured when the event is active. The other dimension is called the *TAUg (global) performance communicator*, and represents a subset of the MPI processes in the application. The notion of the TAUg communicator is that only those processes within the communicator will share TAUg performance views, so as to minimize perturbation of the application.

2.2 Programming Interface

TAUg is designed to be a simple, lightweight mechanism for sharing TAU performance data between MPI processes. The only prerequisites for using TAUg are that the application already be using MPI and TAU. The three methods in the API are designed to be in the same style as MPI methods. These methods are callable from Fortran, C or C++.

An application programmer uses TAUg by first defining the global performance views and communicators. The method `TAU_REGISTER_VIEW` is used to specify a global performance view. This method takes as an input parameter the name of a TAU profiled event, and has an output parameter of an ID for the view. `TAU_REGISTER_VIEW` need only be called by processes that will use the view with TAUg communicators they define.

The method `TAU_REGISTER_COMMUNICATOR` is used to create a global performance communicator. It takes two input parameters; an array of process ranks in `MPI_COMM_WORLD` and the size of the array. The only output parameter is the newly created communicator ID. Because of MPI requirements when creating communicators, `TAU_REGISTER_COMMUNICATOR` must be called by all processes. The following code listing shows an example of how the `TAU_REGISTER_VIEW` and `TAU_REGISTER_COMMUNICATOR` methods would be used in C to create a global performance view of the event `calc()` and a global performance communicator containing all processes.

```
int viewID = 0, commID = 0, numprocs = 0;
TAU_REGISTER_VIEW("calc()", &viewID);
MPI_Comm_size(MPI_COMM_WORLD,&numprocs);
int members[numprocs];
for (int i = 0 ; i < numprocs ; i++) { members[i] = i; }
TAU_REGISTER_COMMUNICATOR(members, numprocs, &commID);
```

Having created all the global performance views and communicators needed to access the global application performance, the application programmer calls the method `TAU_GET_VIEW` to retrieve the data. This method takes a view ID and a communicator ID as input parameters. It also takes a collective communication type as an input parameter. The idea here is to allow TAU communicators to pass profile data between themselves in different ways. The supported communication types are `TAU_ALL_TO_ONE`, `TAU_ONE_TO_ALL` and `TAU_ALL_TO_ALL`. If `TAU_ALL_TO_ONE` or `TAU_ONE_TO_ALL` are used, a processor rank in `MPI_COMM_WORLD` will represent the source or sink for the operation[1]. There are two output parameters which specify an array of doubles and the size of the array. `TAU_GET_VIEW` need only be called by the processes which are contained in the specified TAU global performance communicator. The following code listing shows an example of how the `TAU_GET_VIEW` method would be used in C.

```
double *loopTime;
int size = 0, myid = 0, sink = 0;
MPI_Comm_rank(MPI_COMM_WORLD,&myid);
TAU_GET_VIEW(viewID, commID, TAU_ALL_TO_ONE, sink,
    &loopTime, &size);
if (myid == 0) { /* do something with the result... */ }
```

In summary, this application code is requesting that all processes send performance information for the event `calc()` to the root process. The root process, for example, can then choose to modify the application behavior based on the running total for the specified event.

3   Implementation

TAUg is written in C++, and comprises a public C interface consisting of only the three static methods described in Section 2.2. The complete interface for the API is listed here:

```
void static TAU_REGISTER_VIEW (const char* event_name,
    int* viewID);
void static TAU_REGISTER_COMMUNICATOR (int members[],
    int size, int* commID);
void static TAU_GET_VIEW (int viewID, int commID,
    int type, int sink, double** data, int* outSize);
```

The `TAU_REGISTER_VIEW` method creates a new global performance view structure, and stores it internally. The new view ID is returned to the calling method. The `TAU_REGISTER_COMMUNICATOR` method creates new MPI group and communicator objects which contain the input process ranks, assumed to be relevant in `MPI_COMM_WORLD`. It then stores the MPI communicator ID and all the communicator parameters internally, and returns the new communicator ID (not to be confused with the MPI communicator type) to the calling method.

The `TAU_GET_VIEW` method first looks up the global performance view and communicator in the internal structures. At the same time, the code converts

---

[1] If the `TAU_ALL_TO_ALL` type is specified, the source/sink parameter is ignored.

the source/sink process rank from MPI_COMM_WORLD to its rank in the global performance communicator. The method then accesses TAU to get the profile data for the global performance view. The profile data includes the inclusive and exclusive timer values, number of calls and number of subroutines (events called from this event). This data is then packaged in an MPI type structure and sent to the other processes in the global performance communicator using collective operations. Either MPI_Allgather, MPI_Gather or MPI_Scatter is called, depending on whether the application wants TAU_ALL_TO_ALL, TAU_ALL_TO_ONE or TAU_ONE_TO_ALL behavior, respectively. In the initial implementation, an array of only the exclusive timer values is returned to the user as a *view result*.

4   Experiments

4.1   Application Simulation

TAUg was integrated into a simple simulation program to demonstrate its effectiveness in dynamically load balancing an MPI application. This simulation is intended to replicate general situations where factors external to the application are affecting performance, whether it be hardware differences or other load interference on a shared system. In this experiment, the application program simulates a heterogeneous cluster of n processors, where $n/2$ of the nodes are twice as fast as the other $n/2$ processors.

Initially, each MPI process is designated an equal portion of the work to execute. After each timestep, the application code queries TAUg to get a global view of the application performance. Processes which are slower than the average are given a reduced workload, and the processes which are faster than the average are given an increased workload. This process is iterated 20 times. The application was tested with 5 configurations. Initially, an unbalanced version of the application was tested and compared to a dynamically balanced version. It soon became apparent that different lengths of performance data "decay" are necessary to detect when the load has become balanced, so that the faster nodes are not overburdened simply so that the slower nodes can catch up. Therefore, three more configurations were tested, which used only the previous 1, 2, and 4 timesteps, respectively. Using the unbalanced application as a baseline for the 32 processor simulation, the dynamically balanced simulation is 15.9% faster, and the dynamically balanced simulation which only considers the previous 1 timestep is 26.5% faster. Longer running simulations show similar speedup.

This simple example demonstrates that TAUg can be used to implement the knowledge portion of a load balancing algorithm. In general, load imbalance is reflected in performance properties (execution time and even more detailed behavior), but is caused by and associated with application-specific aspects (such as poor grid partitioning). TAU can be used to measure both performance and application aspects. TAUg then provides an easy-to-use interface for retrieving the information in a portable way.

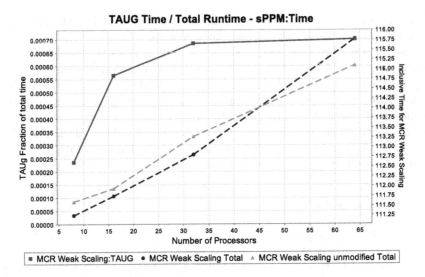

Fig. 2. Comparison of sample execution times from modified and unmodified sPPM, and fraction of time spent in TAUg. Examining the Y-axis on the right to compare total runtime measurements, the application is not significantly affected by the addition of TAUg.

4.2   Overhead and Scalability: sPPM and Sweep3D

The sPPM benchmark[10] solves a 3D gas dynamics problem on a uniform Cartesian mesh using a simplified version of the PPM (Piecewise Parabolic Method) code. We instrumented sPPM with TAU, and TAUg calls were added to get a global performance view for each of 22 subroutines in the application code, for each of 8 equal sized communicators. The sPPM benchmark iterates for 20 double timesteps, and at the end of each double timestep, sPPM was modified to request global performance data for each global performance view / communicator tuple. [2]

We ran sPPM on MCR, a 1,152 node dual P4 2.4-Ghz cluster located at Lawrence Livermore National Laboratory. Using a weak scaling test of up to 64 processors, TAUg total overhead never exceeds 0.1% of the total runtime, and the application is not significantly perturbed. Figure 2 shows the comparison of the modified and unmodified sPPM performance.

ASCI Sweep3D benchmark [11] is a solver for the 3-D, time-independent, neutron particle transport equation on an orthogonal mesh. Sweep3D was instrumented with TAU, and TAUg calls were added to get a global performance view for one of the methods in the application code and one communicator consisting of all processes. The Sweep3D benchmark iterates for 200 timesteps, and at the end of each timestep, Sweep3D was modified to request global performance data for the global performance view / communicator tuple.

---

[2] This resulted in 22 calls to TAU_GET_VIEW since only one subroutine event can be in a view in the current version.

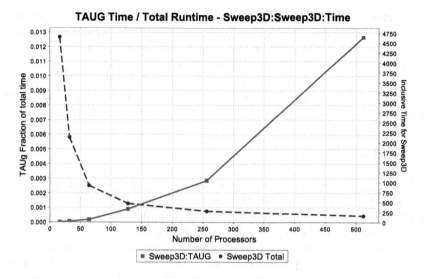

Fig. 3. Fraction of TAUg overhead as measured in Sweep3D in a strong scaling experiment

We ran Sweep3D on ALC, a 960 node dual P4 2.4-Ghz cluster located at Lawrence Livermore National Laboratory. During a strong scaling test of up to 512 processors, TAUg total overhead never exceeded 1.3% of the total runtime, and the application was not significantly perturbed. Figure 3 shows the comparison of the modified Sweep3D performance to the TAUg overhead.

5   Discussion and Future Work

There are several issues to discuss with respect to the current TAUg system as well as areas where we are interested in improving the TAUg design and implementation. Currently, TAUg limits global access to the exclusive value of a single event for a single metric. We will add support for specifying multiple events in a TAUg view and an *all* tag for easily specifying all events. Similarly, hardware performance counter information may be useful to many applications using TAUg, such as floating point operations or cache misses. This information is currently available in TAUg, but only one metric is available at a time. TAU supports tracking multiple concurrent counters, and TAUg will be extended to support this as well. We also will allow TAU_GET_VIEW to be called with an array of views.

The TAUg communication patterns cover what we felt were common use cases. However, they translate into collective MPI operations in TAUg. We believe there will be value to supporting TAUg send and receive operations, to allow more pairwise performance exchange, still within a TAU communicator. This will also allow the opportunity for blocking and non-blocking communication to be used in TAUg. We will also experiment with one-sided communication in MPI-2 to reduce the effects of our current collective operation approach.

Presently, TAUg returns only the raw performance view to the application. We plan to implement TAUg helper functions to compute profile statistics that are typically offered post-mortem (e.g., mean, min, max, and standard deviation for a performance metric). One particularly useful function would take two compatible view results and calculate their difference. This would help to address a problem of calculating incremental global performance data from the last time it was viewed.

6   Related Work

TAUg has similarities to research in online performance analysis and diagnosis. Autopilot [6] uses a distributed system of sensors to collect data about an application's behavior and actuators to make modifications to application variables to change its behavior. Peridot [12] extends this concept with a distributed performance analysis system composed of agents that monitor and evaluate hierarchically-specified "performance properties" at runtime. The Distributed Performance Consultant in Paradyn [5], coupled with MRNet, provides a scalable diagnosis framework to achieve these goals. Active Harmony [13] takes one step further to include a component that automatically tunes an application's performance by adjusting application parameters.

While TAUg can be used to achieve the same purpose of performance diagnosis and online tuning, it focuses as a technology only on the problem of portable access to global performance data. In this way, its use is more general and can be applied more robustly on different platforms.

7   Conclusion

Measurement of parallel program performance is commonly done as part of a performance diagnosis and tuning cycle. However, an application may desire to query its runtime performance state to make decisions that direct how the computation will proceed. Most performance measurement systems provide little support for dynamic performance data access, much less for performance data inspection across all application processes. We developed TAUg as an abstraction layer for parallel MPI-based applications to retrieve performance views from a global performance data space. The TAUg infrastructure is built on top of the TAU performance system which generates performance profiles for each application process. TAUg uses MPI collective operations to provide access to the distributed performance data.

TAUg offers two important benefits to the application developer. First, the TAUg programming interface defines the TAU *communicator* and *view* abstractions that the developer can use to create instances specific to their runtime performance query needs. The `TAU_GET_VIEW` function will return the portion of the global performance profiles selected by the communicator and view parameters. As a result, the developer is insulated from the lower level implementation. Second, the use of MPI in TAUg's implementation affords significant portability, and the scalability of TAUg is only limited by the scalability of the local MPI implementation. Any parallel systems supporting MPI and TAU are candidates for use of TAUg.

It is true that TAUg will necessarily influence the application's operation. We provide some analysis of the overhead generated by TAUg in our benchmark tests. However, the impact of TAUg will depend directly on how the application chooses to use it. This impact is true both of its perturbation of performance as well as its ability to provide the application with performance knowledge for runtime optimization.

References

1. Shende, S., Malony, A.D.: The tau parallel performance system. The International Journal of High Performance Computing Applications (2005) *(to appear)*.
2. KOJAK: Kojak. http://www.fz-jeulick.de/zam/kojak/ (2006)
3. Wismuller, R., Trinitis, J., Ludwig, T.: Ocm – a monitoring system for interoperable tools. In: Proceedings 2nd SIGMETRICS Symposium on Parallel and Distributed Tools (SPDT'98). (1998) 1–9
4. Miller, B., Callaghan, M., Cargille, J., Hollingsworth, J., Irvin, R., Karavanic, K., Kunchithapadam, K., Newhall, T.: The paradyn parallel performance measurement tool. Computer **28**(11) (1995) 37–46
5. Roth, P., Miller, B.: On-line automated performance diagnosis on thousands of processes. In: Proceedings Proc. 11th ACM SIGPLAN Symposium on Principles and Practice of Parallel Programming. (2006) 69–80
6. Ribler, R., Simitci, H., Reed, D.: The Autopilot performance-directed adaptive control system. Future Generation Computer Systems **18**(1) (2001) 175–187
7. Eisenhauer, G., Schwan, K.: An object-based infrastructure for program monitoring and steering. In: Proceedings 2nd SIGMETRICS Symposium on Parallel and Distributed Tools (SPDT'98). (1998) 10–20
8. Gu, W., et al.: Falcon: On-line monitoring and steering of large-scale parallel programs. In: Proceedings of the 5th Symposium of the Frontiers of Massively Parallel Computing. (1995) 422–429
9. Tapus, C., Chung, I.H., Hollingworth, J.: Active harmony: Towards automated performance tuning. In: SC '02: Proceedings of the 2002 ACM/IEEE conference on Supercomputing. (2002)
10. LLNL: The asci sppm benchmark code.
    http://www.llnl.gov/asci/purple/benchmarks/limited/sppm/ (2006)
11. LLNL: The asci sweep3d benchmark.
    http://www.llnl.gov/asci/purple/benchmarks/limited/sweep3d/ (2006)
12. Gerndt, M., Schmidt, A., Schulz, M., Wismuller, R.: Performance analysis for teraflop computers - a distributed automatic approach. In: Euromicro Workshop on Parallel, Distributed, and Network-based Processing (PDP), Canary Islands, Spain (2002) 23–30
13. Hollingsworth, J., Tabatabaee, V., Tiwari, A.: Active harmony. http://www.dyninst.org/harmony/ (2006)

Tracing the MPI-IO Calls' Disk Accesses

Thomas Ludwig, Stephan Krempel, Julian Kunkel,
Frank Panse, and Dulip Withanage

Ruprecht-Karls-Universität Heidelberg
Im Neuenheimer Feld 348, 69120 Heidelberg, Germany
t.ludwig@computer.org
http://pvs.informatik.uni-heidelberg.de/

Abstract. With parallel file I/O we are faced with the situation that we do not have appropriate tools to get an insight into the I/O server behavior depending on the I/O calls in the corresponding parallel MPI program. We present an approach that allows us to also get event traces from the I/O server environment and to merge them with the client trace. Corresponding events will be matched and visualized. We integrate this functionality into the parallel file system PVFS2 and the MPICH2 tool Jumpshot.

Keywords: Performance Analyzer, Parallel I/O, Visualization, Trace-based Tools, PVFS2.

1 Introduction

We now see more and more applications that deploy parallel I/O at different abstraction levels. They use a parallel file system from sequential or parallel programs with regular read/write calls or they already implement parallel I/O with libraries at the level of MPI-IO or even higher. Often, the time spent for file I/O is a significant percentage of the program's execution time and thus optimization is an important issue.

Parallel I/O is a complex concept. It involves client processes which usually form a parallel program using MPI for message passing and MPI-IO for disk access. Every I/O call refers to a set of servers which provide a parallel file system for persistent data storage. Usual concepts deploy a striping scheme (i.e. RAID-0 level) to distribute the data of a logical file onto physical files on several disks. Distribution functions control where the data goes to. Thus, for every I/O call in the program the client library determines on which server the needed byte chunks are located and issues appropriate requests to these servers. As we have parallel client processes and parallel server processes we see a set of message transfers over the network with every I/O activity in the program [4].

The mapping of clients and servers onto the nodes of a cluster is crucial for the overall performance we can get from I/O-bounded applications. In order to optimize this mapping we need to get concise performance measures from our client/server environment. There are tools that visualize the performance of the

B. Mohr et al. (Eds.): PVM/MPI 2006, LNCS 4192, pp. 322–330, 2006.

I/O servers, like e.g. Karma that comes with PVFS2. Karma is an on-line tool that shows the I/O activity during a short sample time period. Unfortunately there is no tracing facility integrated, thus values cannot easily be saved for further investigation.

The main problem that prevents us from getting a deeper insight into our environment's behavior is the fact that there are no tools that can visualize the servers' activities in depence of the clients' I/O calls. We would like to see the relations of the two of them in order to improve certain aspects of parallel I/O. This of course must be supported by a trace-based tool environment as we need the data for analysis after program completion.

What would be the benefits of such a tool? First, the I/O system developers can use it in order to improve internal concepts of their middle-ware. Bottle-necks in the implementation will be visible that would otherwise be hidden. For application programers it will be interesting to see the effects of their I/O calls with respect to server activity. They can change the distribution function of the files in order to balance the load on the servers manually. Future mechanisms could use these results for automatic rebalancing of the file distribution.

What data should the tool present to us? Currently, trace visualizers show relevant events of client processes on so-called time lines. We want to add lines for the I/O servers of our environment and visualize them together with the client processes' time lines. As for MPI the idea behind the I/O API was to make it similar to the message passing API. Thus, reading is like receiving and writing like sending. Regular tools show the relations of sending and receiving e.g. with arrows in a trace visualization tool. We would like to have the same for reading and writing and the corresponding system activities.

At the moment there are no such tools for several reasons: First, there is no tracing of I/O servers. Second, even with traces being provided, we need a correlation to the events in the client processes. This asks for adaptations on both sides. Third, there is no appropriate tool available that can visualize both information sets in an adequate way.

This is where our project starts: We develop and implement a tool environ-ment that can visualize client and server I/O activities at the same time and correlate them. The implementation is based on PVFS2 in combination with MPICH2 and Jumpshot.

The remainder of this paper is structured as follows: Section 2 will describe related work and the state-of-the-art. Section 3 will explain the environment followed by Section 4 that gives an overview over the project. Section 5 will describe the components of our tool environment and Section 6 will show an example.

2   Related Work and State-of-the-Art

With parallel computing we find two classes of tool concepts which are used for different purposes: First there are on-line tools. They use a monitoring system to get data out of the running application and/or instrument this application.

Data is immediately used for several purposes: we can either display them (e.g. with a performance analyzer) or use them for controlling the application (e.g. with a load balancer). Usually data is not stored. A well-known representative of this class is Paradyn [8,5]. Second, we have off-line tools. They present data after program completion (or with a considerable delay during program run). In order to do so the monitoring systems and instrumentations write event traces to files that are used afterwards as data base. Representatives are TAU [11,10], the Intel Trace Analyzer (formerly marketed as Vampir) [2], XMPI [15], and Jumpshot [9]. Our work is for the family of trace-based tools, in particular for performance analysis. So we will concentrate on these aspects here.

With performance analysis tools (on- and off-line) there was considerable progress in the last recent years. In particular we see a focus towards automatic detection of bottlenecks. The working group Apart (Automatic Performance Analysis: Real Tools) [1] has investigated this issue in depth and several tool enhancements were developed over the years. Visualization gets more sophisticated and several tools allow to compare traces from different program runs. We see also other sources of information being integrated into the trace: TAU enters events from performance counters and measures provided by the operating system.

What is still missing are two things: An explicit tracing and visualization of the I/O system's behavior and a correlation of program events and system events. We will refer to these features as multi-source tracing and semantical trace merging.

With trace-based tools we see two typical and crucial problems: Time synchronization of events from sources at different physical locations and total size of the trace files. Both issues are important and there is research available on each of them. However, in our project we postpone these problems as we concentrate first on new categories of functionality.

3   The Parallel I/O Environment

Our environment is composed of two packages: PVFS2 [12], which provides the parallel file system, and MPICH2 [6], which implements MPI and some tools. MPICH2 comprises MPE which includes the tracing environment for MPI programs and the Jumpshot tool for trace visualization. Our enhancements will bring modifications to several of these packages.

To better understand the details of a running MPICH2/PVFS2 system let us have a closer look at PVFS2 itself. PVFS2 has a layered architecture illustrated in Figure 1. Interfaces for the layers use a non-blocking semantics. The user-level interface provides a high abstraction to a PVFS2 file system. Currently, there are integrations with MPI-IO and the kernel VFS available. The system interface API provides functions for the direct manipulation of file system objects and hides internal details from the user. Invoking a request starts a dedicated statemachine processing the operation in small steps. Statemachines break complex requests into several states each representing an atomic operations. Clients

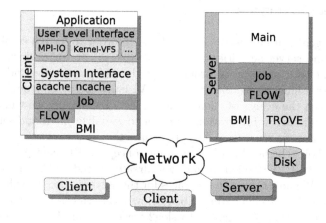

Fig. 1. PVFS2 software architecture

and servers can interlock the execution of these operations to obtain a time-shared processing of different requests. A specific execution order is chosen to ensure that a client crash has no impact on the metadata consistency.

The layers incorporate two caches, which store informations about the directory hierarchy and object attributes to avoid redundant server requests. The job layer consolidates the lower layers into one interface and maintains thread functions for these layers. Data of a larger I/O operation is directly transferred between two endpoints by Flow. An endpoint is one of memory, network, or persistency layer. Flow takes care of the data transmission itself once the endpoints are specified. The Buffered Message Interface (BMI) provides a network independent interface. Clients communicate with the servers by using the request protocol, which defines the message layout for every request. BMI can use different communication methods, currently TCP, Myricom's GM and Infiniband. On the server side a main process decodes incoming requests and starts a new instance of the request's dedicated statemachine. Trove is the persistency layer providing methods for manipulation of key/value pairs (used for metadata) and data streams. BMI, Flow and Trove are modular and the actual implementation can be chosen by the user. Currently, there is only one Trove module available, database plus file (DBPF), which stores metadata in Berkeley databases and data in Unix files. See [13] for details on PVFS2 internals.

Looking at the figure we can identify different processes: the server processes and processes that contain the client library. The latter are the processes of our parallel program and are linked with the MPI-library. As for tracing this has the following consequence: We can get traces from the client processes via the MPE tracing facility of MPICH2. Traces are written in clog2/slog2 format. What we want to have in addition is a trace of the server processes that shows PVFS2 activities at the module level, i.e. in particular of the Trove module, as it does the real file I/O to disk, but also the other modules are of interest. The next section will give details on this.

4   Project Overview

The project goal can be described as follows: In order to have a combined view onto client and server activities we must be able to get traces during program runtime from both of them. In order to see the server activities induced by an MPI-IO call we must relate the two traces. Visually this will be done by adding arrows between the events of clients and servers. Technically we will merge the two traces before arrow integration. In order to visualize client and server traces at the same time we must have an appropriate tool.

Figure 2 shows the block structure of our concept. MPI client processes which are linked to the PVFS2 client library get already traced by the MPICH2/MPE environment. We now need to have traces for the servers, too. We will use a similar approach as for the clients (described below). In both case we get at first clog2 node local traces that are merged to slog2 traces for all clients and all servers respectively. These two traces now get merged and visualized concurrently in the Jumpshot tool.

The list of issues to be covered in order to visualize this double trace with additional arrows comprises the following points:

  – Generate a single trace from the server processes.
  – Forward information from the clients to the servers that allow to find corresponding MPI-IO and Trove calls later on.
  – Merge the two traces from all clients and all servers.
  – Add arrow elements to the combined trace.
  – Distribute client and server activities over time lines in the visualizer.

The implementation of our enhancements takes places at various locations: We need to instrument the PVFS2 server code and add own code to the client library. Server trace generation will be described later. For the two slog2 trace files we have several tools implemented to manipulate them. They are included

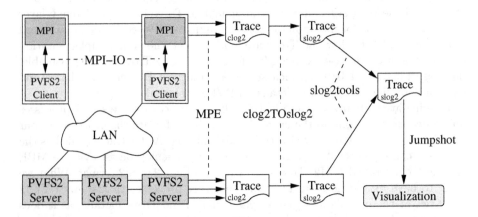

Fig. 2. Architecture of the tool environment

in our package *slog2tools* that will be released in summer 2006. As Jumpshot just visualizes time lines without knowing anything about their meaning and context we could even live without any modifications of this tool. However, for better adaptation to our project goals we slightly enhance the GUI. As this is work in progress it will not be described here.

5   Tool Components

This section will give a more detailed overview over implemented components and enhancements.

Double Trace Generation. We need two event traces, one from the parallel program, the other from the parallel file system. For the client processes we use the MPICH2/MPE environment that provides trace generation and visualization with Jumpshot. As for the servers we use a small trick: we start the PVFS2 server environment as an MPI program. This gives us the potential to generate traces from the server pocesses in clog2 format and merge them to a single slog2 format trace after program completion. A small patch allows us to perform this execution via MPI. As the server environment runs permanently, the client program however does not, we still need a small patch to start and stop the server tracing and trigger the final trace conversion. The server trace has at least the same time segment as the client trace.

I/O Server Events. The I/O servers have a complex internal structure and there are several places in the code where a tracing could be advantageous. From all the software layers in PVFS2 we will at the moment concentrate on the Trove layer. It is responsible for the disk access and thus is related to the load of physical I/O. The tricky issue here is that one client request causes several activities in several servers that will overlap with other activities from other requests. The orders of execution might be changed and the requests trigger different events in different layers. It is of extreme importance to always keep track of what event belongs to which request. We instrument the server environment to write relevant events from all major modules to the trace. For technical details refer to [7].

Trace Merging. Both traces have now to be merged. The program trace is limited in its extend whereas for the servers' trace we have to provide a mechanism to select the correct section of all traced events. Remember that they act as daemons, i.e. have a start and stop time that is not related to that of the MPI program. The two corresponding parts found we have to merge them into a single trace. In addition we need some adaptation of the time lines of the clients' and the servers' trace. We end up with a single trace that has as many horizontal time lines as we had client and server processes.

This combined trace can already be visualized by Jumpshot. The tool shows just the horizontal line with their events and does not distinguish between lines that belong to user processes and to I/O server processes. No connecting arrows between client and server events are included in the trace. Due to the asynchronous request processing in the servers we will see overlapping event visualizations that prevent us from getting a clear view onto the server behavior.

Relate Abstraction Layers. We want to see the relation between program calls and server activities by drawing arrows between the corresponding events. For this to be possible we need the appropriate information to be added into the traces. The concept is as follows: For every MPI-IO call issued we generate a unique call ID and make sure that this information is also written to the client processes' trace. When passing requests from a client to the servers these IDs have to be transferred, too. Normally the server is not interested in the MPI-IO call the request was generated from, however, now the situation changes. For every server activity of interest that gets recorded in the trace we need to know to which MPI-IO call it belongs. Thus, the call IDs need to be written to the server trace together with the regular event informations. After having the traces merged we can now easily detect corresponding events from clients and servers. Conceptionally this is simple, however, the modification of two environments like MPICH/MPE and PVFS2 is a challenge. Obviously the changes would be easy to do while doing a redesign of at least PVFS2 together with its client server request protocol. For technical details refer to [3].

Beautifying the Output. Finally we see client and server events and many arrows in our trace visualization. Due to the asynchronous nature of the servers we have many overlaps with the server events. We added another tool that distributes the overlapping events from one timeline onto as many timelines as are needed in order not to have any more overlaps. This results in a nice visualization. For technical details refer to [14].

Trace merging, arrow generation, and the distribution of overlapping events onto several timelines are performed at the level of the slog2 trace files. Thus, no code changes needed to be done with Jumpshot. In order to get trace information and to forward call IDs from clients to server we had to modify the PVFS2 implementation on many places. There is however no modification at the MPICH2 level.

6   A First Example

A first example is shown in the screen dump in Figure 3. The program that produced these events is just a simple MPI program running in two processes and doing write calls. On the left side you see the information that Jumpshot displays in a pop-up window when you click on the MPI call and the corresponding arrows. The right window shows the Trove events (i.e. disk write activities) triggered by an MPI call in another process. We display here only write events on the client and server side. All other events are supressed. We can see clearly, that a single MPI-IO call results in a set of Trove calls being triggered. Other PFVS2 server activities are recorded, too and give a deeper insight into the servers' activities. For clearness they are not shown here.

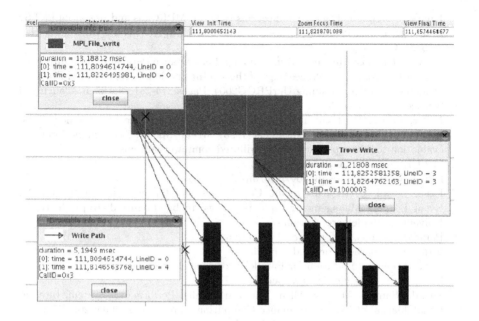

Fig. 3. Two MPI processes invoke write calls (upper two lines) and thus induce trove acticities in the PVFS2 layers (lower two lines). Arrows show the relations between these events.

7   Conclusion and Future Work

The paper presents a new approach to analyze the I/O related performance aspects in an MPICH2/PVFS2 environment. We design and implement an enhancement to PVFS2 that allows us to get meaningful trace information out of the server processes. Server traces get merged with client traces and additional information is added to the trace: We connect corresponding MPI-IO calls and events in the persistency layer of PVFS2 (Trove). Thus, we can observe which calls at program level trigger which activities at system level.

The complete environment is already functional. All components are implemented and interact, however, we still need some time for tuning the components and intensive tests with applications.

In a next step we will integrate the Karma on-line performance values into the server trace such that with the visualization of the events we can afterwards also review important I/O performance values.

References

1. APART (Homepage). http://www.kfa-juelich.de/apart/
2. Intel Trace Analyzer & Collector (Home page).
   http://www.intel.com/cd/software/products/asmo-na/eng/cluster/tanalyzer/index.htm

3. Krempel, Stephan: Tracing Connections Between MPI Calls and Resulting PVFS2 Disk Operations, Bachelor's Thesis, March 2006, Ruprecht-Karls-Universität Heidelberg, Germany.

4. Ludwig, Thomas:. Research Trends in High Performance Parallel Input/Output for Cluster Environments, Proceedings of the 4th International Scientific and Practical Conference on Programming UkrPROG2004, Pages 274-281, National Academy of Sciences of Ukraine, Kiev, Ukraine, 2004.

5. Miller, Barton P. et al.: The Paradyn Parallel Performance Measurement Tool, IEEE Computer 28, 11, (November 1995): 37-46. Special issue on performance evaluation tools for parallel and distributed computer systems.

6. MPICH2 home page (Home page).
   http://www-unix.mcs.anl.gov/mpi/mpich2/index.htm

7. Panse, Frank: Extended Tracing Capabilities and Optimization of the PVFS2 Event Logging Management, Diploma Thesis, to be submitted, Ruprecht-Karls-Universität Heidelberg, Germany.

8. Paradyn Parallel Performance Tools (Home page).
   http://www.paradyn.org/index.html.

9. Performance Visualization for Parallel Programs (Home page).
   http://www-unix.mcs.anl.gov/perfvis/

10. Shende, Sameer; Malony, Allen: The Tau Parallel Performance System, International Journal of High Performance Computing Applications, 2006; 20: 287-311

11. TAU – Tuning and Analysis Utilities (Home page).
    http://www.cs.uoregon.edu/research/tau/

12. The Parallel Virtual File System – Version 2 (Home page).
    http://www.pvfs.org/pvfs2/

13. The PVFS2 Development Team: PVFS2 Internal Documentation included in the source code package (2006).

14. Withanage, Dulip: Performance Visualization for the PVFS2 Environment, Bachelor's Thesis, November 2005, Ruprecht-Karls-Universität Heidelberg, Germany.

15. XMPI – A Run/Debug GUI for MPI (Home page).
    http://www.lam-mpi.org/software/xmpi/

Measuring MPI Send and Receive Overhead and Application Availability in High Performance Network Interfaces

Douglas Doerfler and Ron Brightwell

Center for Computation, Computers, Information and Math
Sandia National Laboratories*
Albuquerque, NM 87185-0817
{dwdoerf, rbbrigh}@sandia.gov

Abstract. In evaluating new high-speed network interfaces, the usual metrics of latency and bandwidth are commonly measured and reported. There are numerous other message passing characteristics that can have a dramatic effect on application performance that should be analyzed when evaluating a new interconnect. One such metric is overhead, which dictates the networks ability to allow the application to perform non-message passing work while a transfer is taking place. A method for measuring overhead, and hence calculating application availability, is presented. Results for several next-generation network interfaces are also presented.

Keywords: MPI, Overhead, Availability, High Performance Computing, High Speed Networks.

1 Introduction

Scaling efficiency of parallel applications in many instances depends on the ability to overlap communication with computation. If there is sufficient computation to overlap with communication, the application becomes insensitive to the bandwidth provided by the network. Overlap is also beneficial for inherently communication bound codes. In this instance the overhead of preparing the next messages can be overlapped with the transmission of the messages already in the send queue. In MPI application codes, the non-blocking send and receive calls are the primary means of achieving overlap. Unlike other MPI communication metrics, e.g. latency and bandwidth, there is a lack of readily available open-source micro-benchmarks that measure MPI overhead for non-blocking calls. This paper presents a method for measuring overhead and application availability and then applies this method to several current state-of-the-art high-performance network interfaces. It is not within the scope of this paper to explain why some interconnects and protocols provide low overhead and high availability.

---

* Sandia is a multiprogram laboratory operated by Sandia Corporation, a Lockheed Martin Company, for the United States Department of Energy's National Nuclear Security Administration under contract DE-AC04- 94AL85000.

B. Mohr et al. (Eds.): PVM/MPI 2006, LNCS 4192, pp. 331 – 338, 2006.

2 Method

There are multiple methods an application can use to overlap computation and communication using MPI. The method assumed by this paper is the post-work-wait loop using the MPI non-blocking send and receive calls, MPI_Isend() and MPI_Irecv(), to initiate the respective transfer, perform some work, and then wait for the transfer to complete using MPI_Wait(). This method is typical of most applications, and hence makes for the most realistic measure of a microbenchmark. Periodic polling methods have also been analyzed [1], but that particular method only makes sense if the application knows that progress will not be made without periodic MPI calls during the transfer. Overhead is defined to be [2]:

> *... the overhead, defined as the length of time that a processor is engaged in the transmission or reception of each message; during this time, the processor cannot perform other operations.*

Application availability is defined to be the fraction of total transfer time[1] that the application is free to perform non-MPI related work.

$$\text{Application Availability} = 1 - (\text{overhead} / \text{transfer time}) \tag{1}$$

Figure 1 illustrates the method used for determining the overhead time and the message transfer time. For each iteration of the post-work-wait loop the amount of work performed (work_t), which is overlapped in time with the message transfer, increases and the total amount of time for the loop to complete (iter_t) is measured. If the work interval is small, it completes before the message transfer is complete. At some point the work interval is greater than the message transfer time and the message transfer completes first. At this point, the loop time becomes the amount of time required to perform the work plus the overhead time required by the host processor to complete the transfer. The overhead can then be calculated by measuring the amount of time used to perform the same amount of work without overlapping a message transfer and subtracting this value from the loop time.

The message transfer time is equal to the loop time before the work interval becomes the dominant factor. In order to get an accurate estimate of the transfer time, the loop time values are accumulated and averaged, but only those values measured before the work interval starts to contribute to the loop time. These values used in the average calculation are determined by comparing the iteration time to a given threshold (base_t). This threshold must be set sufficiently high to avoid a pre-mature stop in the accumulation of the values used for the average calculation, but not so high as to use values measured after the work becomes a factor. The method does not automatically determine the threshold value. It is best to determine it empirically for a given system by trying different values and observing the results in verbose mode. A typical value is 1.02 to 1.05 times the message transfer time.

---

[1] Per the MPI non-blocking call definitions, the MPI_Wait() call only signifies that for a send the buffer can be reused and for a receive the data can be accessed in the receive buffer [3].

Figure 1 also shows an iteration loop stop threshold (iter_t). This threshold is not critical and can be of any value as long as it is ensured that the total loop time is significantly larger than the transfer time. A typical value is 1.5 to 2 times the transfer time. In theory, the method could stop when the base_t threshold is exceeded, but in practice it has been found that this point can be too close to the knee of the curve to provide a reliable measurement. In addition, it is not necessary to calculate the work interval without messaging until the final sample has been taken.

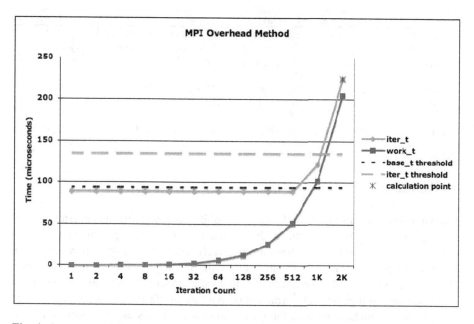

Fig. 1. A conceptual illustration of the post-work-wait loop time (iter_t) of a given message size for each iteration of the algorithm, with the work performed (work_t) increasing for each iteration. The message transfer time calculation threshold (base_t) and the iteration stop threshold (iter_t) are also shown along with the point at which the overhead calculation is taken.

3 Platforms

Overhead and availability was measured on a variety of platforms, summarized in Table 1. All of the platforms except Red Storm are Linux clusters using the respective vendor's commercial software stacks. The Thunderbird cluster's MPI software stack has been modified and parameters have been set to reduce the memory required by the MPI stack at a scale of several hundred to a thousand processes. These modifications do affect the real-world application performance, but it is unknown how those modifications affect the MPI overhead microbenchmark used in this analysis. The Red Storm platform uses the Catamount lightweight kernel [4], with low-level communications implemented using the Portals API [5]. All of the platforms use MPICH 1.x for their implementation of MPI, although several of these implementations have been optimized for their respective network interface. In particular, many vendors have optimized the collective communication routines. The Quadrics software stack uses a patched kernel, which allows optimizations benefiting overhead and host availability performance.

Table 1. Overview of Test Platforms

	Red Storm	*Thunderbird*	*CBC-B*	*Odin*	*Red Squall*
Interconnect	Seastar 1.2	InfiniBand	InfiniBand	Myrinet 10G	QsNetII
Manufacturer	Cray	Cisco/Topspin	PathScale	Myricom	Quadrics
Adaptor	Custom	PCI-Express HCA	InfiniPath	Myri-10G	Elan4
Host Interface	HT 1.0	PCI-Express	HT 1.0	PCI-Express	PCI-X
Programmable coprocessor	Yes	No	No	Yes	Yes
MPI	MPICH-1	MVAPICH	InfiniPath	MPICH-MX	MPICH QsNet

4 Results

From a practical perspective, application availability is usually not a concern for small message sizes, as there is little to be gained trying to overlap computation with communication when transfer times are relatively small. Most applications will only try to overlap computation when they know the message size is sufficiently large. However, as an academic exercise, it still may be interesting to view availability for a small message as it provides information on how an interface's characteristics change at a protocol boundary, such as the switch from a short message protocol to a large message protocol. If an application writer is trying to optimize to a given platform, he/she may want to know where the protocol boundaries are and modify the code to better suit the platform. Overlap may also be beneficial to codes that need to send multiple small messages at a time. In this case, overlap allows preparation of the next message to be put in the queue while the messages already in the queue are being transmitted. However, this is the message throughput metric and is not within the scope of this study.

Figure 2 illustrates the MPI_Isend() overhead as a function of message size for the platforms tested[2]. Figure 3 shows application availability. The overhead for the Red Storm, Odin (Myri-10G) and Red Squall (Elan4) interconnects is relatively constant for all message sizes. As such, application availability increases with message size until it is nearly 100% for large message transfers. The Thunderbird (InfiniBand) and CBC (InfiniPath) interconnects show a high overhead for large message transfers, with a corresponding drop in application availability. It should be noted that the InfiniPath network has a relatively low overhead for small transfers, which allows for that interconnect to achieve its high, advertised message throughput rate.

---

[2] Note that this figure uses a logarithmic axis for overhead.

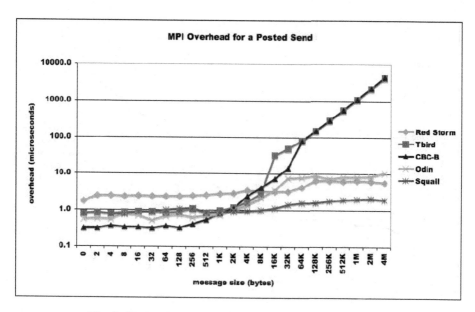

Fig. 2. Overhead as a function of message size for MPI_Isend()

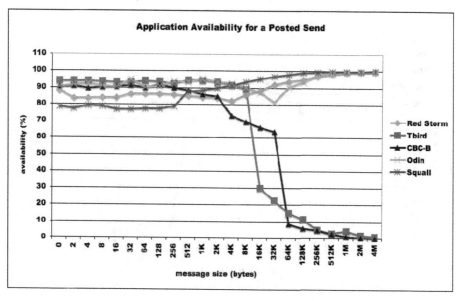

Fig. 3. Application availability as a function of message size for MPI_Isend()

MPI receive performance is charted in Figures 4 and 5. In general, receive performance is similar to the send performance for all of the interconnects tested. The Odin (Myri-10G) cluster does exhibit a more noticeable drop in application availability until the 32K byte message size, which is presumably a protocol boundary. After this point availability increases to an asymptotic value of 100%.

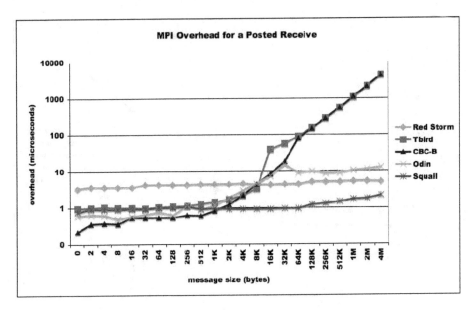

Fig. 4. Overhead as a function of message size for MPI_Irecv()

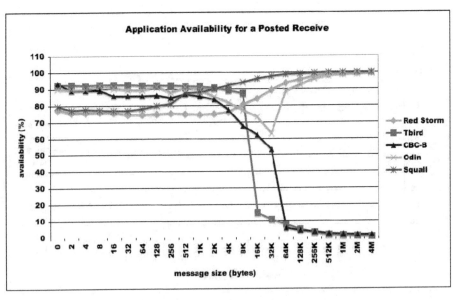

Fig. 5. Application availability as a function of message size for MPI_Irecv()

5 Related Work

A significant amount of prior work has been done to measure and study the effect of overhead on application performance [1], [6], [7], [8] and [9]. Lawry [1] analyzes application availability, but the analysis and results are for a fixed message size and the results are a function of the polling interval. The other previous work does not quantify the overhead as a function of message size, but rather looks at its effect on application performance. An additional contribution of this paper is a comparison of overhead results for relatively new networking technologies, such as Red Storm's SeaStar, Pathscale's InfiniPath, and Myricom's Myri-10G.

6 Conclusion

Simple ping-pong micro-benchmarks do not accurately capture all of the capabilities of a high-performance network. Host overhead and the ability to overlap computation with communication are important performance characteristics that can have a direct impact on an application's scalability. Two networks that have similar latency and bandwidth performance can vary significantly in their ability to provide overlap.

This paper presented a method for measuring overhead and application availability for high-speed networks using MPI and then applied the method to five test platforms, each with a different network interface. Performance for MPI send and MPI receive operations was presented. In general, the send and receive characteristics for a given interconnect were similar. The Red Storm, Odin (Myri-10G) and Red Squall (Elan4) platforms demonstrated a relatively small overhead as a function of message size, and thus showed high application availability for all message sizes. The CBC (InfiniPath) platform demonstrated excellent small message overhead, but for large messages overhead increased linearly with message size and application availability was very low. The Thunderbird (InfiniBand) cluster demonstrated good small message overhead, but like the CBC cluster large message overhead is high and application availability is low.

7 Future Work

It is the intent of the authors to make the source to the code used in this study generally available and downloadable from an open web site, with the hope that this will allow overhead and application availability to become a common micro-benchmark used in the evaluation of interconnects. We also expect that this will encourage contributions from the community to make the code more robust and accurate.

References

1. W. Lawry, C. Wilson, A. Maccabe, R. Brightwell. COMB: A Portable Benchmark Suite for Assessing MPI Overlap. In *Proceedings of the IEEE International Conference on Cluster Computing (CLUSTER 2002)*, p. 472, 2002.
2. D. Culler, R. Karp, D. Patterson, A. Sahay, K. E. Schauser, E. Santos, R. Subramonian and T. von Eicken. LogP: Towards a Realistic Model of Parallel Computation. In *Fourth ACM SIGPLAN symposium on Principles and Practice of Parllel Programming*, pp. 262-273, 1993.

3. M. Snir, S. W. Otto, S. Huss-Lederman, D. W. Walker, J. Dongara. MPI: The Complete Reference. p. 52, The MIT Press, Cambridge, Massachusetts, 1996.
4. S. Kelly, R. Brightwell. Software Architecture of the Light Weight Kernel, Catamount. In *Proceeding of the 47th Cray User Group (CUG 2005)*, 2005.
5. *Portals API*, http://www.cs.sandia.gov/Portals.
6. R. Martin, A. M. Vahdat, D. E. Culler, T. E. Anderson. The Effects of Communication Latency, Overhead, and Bandwidth in a Cluster Architecture. In *Proceedings of the International Symposium on Computer Architecture,* 1997.
7. D. Culler, L. T. Liu, R. P. Martin, C. O. Yoshikawa. Assessing Fast Network Interfaces. *IEEE Micro*, pp. 35-43, Feb., 1996.
8. C. Bell, D. Bonachea, Y. Cote, J. Duell, P. Hargrove, P. Husbands, C. Iancu, M. Welcome, K. Yelick. An Evaluation of Current High-Performance Networks. In *Proceedings IEEE International Parallel & Distributed Processing Symposium (IPDPS '03)*, 2003.
9. R. Brightwell, D. Doerfler, K. D. Underwood. A Preliminary Analysis of the InfiniPath and XD1 Interfaces. In *Proceedings IEEE International Parallel & Distributed Processing Symposium (IPDPS '06)*, 2006.

Challenges and Issues in Benchmarking MPI

Keith D. Underwood*

Sandia National Laboratries
P.O. Box 5800, MS-1110
Albuquerque, NM, 87185-1110
kdunder@sandia.gov

Abstract. Benchmarking MPI is a contentious subject at best. Micro-benchmarks are used because they are easy to port and, hypothetically, measure an important system characteristic in isolation. The unfortunate reality is that it is remarkably difficult to create a benchmark that is a fair measurement in the context of modern system. Software optimizations and modern processor architecture perform extremely efficiently on benchmarks, where it would not in an application context. This paper explores the challenges faced when benchmarking the network in a modern microprocessor climate and the remarkable impacts on the results that are obtained.

1 Introduction

Accurately measuring the MPI performance of a network is challenging. Application codes are notorious for having poor portability. More importantly, full applications are dependent on so many system factors that they obscure any attempt to assess a single system component like the network. As a result, many people turn to microbenchmarks to compare networks. The most widely quoted benchmark — ping-pong latency and bandwidth — are known to bear little resemblance to application codes, and yet they are widely quoted anyway. Other attempts to measure MPI characteristics, such as the Pallas Micro-Benchmark suite (PMB)[2] and the OSU streaming bandwidth test, attempt to address different sets of application characteristics.

The universal problem with microbenchmarks is that they do not account for interference with or interference from the application. A perfect example is found in the collective benchmarks in PMB: hundreds of collective operations are called consecutively. In real applications, there is load imbalance that means that every node does not arrive at the collective at the same time, much as with the Rogue OS effect[9]. Furthermore, applications tend to call collectives one at a time with work between the calls; thus, the code and MPI data structures will not be in cache.

---

* Sandia is a multiprogram laboratory operated by Sandia Corporation, a Lockheed Martin Company, for the United States Department of Energy's National Nuclear Security Administration under contract DE-AC04-94AL85000.

B. Mohr et al. (Eds.): PVM/MPI 2006, LNCS 4192, pp. 339–346, 2006.

This paper, however, is not about a new benchmark. Instead, it discusses system level aspects that make it challenging to write a new benchmark that actually captures the properties of interest. The issues discussed include compiler optimizations, software optimizations, caching effects, and microprocessor optimizations. Quantitative data from previous benchmarks written by the author demonstrate the challenges of getting the benchmark right and creating a "fair" comparison.

2   Related Work

Many benchmarks and metrics of network performance exist. NetPIPE[11] and Netperf[1] are commonly used to measure ping-pong latency and streaming performances, but it is almost as common for individuals (or network vendors) to write their own. With the advent of MPI, there has been surprisingly little published research on more realistic latency measurements. Preliminary work in [10] presented a new latency micro-benchmark that includes the variance in transmission time, which the standard ping-pong benchmark does not reveal. Another previous work[13] considered the impacts of the message queue lengths on message latency.

There have been some attempts at providing a more complete set of micro-benchmarks that better characterize the behavior of real applications and/or expose potential performance advantages that applications may leverage. Microbenchmark suites now attempt to measure the potential for overlap[6] as well as overhead, the impact of buffer re-use, and memory consumption[7]. There have been efforts to model these parameters (LogP[5]) and assess their impact on applications [8]. Other work has attempted to measure the LogGP[3] parameters of modern networks[4]. While all of these efforts are moving toward a more complete picture of network performance, they all have a common, previously unquantified challenge: the system level optimizations that help MPI performance more in the benchmark case than the application case.

3   Benchmarks and Platforms

Table 1 highlights the platforms used in this evaluation. The benchmarks used come from recent efforts to benchmark MPI in a more realistic scenario[13]. The *preposted latency* benchmark[13] is a modification of the classic ping-pong latency benchmark to examine latencies in the presence of preposted non-blocking receives. It builds a posted receive queue of a specified length and inserts a target receive such that when the target message arrives it must traverse a specified percentage of the posted receive queue. The *message rate* benchmark is a variant on the concept of a streaming bandwidth benchmarks. Where traditional streaming bandwidth benchmarks post a long set of receives and then always send a message to the item at the head of the queue, this benchmark recognizes that any application that posts a significant number of receives posts them for

Table 1. Overview of Test Platforms

	Infinipath	**Red Squall**	**Thunderbird**	**Liberty**
Interconnect	4x InfiniPath	Elan-4	4x InfiniBand	Myrinet-2000
Host Interface	HyperTransport	PCI-X	x8 PCI-Express	PCI-X
Link BW	2 GB/s	2.133 GB/s	2 GB/s	500 MB/s
Host BW	6.4 GB/s	1.0 GB/s	4 GB/s	1.0 GB/s
Host Proc.	2.6GHz Opteron	2.2GHz Opteron	3.4GHz Xeon	3.06GHz Xeon
Mem. Speed	dual DDR-400	dual DDR-333	dual DDR-400	dual DDR-266
OS	Fedora 3	SUSE 9.1 Pro	SUSE 9.1 Pro	RHEL 3
Compiler	PathScale 2.3.1	PathScale 2.1	PathScale 2.1	Intel 8.1
MPI Software	InfiniPath 1.3	MPICH 1.24-43	MVAPICH 0.92	MPICH 1.2.6

more than one neighbor. Messages then arrive in an arbitrarily interleaved fashion from those neighbors. To make this happen on two nodes, MPI tags are used to force partial traversal of the list.

4 Analysis and Results

Challenges to writing a good benchmark that operates in a realistic application like environment abound, but they fall into four basic categories: the software, the hardware, the interaction of hardware and software, and the compiler. Basically, they fall in every aspect of the system.

4.1 Software Optimizations

An example of a well meaning software optimization that interacts badly with the benchmark writer comes from the Quadrics Elan4 MPI stack. For the Elan4, Quadrics partitioned the local and remote posted receive queues. This is the "right thing to do" in that, while the posted receive queue for remote messages lives on the NIC, it doesn't make sense to make a trip to the NIC to traverse the posted receive queue for strictly local messages.

How does this interact with the benchmark writer? Consider the *preposted latency* benchmark described in Section 3. The goal is to measure the impact of a long posted receive queue; thus, there are several posted receives that must be cleared (preferably quickly) on each iteration of the benchmark. Since they are zero length messages, it is actually far quicker on many platforms to clear them locally (without having to go out over the wire). Thus, they are posted from the local node, which triggers the Elan4 optimization. This defeats the point of the benchmark, but clearing the extra posted receives remotely takes significant extra time. To work around this, the messages were posted with MPI_ANY_SOURCE and differentiated by tag. This is not a particularly uncommon application characteristic, since some applications do not know who to expect messages from or how many to expect. The impact of this change is seen in Figure 1, where traversing a queue looks "free" in one case to taking approximately 150 ns when the processor on the NIC actually has to traverse the messages.

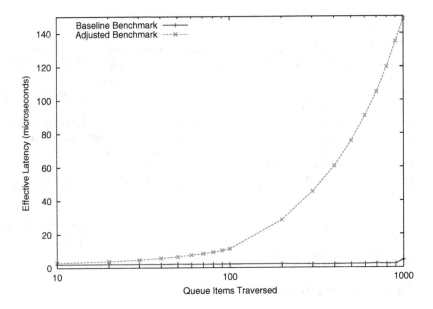

Fig. 1. Impact of the Elan 4 optimization on the preposted receive benchmark

4.2 Microprocessor Optimizations

While some modern networks (e.g. Quadrics Elan4) handle the MPI posted receive queue on the network interface, most of them handle it on the microprocessor. Some even go so far as to do many of the data movement operations using the host processor (PathScale Infinipath). That means that the performance of the microprocessor's memory hierarchy (and particularly the cache) matters a lot.

Flushing the cache is slightly tricky because of compiler optimizations (discussed in Section 4.4), but it also requires care in message passing. Because a cache flush operation is inherently a variable time operation, a custom "barrier" operation is used so that node 1 is guaranteed to exit the barrier before node 0 (the initiator of the ping-pong test).

When the cache is flushed after the receives are posted, the performance of microprocessors in the preposted receive benchmark drops dramatically, as shown in Figure 2. The impact can be as large as 4× for networks like Infinipath that rely on the processor to obtain performance. It also suggests that many of the modern network performance claims for ping-pong latency (far left of the graph) are heavily based on caching performance that is unlikely to be obtained.

Figure 2 shows similar results using the message rate benchmark with 6 simulated neighbors with 64 posted receives each. That results in 6 sets of 64 posted receives where tags cause the message target to be rotated among the sets. The differences are less dramatic, but approach 25% in many cases.

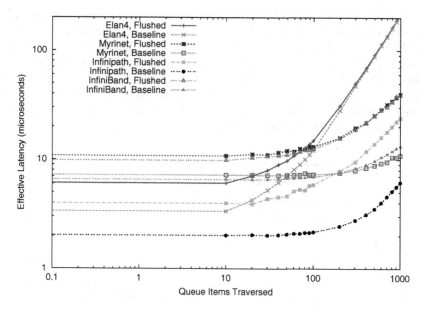

Fig. 2. Impact of flushing the cache on the preposted receive benchmark

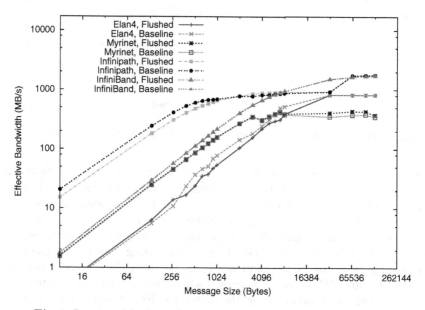

Fig. 3. Impact of flushing the cache on the message rate benchmark

4.3   Hardware and Software Interactions

Things get even more interesting when hardware and software interact. Take, for example, the slab cache optimization in MPICH. Rather than dynamically allocate queue entries, MPICH manages its own memory for these structures. Requests are taken from a queue and placed back on the queue in the order they are used. Because the slab cache starts from a contiguous, in-order region, a curious interaction occurs with optimizations in the microprocessor. Modern microprocessors include a hardware prefetcher, which detect the effectively stride-N access that occurs when traversing a linked list that was sequentially allocated from a contiguous region. This results in list traversals that are remarkably fast relative to the memory latencies of the machine.

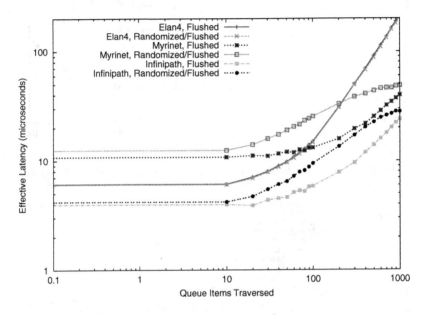

Fig. 4. Impact of the hardware prefetcher on the preposted receive benchmark

In a *real* application, list items are cleared (and thus returned to the free list) in a more random order. To mimic this behavior requires randomizing the list entries, so a slab cache randomization function (one that preposts a large number of receives and clears them in a random order) was added. Figure 4 shows the impact of this change on the preposted receive benchmark. As the length of the list grows long, the effect becomes dramatic. Another interesting note is that the impact is less dramatic for Infinipath than for Myrinet because the Infinipath network is connected to an AMD Opteron processor, which has a lower memory latency. The Elan4 network, in contract, stores the posted receive list on the NIC, where the memory is not nearly so impacted by caching behavior.

4.4 Compiler Optimizations

Benchmarks should generally be compiled with optimization. Otherwise, compilers have been known to do things that are just dumb. Unfortunately, compilers can be quite aggressive in determining what code is "dead" when they are doing dead code elimination. Almost everyone has encountered this when writing a timing loop. You want to measure the time of operation X, so you write a loop to do it N times. With optimization, the loop takes zero time because the compiler figured out you never used the result variable.

Knowing this, an early version of the benchmarks described above had an option to flush the cache. On every benchmark iteration (but outside the timing portion), this option summed a large array of random numbers generated at run-time and put the result into the data being transmitted as part of the benchmark. Normally, one might expect the compiler to leave this alone; however, to prevent excessive execution times, the array was only initialized once. When optimization was turned on, the compiler detected this and the results with cache flushing were no different from the results without it. In the C language, the solution was to mark the target data location as volatile so that the compiler would leave it alone. While this has negative implications for what the compiler will do with that particular piece of code, it is not in the timing loop, so that does not matter, and it is still faster than generating 1,000,000 random numbers on each iteration.

5   Conclusions

The two most important characteristics of a benchmark are that it measure properties of interest to applications and that it measure them in an application context. While it can be conceptually difficult to define properties of interest to applications, measuring them in the context of applications is unduly complicated by important system level optimizations. This paper discusses five optimizations spread across the hardware and software stack that generally work extremely well in the benchmark context, but are unlikely to provide such performance in a real application context. Techniques are presented for reducing these impacts along with quantitative data on the magnitude of their impact. In extreme cases, the difference can be 4×.

References

1. *Netperf.* http://www.netperf.org.
2. *Pallas MPI Benchmarks.* http://http://www.pallas.com/e/products/pmb/index. htm.
3. A. Alexandrov, M. F. Ionescu, K. E. Schauser, and C. Sheiman. LogGP: Incorporating long messages into the LogP model. *Journal of Parallel and Distributed Computing*, 44(1):71–79, 1997.
4. C. Bell, D. Bonachea, Y. Cote, J. Duell, P. Hargrove, P. Husbands, C. Iancu, M. Welcome, and K. Yelick. An evaluation of current high-performance networks. In *17th International Parallel and Distributed Processing Symposium (IPDPS'03)*, Apr. 2003.
5. D. E. Culler, R. M. Karp, D. A. Patterson, A. Sahay, K. E. Schauser, E. Santos, R. Subramonian, and T. von Eicken. LogP: Towards a realistic model of parallel computation. In *Proceedings 4th ACM SIGPLAN Symposium on Principles and Practice of Parallel Programming*, pages 1–12, 1993.

6. W. Lawry, C. Wilson, A. B. Maccabe, and R. Brightwell. COMB: A portable benchmark suite for assessing MPI overlap. In *IEEE International Conference on Cluster Computing*, September 2002. Poster paper.

7. J. Liu, B. Chandrasekaran, J. Wu, W. Jiang, S. Kini, W. Yu, D. Buntinas, P. Wyckoff, and D. K. Panda. Performance comparison of MPI implementations over InfiniBand, Myrinet and Quadrics. In *The International Conference for High Performance Computing and Communications (SC2003)*, November 2003.

8. R. P. Martin, A. M. Vahdat, D. E. Culler, and T. E. Anderson. Effects of communication latency, overhead, and bandwidth in a cluster architecture. In *Proceedings of the 24th Annual International Symposium on Computer Architecture*, June 1997.

9. F. Petrini, D. J. Kerbyson, and S. Pakin. The case of the missing supercomputer performance: Identifying and eliminating the performance variability on the ASCI Q machine. In *Proceedings of the 2003 Conference on High Performance Networking and Computing*, November 2003.

10. R. Riesen, R. Brightwell, and A. B. Maccabe. Measuring MPI latency variance. In J. Dongarra, D. Laforenza, and S. Orlando, editors, *Recent Advances in Parallel Virtual Machine and Message Passing Interface: 10th European PVM/MPI Users' Group Meeting, Venice, Italy, September/October 2003 Proceedings*, volume 2840 of *Lecture Notes in Computer Science*, pages 112–116. Springer-Verlag, 2003.

11. Q. O. Snell, A. Mikler, and J. L. Gustafson. NetPIPE: A network protocol independent performance evaluator. In *Proceedings of the IASTED International Conference on Intelligent Information Management and Systems*, June 1996.

12. K. D. Underwood. The impacts of message rate on applications programming. In *submitted*.

13. K. D. Underwood and R. Brightwell. The impact of MPI queue usage on message latency. In *Proceedings of the International Conference on Parallel Processing (ICPP)*, Montreal, Canada, August 2004.

Implementation and Usage of the PERUSE-Interface in Open MPI

Rainer Keller[1], George Bosilca[2], Graham Fagg[2],
Michael Resch[1], and Jack J. Dongarra[2]

[1] High-Performance Computing Center, University of Stuttgart,
{keller, resch}@hlrs.de
[2] Innovative Computing Laboratory, University of Tennessee
{bosilca, fagg, dongarra}@cs.utk.edu

Abstract. This paper describes the implementation, usage and experience with the MPI performance revealing extension interface (Peruse) into the Open MPI implementation. While the PMPI-interface allows timing MPI-functions through wrappers, it can not provide MPI-internal information on MPI-states and lower-level network performance. We introduce the general design criteria of the interface implementation and analyze the overhead generated by this functionality. To support performance evaluation of large-scale applications, tools for visualization are imperative. We extend the tracing library of the Paraver-toolkit to support tracing Peruse-events and show how this helps detecting performance bottlenecks. A test-suite and a real-world application are traced and visualized using Paraver.

1   Introduction

The Message Passing Interface (MPI) [7,8] is the standard for distributed memory parallelization. Many scientific and industrial applications have been parallelized and ported on top of this parallel paradigm. From the very beginning the MPI standard offered a way for performance evaluation of all provided functions including the communication routines with the so-called Profiling-Interface (PMPI). Thereby all MPI-function calls are accessible through the prefix PMPI_, allowing wrapper-functions, which mark the time at entry and exit. The tracing libraries of performance analysis tools, such as Vampir[3], Paraver[5] and Tau[9] are build upon the PMPI-Interface. However, the information gathered using this interface has a limited impact, as it can only provide high level details about any communications (such as starting and ending time), rather than more interesting internal implementation and networking activities triggered by the MPI calls.

   In order to know the internals of how the communication between two processes proceeds and where possible bottlenecks are located, a more in-depth and finer-grained knowledge is required than is available from the PMPI-interface level. The Peruse-interface [2], a multi-institution effort driven by LLNL which

B. Mohr et al. (Eds.): PVM/MPI 2006, LNCS 4192, pp. 347–355, 2006.
© Springer-Verlag Berlin Heidelberg 2006

gained larger audience at a BoF at SC2002, proposes a standard way for applications and libraries to gather this information from a Peruse-enabled MPI-library. Especially with more diverse hardware, such as multi-core chips using shared-memory and many hierarchies in large-scale clusters, this performance evaluation becomes essential for in-depth analysis.

This paper introduces an implementation of Peruse in the Open MPI [4] implementation. In section 2 we describe the general design and implementation of the Peruse-interface within Open MPI, and state the impact on communication performance degradation, while section 3 shows the performance metrics gathered. Section 4 illustrates a possible method to evaluate the communication performance by extending the `mpitrace`-library of the Paraver-toolkit. In section 5 a real-world application is traced and visualized with Paraver. Finally, the last section gives a conclusion and an outlook on future developments.

2   Design and Implementation

The Open MPI implementation uses the so-called modular component architecture (MCA) to support several component implementations offering a specific functionality [10]. In this paper, we will consider only the frameworks and components used for communication purposes, i.e., the Point-to-Point management layer (PML), the recursively named BML management layer (BML) and the Bit-transport layer (BTL). These frameworks are stacked, as may be seen in figure 1. MPI communication calls are passed on to the PML, which uses the BML to select the best possible BTL, and then passes the message (possibly in multiple fragments depending on length) to the BTL for transmission.

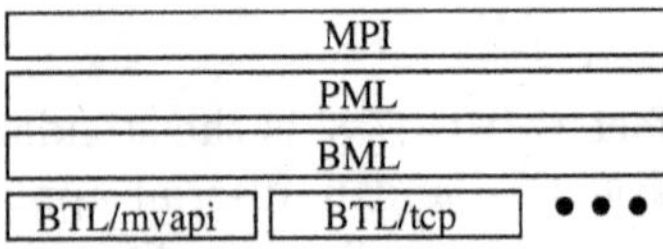

Fig. 1. Open MPI stack of frameworks and modules for communication

The Peruse interface allows an application or performance measurement library to gather information on state-changes within the MPI library. For this, user-level callbacks have to be registered with the Peruse interface, which are subsequently invoked upon the triggering of corresponding events. The interface allows a single callback function to be registered for multiple events, as well as multiple callback functions for one event (which covers the rare instance of an application and one or more libraries wanting to gather statistics on a single event simultaneously). Peruse does not impose any particular message passing method and recommends not supporting a particular event, if this would burden or slow down the MPI implementation. The interface is portable in design, by allowing applications or performance tracing libraries to query for supported events using defined ASCII strings. The tracing library may then register for an event, supplying a callback function, which is invoked upon triggering a particular event, e. g. `PERUSE_COMM_REQ_XFER_BEGIN` when the first data transfer of a request is scheduled. Registration then returns an event-handle. Events implemented in Open MPI are presented in sec. 3.

Prior experience with the implementation of Peruse-functionality was gained with PACX-MPI [6]. Special care was taken not to slow down the critical fast path of the Open MPI library. The actual test for an active handle and the immediate invocation of the callback function is implemented as a macro, which the preprocessor optimizes away in a default build of the library. When building with the configure parameter `--enable-peruse`, the actual test for an active handle involves at most two additional `if`-statements: whether any handles are set on this communicator and whether the particular one is set and active.

Although most of the events are pertaining to messages being sent and received, the actual calls to the callback functions are performed in the PML-layer, as it has all the necessary information regarding requests and fragments being sent. Currently, only one major PML-module exists (`ob1`), in contrast to the six major BTLs (`sm`, `tcp`, `mvapi`, `openib`, `gm`, `mx`), which would have each required modifications for every possible Peruse event.

Additionally, this initial implementation only allows a single callback function per event. As handles are stored per communicator (`PERUSE_PER_COMM`) as array of `ompi_peruse_handle_t`-pointer, allowing more callbacks per event or worse case multiple handles (instances) per event would have required iterating over all the registered and active handles in the communicator-storage, greatly increasing the overall overhead.

Table 1. Configuration of clusters for the Peruse overhead evaluation

Cluster	cacau	strider
Processor	Dual Intel Xeon EM64T, 3.2 GHz	Dual AMD Opteron 246, 2GHz
Interconnect	Infiniband	Myrinet 2000
Interface	`mvapi-4.1.0`	`gm-2.0.8`
Compiler	Intel compiler 9.0	PGI compiler 6.1.3
Open MPI	no debug, static build	no debug, dynamic build
Native MPI	Voltaire MPIch-1.2.6	MPIch-1.2.6

For performance comparison with and without the Peruse-interface implementation, several measurements were conducted on the clusters given in table 1. We compare the latency induced by the additional overhead by using a build without any Peruse-support and two versions with Peruse-support: one without any callbacks and one with callbacks attached for all possible events. Additionally this is compared to the latency of the native MPI-implementation provided on each cluster.

Table 2. Latency (in μs) of zero-byte messages using IMB-2.3 with `PingPong`

	cacau			strider		
	native	mvapi	sm	native	gm	sm
No Peruse	4.13	4.69	1.02	7.16	7.16	1.33
Peruse, no callbacks		4.67	1.06		7.26	1.71
Peruse, no-op callbacks		4.77	1.19		7.49	1.84

Table 2 shows the measurements done with the Intel-MPI Benchmark using the zero Byte PingPong-test. The `IMB_settings.h` was changed to perform each test for 10000 iterations with ten warm-up phases. For the native MPI, the optimized vendor's version on the cluster was used as listed in table 1.

In comparison with the cluster's native MPI, the Open MPI's BTLs `mvapi` and `gm` only show marginal difference in latency being 1.7% and 4.6% respectively. Therefore, a much more sensitive test using the shared-memory BTL `sm` was performed. Here, one experiences a degradation in latency – but even with all 16 communication events registered, the increase is 16% and 38% respectively for the two target systems. For larger message sizes, the overhead compared to the bandwidth without any Peruse-support is shown in Fig. 2.

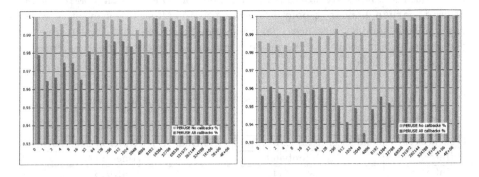

Fig. 2. Percentage of achieved bandwidth on cacau (left) and strider (right) compared to Open MPI without Peruse

3   Performance Metrics Gathered

The current implementation in Open MPI supports all events stated in the current Peruse-2.0 specification [2]. Orthogonal to the `PERUSE_COMM_REQ_XFER_BEGIN/_END` Open MPI implements the `PERUSE_COMM_REQ_XFER_CONTINUE` notifying of new fragments arriving for this request. This event is only issued in the case of long messages not using the `eager` protocol. The sequence of callbacks that may be generated on the way for sending / receiving a message are given in Fig. 3.

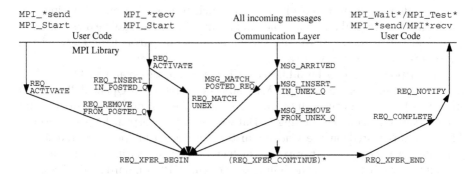

Fig. 3. Sequence of Peruse events implemented in Open MPI

The following example shows the callback sequence when sending a message from rank zero to rank one, which nicely corresponds to figure 3. Here, we have imposed an early receiver by delaying the sender by one second. One may note the early activation of the request, searching in the unexpected receive queue, insertion into the expected receive queue, and finally the arriving message with the subsequent start of transfer of messages events (edited):

```
PERUSE_COMM_REQ_ACTIVATE at 0.00229096 count:10000 ddt:MPI_INT
PERUSE_COMM_SEARCH_UNEX_Q_BEGIN at 0.00229597 count:10000 ddt:MPI_INT
PERUSE_COMM_SEARCH_UNEX_Q_END at 0.00230002 count:10000 ddt:MPI_INT
PERUSE_COMM_REQ_INSERT_IN_POSTED_Q at 0.00230312 count:10000 ddt:MPI_INT
PERUSE_COMM_MSG_ARRIVED at 1.00425 count:0 ddt:0x4012bbc0
PERUSE_COMM_SEARCH_POSTED_Q_BEGIN at 1.00426 count:0 ddt:0x4012bbc0
PERUSE_COMM_SEARCH_POSTED_Q_END at 1.00426 count:0 ddt:0x4012bbc0
PERUSE_COMM_MSG_MATCH_POSTED_REQ at 1.00426 count:10000 ddt:MPI_INT
PERUSE_COMM_REQ_XFER_BEGIN at 1.00427 count:10000 ddt:MPI_INT
PERUSE_COMM_REQ_XFER_CONTINUE at 1.0043 count:10000 ddt:MPI_INT
  -- subsequent XFER_CONTINUES deleted --
PERUSE_COMM_REQ_XFER_CONTINUE at 1.00452 count:10000 ddt:MPI_INT
PERUSE_COMM_REQ_XFER_END at 1.00452 count:10000 ddt:MPI_INT
PERUSE_COMM_REQ_COMPLETE at 1.00453 count:10000 ddt:MPI_INT
PERUSE_COMM_REQ_NOTIFY at 1.00453 count:10000 ddt:MPI_INT
```

Collecting the output of the callbacks from a late sender:

```
PERUSE_COMM_REQ_ACTIVATE at 1.00298 count:10000 ddt:MPI_INT
PERUSE_COMM_REQ_XFER_BEGIN at 1.0031 count:10000 ddt:MPI_INT
PERUSE_COMM_REQ_XFER_CONTINUE at 1.00313 count:10000 ddt:MPI_INT
  -- subsequent XFER_CONTINUEs deleted --
PERUSE_COMM_REQ_XFER_CONTINUE at 1.00322 count:10000 ddt:MPI_INT
PERUSE_COMM_REQ_XFER_END at 1.00327 count:10000 ddt:MPI_INT
PERUSE_COMM_REQ_COMPLETE at 1.00328 count:10000 ddt:MPI_INT
PERUSE_COMM_REQ_NOTIFY at 1.00328 count:10000 ddt:MPI_INT
```

4   Trace-File Generation

To cope with the information provided by Peruse's functionality, one needs tools to visualize the output generated. We have ported the `mpitrace`-library of the Paraver-toolkit [1] to Open MPI. Paraver is a powerful performance analysis and visualization tool developed at CEPBA/BSC. Similar to Vampir, a trace is a time-dependant function of values for each process. Through filtering and combination of several functions, meaningful investigations may be deduced even within large traces, e. g. searching and highlighting of parts of the trace with a GFlop-rate below a specified value.

Several points had to be addressed when porting `mpitrace` to Open MPI: removing assumptions on opaque MPI-objects (pointers to Open MPI internal structures) being integer values and separating helper functions into C- and Fortran-versions to avoid passing C-Datatypes to the Fortran PMPI-Interface. The port was tested on the Cacau-Cluster (having 64-bit pointers and 32-bit integers) with the `mpi_test_suite`, which employs combinations of simple functionality to stretch tests to the boundaries of the MPI-standard's definition.

For tracing, an application needs re-linking with the Peruse-enabled `mpitrace`-library. Peruse-events to be tested for are specified by the environment variable `MPITRACE_PERUSE_EVENTS`, separated by colons. Figure 4 shows the Paraver-window of an exemplary trace of ten sends, each of 10MB-size messages from rank zero to rank one with four Peruse-events attached[1]. Clearly, the initialization of the buffer on rank zero is visible as running time, while rank one awaits the message in the first `MPI_Recv`. Only with the Peruse-events (shown in gray), can the actual transfer be seen as the small green flags for each transmitted data fragment. By clicking into the trace-window, one may get further information on the Peruse-Events of the trace.

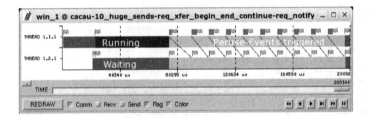

Fig. 4. Paraver visualization of 10 large msgs sent from rank zero to one (edited)

5   Application Measurement

To demonstrate the suitability of Peruse-events tracing with the Paraver-toolkit, we show the tracing of the large molecular-dynamics package IMD with a benchmark test (`bench_cu3au_1048k.param`). The overall trace with 32 processes on cacau is shown in Fig. 5. One may note the long data distribution done using a linear send, followed by a collective routine during the initialization at the beginning of the execution. The overall run shows ten iterations and a final collection phase. The right-hand window of Fig. 5 shows the achieved bandwidth, here ranging from 101 to 612 MB/s.

---

[1] `MPITRACE_PERUSE_EVENTS=PERUSE_COMM_REQ_XFER_BEGIN:PERUSE_COMM_REQ_XFER_END: PERUSE_COMM_REQ_XFER_CONTINUE:PERUSE_COMM_REQ_NOTIFY`

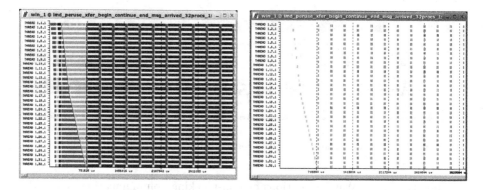

Fig. 5. Trace of IMD with 32 processes – overall run (left) and bandwidth (right)

Figure 6 zooms into one communication step of the run (left) with the corresponding bandwidth graph on the right hand side with up to 611 MB/s using on average 524 kB-sized messages. In order to appreciate the additional information Peruse-events give to the performance analyst, the actual time between message fragments arriving with the `PERUSE_COMM_REQ_XFER_CONTINUE` event are shown at the top-right of Fig. 6. Here, one may see how the in-flow rate of messages changes over time, ranging from $46\mu s$ to $224\mu s$ between fragments, corresponding to 4464 fragments/s up to 21739 fragments/s.

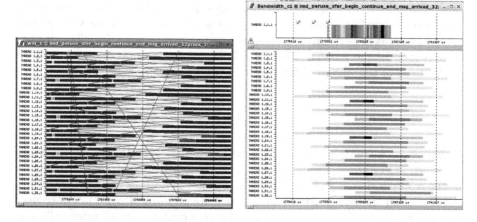

Fig. 6. Zoom into one communication step; bandwidth(right) and interval between fragments(right-top)

While with PMPI-based tracing it is possible to detect performance problems, such as "Late Sender", or "Late Receiver", the actual transferral of the message can not be seen. Particularly, for eager sends (small message) sends, the actual logical transferral of the message is far longer than the physical. This may be detected only with a corresponding `PERUSE_COMM_MSG_ARRIVED`-event on the receiver side.

Fig. 7. Detection of late wait situation with Peruse on Open MPI (edited)

Similarly, "Late Wait" situations of non-blocking communication cannot be detected through PMPI, as the communication will only be considered finished upon the corresponding `MPI_Wait`/`MPI_Wait`; here the `PERUSE_COMM_REQ_COMPLETE`-event notifies of the completion. Figure 7 shows a trace of such a situation. Process zero again sends a small message with eager protocol to process one, using non-blocking send and receive, respectively. The recv's `MPI_Wait` however is delayed by roughly 1.6ms. While the PMPI-based tracing considers the logical communication to finish within the `MPI_Wait` only, with Peruse one receives the early `PERUSE_COMM_REQ_COMPLETE`.

Furthermore, together with PMPI-wrapper, the tracing-library may additionally uncover book-keeping work commonly done by MPI-implementations before returning to the application, e. g. running event-handlers to progress other communication.

Additionally to message send/arrival times, Peruse allows information on the timing of internal traversal of message queues, which may be used to distinguish low network performance from slow queue management. Finally, with the introduction of the `PERUSE_COMM_REQ_XFER_CONTINUE`-event one may uncover fluctuations of the stream of fragments in case of network congestion.

6   Conclusion

In this paper we have described the implementation of the Peruse-interface into the Open MPI library. The integration into Open MPI was straightforward due to the modular design and the target platform. The authors are however aware, that for other implementations, the current design of the Peruse interface may not be feasible due to MPI running in a different context, not allowing callbacks or due to the overhead introduced.

In the future, the authors would like to extend the Open MPI Peruse system with additional events yet to be defined in the current Peruse specification, e.g., collective routines and/or one-sided operations. Additionally the functionality for very low level events such as those defined within networking devices is also envisaged.

We would like to thank BSC for making `mpitrace` available. This work was made possible by funding of the EU-project HPC-Europa (Contract No. 506079), and also by the "Los Alamos Computer Science Institute (LACSI)", funded by Rice University Subcontract No. R7B127 under Regents of the University Subcontract No. 12783-001-05 49.

References

1. Paraver Homepage. WWW, May 2006. http://www.cepba.upc.es/paraver.
2. Peruse specification. WWW, May 2006. http://www.mpi-peruse.org.
3. Holger Brunst, Manuela Winkler, Wolfgang E. Nagel, and Hans-Christian Hoppe. Performance Optimization for Large Scale Computing: The scalable VAMPIR approach. In V.N. Alexandrov et al., editors, *Computational Science (ICCS'01)*, volume 2, pages 751–760. Springer, May 2001.
4. Edgar Gabriel, Graham E. Fagg, George Bosilca, Thara Angskun, Jack J. Dongarra, and Jeffrey M. Squyres et al. Open MPI: Goals, Concept, and Design of a Next Generation MPI Implementation. In D. Kranzlmüller, P. Kacsuk, and J.J. Dongarra, editors, *Recent Advances in Parallel Virtual Machine and Message Passing Interface*, volume 3241, pages 97–104, Budapest, Hungary, September 2004. Springer.
5. Gabriele Jost, Haoquian Jin, Jesus Labarta, Judit Gimenez, and Jordi Caubet. Performance analysis of multilevel parallel applications on shared memory architectures. In *International Parallel and Distributed Processing Symposium (IPDPS 2003)*, volume 00, page 80b, April 2003. Nice, France.
6. Rainer Keller, Edgar Gabriel, Bettina Krammer, Matthias S. Müller, and Michael M. Resch. Towards efficient execution of MPI applications on the Grid: Porting and Optimization issues. *Journal of Grid Computing*, 1(2):133–149, 2003.
7. Message Passing Interface Forum. *MPI: A Message Passing Interface Standard*, June 1995. http://www.mpi-forum.org.
8. Message Passing Interface Forum. *MPI-2: Extensions to the Message-Passing Interface*, July 1997. http://www.mpi-forum.org.
9. Sameer Shende and Allen D. Malony. TAU: The TAU Parallel Performance System. 2005.
10. T.S. Woodall, R.L. Graham, R.H. Castain, D.J. Daniel, M.W. Sukalski, G.E. Fagg, E. Gabriel, G. Bosilca, T. Angskun, J.J. Dongarra, J.M. Squyres, V. Sahay, P. Kambadur, B. Barrett, and A. Lumsdaine. Open MPI's TEG Point-to-Point Communications Methodology: Comparison to Existing Implementations. In *Recent Advances in Parallel Virtual Machine and Message Passing Interface*, volume 3241, pages 105–111, Budapest, Hungary, September 2004. Springer.

5th International Special Session on

Current Trends in Numerical Simulation for Parallel Engineering Environments*

New Directions and Work-in-Progress

ParSim 2006

In today's world, the use of parallel programming and architectures is essential for simulating practical problems in engineering and related disciplines. Remarkable progress in CPU architecture, system scalability, and interconnect technology continues to provide new opportunities, as well as new challenges for both system architects and software developers. These trends are paralleled by progress in parallel algorithms, simulation techniques, and software integration from multiple disciplines.

ParSim brings together researchers from both application disciplines and computer science and aims at fostering closer cooperations between these fields. Since its successful introduction in 2002, ParSim has established itself as an integral part of the EuroPVM/MPI conference series. In contrast to traditional conferences, emphasis is put on the presentation of up-to-date results with a short turn-around time. This offers a unique opportunity to present new aspects in this dynamic field and discuss them with a wide, interdisciplinary audience. The EuroPVM/MPI conference series, as one of the prime events in parallel computation, serves as an ideal surrounding for ParSim. This combination enables the participants to present and discuss their work within the scope of both the session and the host conference.

This year, eleven papers from authors in nine countries were submitted to ParSim, and we selected five of them. They cover a wide range of different application fields including gasflow simulations, thermo-mechanical processes in nuclear waste storage, and cosmological simulations. At the same time, the selected contributions also address the computer science side of their codes and discuss different parallelization strategies, programming models and languages, as well as the use nonblocking collective operations in MPI. We are confident that this provides an attractive program and that ParSim will be an informal setting for lively discussions and for fostering new collaborations.

Several people contributed to this event. Thanks go to Jack Dongarra, the EuroPVM/MPI general chair, and to Bernd Mohr, Jesper Larsson Träff, and Joachim Worringen, the PC chairs, for their encouragement and support to continue the ParSim series at EuroPVM/MPI 2006. We would also like to thank the

* Part of this work was performed under the auspices of the U.S. Department of Energy by University of California Lawrence Livermore National Laboratory under contract No. W-7405-Eng-48. UCRL-PROC-222517.

B. Mohr et al. (Eds.): PVM/MPI 2006, LNCS 4192, pp. 356–357, 2006.

numerous reviewers, who provided us with their reviews in such a short amount of time (in most cases in just a few days) and thereby helped us to maintain the tight schedule. Last, but certainly not least, we would like to thank all those who took the time to submit papers and hence made this event possible in the first place.

We hope this session will fulfill its purpose to provide new insights from both the engineering and the computer science side and encourages interdisciplinary exchange of ideas and cooperations. We hope that this will continue ParSim's tradition at EuroPVM/MPI.

Carsten Trinitis
Lehrstuhl für Rechnertechnik und Rechnerorganisation (LRR)
Institut für Informatik
Technische Universit"at M"unchen, Germany
Carsten.Trinitis@in.tum.de

Martin Schulz
Center for Applied Scientific Computing
Lawrence Livermore National Laboratory
Livermore, CA, USA
schulzm@llnl.gov

MPJ Express Meets Gadget: Towards a Java Code for Cosmological Simulations

Mark Baker[1], Bryan Carpenter[2], and Aamir Shafi[3]

[1] School of Systems Engineering, University of Reading
[2] Open Middleware Infrastructure Institute, University of Southampton
[3] Distributed Systems Group, University of Portsmouth

Abstract. Gadget-2 is a massively parallel structure formation code for cosmological simulations. In this paper, we present a Java version of Gadget-2. We evaluated the performance of the Java version by running colliding galaxies simulation and found that it can achieve around 70% of C Gadget-2's performance.

1 Introduction

Various computer scientists have argued that Java could make an excellent language for developing scientific codes. To date this argument has not convinced too many practising computational scientists. The scarcity of high-profile number-crunching codes implemented in Java does not help the case.

We have recently released MPJ Express [1], a thread-safe, production quality Java messaging system for high performance computing. To help establish the practicality of real scientific computing using message passing Java we have ported the parallel cosmological simulation code, Gadget-2, from C to Java, using MPJ Express. Gadget-2 [7] is a massively parallel structure formation code developed by Volker Springel at the Max Planck Institute for Astrophysics. Versions of Gadget-2 have been used in various research papers in astrophysics literature, including the noteworthy "Millennium Simulation" [8]—the largest ever model of the Universe.

Producing a Java version of Gadget is an experiment that helps us to understand where Java stands in comparison to C—an already established HPC language. Concerns about Java's performance have stopped many computational scientists from seriously considering it. But constant improvements in JIT (Just In Time) compilers, which translate bytecode into the native machine code at runtime, have improved the computational performance.

Exploitation of Java for simulation projects has been ongoing for some years. JWarp [3] is a Java library for discrete-event parallel simulations. MONARC [4] is a simulation framework for large scale computing resources. It has been deployed on an inter-continental testbed to verify simulation results with success. CartaBlanca [5], from Los Alamos National Lab, is a general purpose non-linear solver environment for physics computations on non-linear grids. It employs an object-oriented component design, and is pure Java. These projects suggest

B. Mohr et al. (Eds.): PVM/MPI 2006, LNCS 4192, pp. 358–365, 2006.

that Java has already made its mark on a range of projects involved in parallel simulations, or scientific computing in general.

Section 2 of this paper presents an overview of Gadget-2. We discuss our experiences in porting Gadget-2 to Java in Section 3. We evaluate the performance of the Java version in section 4 and also compare it with the original C version. We conclude and discuss future work in Section 5.

2   Overview of Gadget-2

Gadget-2 is a free production code for cosmological N-body and hydrodynamic simulations. The code is written in the C language and parallelized using MPI. It simulates the evolution of very large, cosmological-scale systems under the influence of gravitational and hydrodynamic forces. The universe is modelled by a sufficiently large number of test particles, which may represent ordinary matter or dark matter.

We are particularly interested in the parallelization strategy, which is based on an irregular and dynamically adjusted domain decomposition, with copious communication between processors.

To give some feeling for the scale of interesting problems, consider the so-called "Millennium Simulation" [8]. This simulation follows the evolution of 10^{10} dark matter particles from the early Universe to the current day. It was performed on 512 processors and used 1 Terabytes of distributed memory. The simulation used 350,000 CPU hours over 28 days of elapsed time.

2.1   Computing Gravitational Forces

One of the main tasks of a structure formation code is to calculate gravitational forces exerted on a particle.

In a N-body cosmological simulation, every particle exerts gravitational force on every other particle. The reason is that gravity is a long range force. Thus, calculating gravitational force in such simulations can be computationally intensive—the total cost is $O(N^2)$ for the naive summation approach. This is not feasible for the scale of problems that Gadget-2 aims to solve.

Thus, Gadget-2 can use either of two efficient algorithms to calculate gravitational forces. The first is Barnes-Hut (BH) [2] oct tree, and the second is a hybrid of BH tree and Particle-Mesh (PM) method called *TreePM*. In this paper, we restrict our attention to the pure BH tree algorithm.

Barnes-Hut Tree Algorithm. The cubical region of 3D space is divided into eight sub-regions by halving each dimension. Every sub-region that contains any particles is recursively divided until each region has at most one particle. The root of the Barnes-Hut tree corresponds directly to the whole 3D space. The first division of space results in eight sub-regions that become the daughter nodes of the root. This process continues until each node of the tree contains one particle.

The reason for arranging the particles in a tree data-structure is that it allows efficient calculation of gravitational forces. The tree is traversed from root to compute the force, for example on a particle i. If a node n is *distant from* particle i, the contribution of node n is added to force on i from the center of mass of n. In this case, there is no need to to visit the daughter nodes of n. The daughter nodes of node n are visited recursively if it is *close to i*.

The definition of *distant from* or *close to* depends on an opening criterion. The basic idea is that a node representing some region in space is *distant from* a particle i if the angle it subtends is smaller than a threshold opening angle. Otherwise, a node is considered *close to* particle i.

Using this approach, it is possible to calculate the gravitational force for each particle in $O(\log N)$ steps. For the range of N of practical interest this is clearly a huge win over the summation approach that results in $O(N)$ steps.

2.2   Domain Decomposition

Being a massively parallel code, Gadget-2 needs to divide space or particle set into domains, where each domain is handled by a single processor. It is particularly challenging in Gadget-2 because it is not practical to divide space evenly. This would result in poor load balancing because some regions have more particles than the others. Conversely, it is also not possible to divide particles evenly in a fixed way because they move throughout space and it is desirable to keep physically close particles on the same processor.

To solve this, Gadget-2 uses a space-filling *Peano-Hilbert curve* originally suggested by Warren and Salmon [6]. Gadget-2 applies the standard recursion for constructing the curve 20 times, logically dividing space into up to $2^{20} \times 2^{20} \times 2^{20}$ cells on the Peano-Hilbert curve. Each cell is labelled by its location along the Peano-Hilbert curve—2^{60} possible locations. The information about the location of each cell can be stored in a `long` word called the *Peano-Hilbert key*. These Peano-Hilbert keys play an important role during domain decomposition. Because the total number of cells is far greater than total number of particles, points of the discrete linear Peano-Hilbert curve are sparsely populated with particles. To establish the domain decomposition, one sorts particles by their Peano-Hilbert keys and then divides them evenly into P sections, where P is the total number of processors.

This technique implements an efficient domain decomposition. It provides good load balancing. The domains are simply connected and quite "compact" in real space, because particles that are close along Peano-Hilbert curve are close in real space (the converse is often but not always true). An added advantage is that the Peano-Hilbert curves provide simple mapping to Barnes-Hut tree nodes.

Distributed Representation of Tree . The BH tree is implemented as a distributed data structure. Nodes of the tree can be classified according to whether all particles in the node belong to one processor, or the node contains particles from multiple processors. Nodes in the first category are stored locally on the relevant processors. All nodes in the second category—this typically means higher nodes in the tree—are replicated over all processors.

So every processor holds a copy of the root nodes and all daughter nodes down to the point where all particles of a node are held on a single processor. Where this is a remote processor the corresponding node is called a *pseudo-particle*. To compute the force on a single local target particle, the tree is traversed starting from root as usual accumulating force contributions from locally held particles.

2.3   Communication

The original Gadget-2 is parallelized following the standard MPI specifications. As part of the parallel tree-force computation, a processor walks the tree for every locally held particle accumulating force contributions. These contributions may come from local particles or *pseudo-particles*. If the daughters of a node representing *pseudo-particles* need to be traversed, the locally held particles are marked for export to the processor that owns the *pseudo-particle* in question. After the tree-walk, all particles marked for export are communicated to remote hosts. These hosts calculate the force contributions and communicate them back. Also, there is some communication involved during domain decomposition for distributed sorting of the particle list.

3   Porting Gadget-2 to Java

Gadget-2 was manually translated to the Java language. We deliberately kept similar data structures in the translated version so that we could cross reference the original source code for debugging. Currently there are some functional limitations compared with the C code. For example, the Java version only provides the option of using BH oct tree for calculating gravitational forces.

There are three dependencies for Gadget-2; GNU Scientific Library (GSL), parallel version of Fastest Fourier Transforms in the West (FFTW), and of course a MPI library. Gadget-2 only uses a handful of GSL functions—we manually translated these to Java. FFTW would be required for the *TreePM* algorithm, and for this reason we use BH tree algorithm for calculating gravitational forces in the current Java version. For communication, we use MPJ Express, our own thread-safe implementation of MPI-like bindings for the Java language.

The main simulation loop increments timesteps and drift the particles to the next timestep. This involves calculating gravitational forces for each particle in the simulation and updating their accelerations. The BH tree could either be dynamically updated or redrawn to depict the new state of the system. Calculating the gravitational forces, or in other words, walking the tree is the most compute intensive task in the simulation.

3.1   Test Cases for Java Version

The source distribution of the original Gadget-2 code comes with some initial conditions files including *Colliding Galaxies* and *Cluster Formation*. The Gadget-2 code produces snapshot files at regular intervals during the simulation which

can be used to plot the state of the system. The distribution also provides some
IDL (software for data visualisation and analysis) scripts to view the system.
We used these scripts along with the snapshot files to generate visual output,
which are indistinguishable for the two versions. This provides us with a very
high degree of confidence in correctness of the translated code. We perform this
comparison to ensure no bug has been introduced in the Java version.

3.2   Initial Java Optimizations

The performance evaluation of the initial Java version revealed that the performance was approximately three times slower than the C version. We now
describe the principal optimizations applied to improve performance.

Custom Serialization and Deserialization. Initial versions of Java Gadget-
2 communicated Java objects, which was made possible by exploiting the JDK
default serialization and de-serialization mechanism in MPJ Express. The object
serialization and de-serialization is the process of converting Java objects to a
byte array and vice versa. It can have detrimental effects on the performance of
a parallel application. Thus, we decided to replace Java object communication
in Java Gadget-2 with primitive datatypes.

In the original C Gadget-2, initial conditions are read into an array of C
`structs` called `ParticleData`. In the Java version, this array of `structs` is
replaced by an object array called `ParticleData`. Particles that need to be
exported are copied to a contiguous memory region called `CommBuffer` in the
original C version. We replaced this with `CommBuffer` object, which contained
object arrays. Before the communication operation, the data was copied from
`ParticleData` array onto a related object array in `CommBuffer` object and
communicated.

In the optimized version of Java Gadget-2, this `CommBuffer` object is replaced
by a contiguous memory region, which is an instance of `ByteBuffer` class. Before the actual communication, we copy primitive data from each element of
`ParticleData` array to `CommBuffer`. Once all the data has been packed onto
this `ByteBuffer`, it is communicated to the receiver process. The receiver process receives the data in `CommBuffer`, and unpacks it onto the `ParticleData`
object array. This technique helped us not only to avoid the Java object serialization overhead, but also reduced the memory footprint of the JVM (Java
Virtual Machine) by 60%.

Maintaining Memory Locality. It is hard to maintain memory locality for
Java HPC applications. The reason is that native machine architecture is not
aware of Java objects that might be involved in computationally intensive sections of the code. This might result in poor usage of processor cache and page
faults. The authors in [9] have identified this problem and proposed an object-
aware memory architecture.

In the Java version of Gadget-2, we maintained memory locality by *flattening*
sensitive data structures. Using this technique, we replaced Java object arrays

with primitive datatype arrays. For example, BH tree nodes are stored in an array of Java objects called Nodes_base. Each element of this array has members like an array of doubles called center and a double called len, that represents the side length of a tree node. In the Java version, these two members center and len are stored in a doubles array. This ensures that when a particular tree node is accessed, all the members of particular object element in Nodes_base array are in close vicinity in the memory.

We also *flattened* the ParticleData array, where each object has attributes like a three element array of pos and vel representing position and velocities in three dimensions. In addition, we also *flattened* the TopNodes array.

4 Performance Evaluation

In this section, we evaluate the performance of the Java version against the C Gadget-2 code. We used the *Colliding Galaxies* simulation for comparison. Note that the C version of Gadget-2 is meant to be a massively parallel code. The *Colliding Galaxies* simulation is too small to utilize its full potential. Nevertheless, it gives us a starting point for evaluating the performance of Java Gadget-2.

We conducted these tests on a cluster called *StarBug* at the DSG. This cluster consists of 8 dual Intel Xeon 2.8 GHz processors. The PCs were equipped with 2 Gigabytes of ECC RAM with 533 MHz Front Side Bus (FSB). The PCs were running the Debian GNU/Linux with the 2.4.32 Linux kernel. The C compiler on this cluster was GNU GCC 3.3.5. There is an option to use Myrinet or Fast Ethernet for communication. We used MPJ Express (version 0.23) with Sun JDK 1.5 (Update 6) to run the Java version of Gadget-2. The original C Gadget-2 code used MPICH (version 1.2.5.2) on Fast Ethernet and MPICH-MX (version 1.2.6..0.94) using Myrinet.

Figure 1 shows execution time of C and Java Gadget-2 on 1, 2, 4, and 8 processors using Fast Ethernet. Note that one MPI or Java process is running on a dual CPU node using one of the two available processors. A similar comparison of execution time on Myrinet is shown in Figure 2. The Java version is almost 30% slower than the C version.

Figure 3 shows tree-walk time of C and Java Gadget-2. The presented tree-walk is the average of all processors for more than one processor case. The Java version is approximately 30% slower in calculating gravitational force than the C version.

The speed-up for C and Java version is modest because of small problem size but the focus of this paper is the performance comparison of the two versions.

5 Conclusions and Future Work

In this paper, we have presented a Java version of Gadget-2. The performance evaluation of the Java version revealed that it can achieve around 70% of C Gadget-2's performance. This is understandable given that Java has extra run-time safety features. Also, it should be noted that the comparison is between a production quality C code against a Java code that could be further optimized.

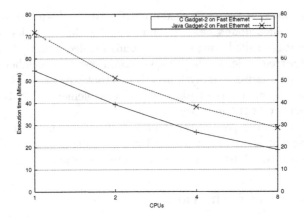

Fig. 1. Execution Time Comparison on Fast Ethernet

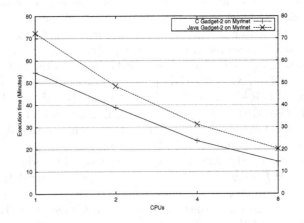

Fig. 2. Execution Time Comparison on Myrinet

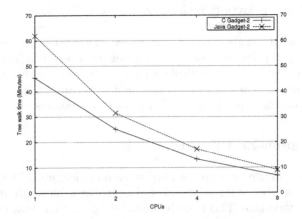

Fig. 3. Tree Walk Time Comparison

The performance of Java Gadget-2 shows that with careful programming, it is possible to achieve performance in the same general ballpark as C code. It could be argued that Java is an acceptable choice for HPC applications, especially the ones that require high reliability. Java ensures reliability by providing extra safety features including array bounds checking. For example, we discovered a scenario in the original C Gadget-2 where seventh element of a six element array was accessed. The Java Gadget-2 helped identify this scenario by throwing a `ArrayOutOfBound` exception. We have informed the developer of C Gadget-2, who has fixed this problem in the distribution.

In general, Java encourages better software engineering by being an object oriented language and is more portable than its precursors.

We plan to continue working on the Java Gadget-2 software and make a public release in the future. Our MPI-like messaging software MPJ Express is publicly available from http://dsg.port.ac.uk/projects/mpg.

References

1. Mark Baker, Bryan Carpenter, and Aamir Shafi. An Approach to Buffer Management in Java HPC Messaging. In V. Alexandrov, D. van Albada, P. Sloot, and J. Dongarra, editors, *International Conference on Computational Science (ICCS 2006)*, LNCS. Springer, 2006.
2. J. Barnes and P. Hut. A Hierarchical O(N log N) Force-calculation Algorithm . *Nature*, 324(4):446–449, 1986.
3. Pedro Bizarro, Luís Moura Silva, and João Gabriel Silva. JWarp: A Java Library for Parallel Discrete-Event Simulations. *Concurrency: Practice and Experience*, 10(11-13):999–1005, 1998.
4. The MONARC project. www.cern.ch/MONARC.
5. N. T. Padial-Collins, W. B. VanderHeyden, D. Z. Zhang, E. D. Dendy, and D. Livescu. Parallel operation of CartaBlanca on shared and distributed memory computers. *Concurrency and Computation: Practice and Experience*, 16(1):61–77, 2004.
6. John K. Salmon and Michael S. Warren. Skeletons from the Treecode Closet. *J. Comput. Phys.*, 111(1):136–155, 1994.
7. Volker Springel. The cosmological simulation code GADGET-2. *Monthly Notices of the Royal Astronomical Society*, 364:1105, 2005.
8. Volker Springel, Simon D. M. White, Adrian Jenkins, Carlos S. Frenk, Naoki Yoshida, Liang Gao, Julio Navarro, Robert Thacker, Darren Croton, John Helly, John A. Peacock, Shaun Cole, Peter Thomas, Hugh Couchman, August Evrard, Joerg Colberg, and Frazer Pearce. Simulating the joint evolution of quasars, galaxies and their large-scale distribution. *Nature*, 435:629, 2005.
9. Greg Wright, Matthew L. Seidl, and Mario Wolczko. An object-aware memory architecture. Technical Report TR-2005-143, Sun Microsystems, February 2005. http://research.sun.com/techrep/2005/abstract-143.html.

An Approach for Parallel Fluid-Structure Interaction on Unstructured Meshes

Ulrich Küttler and Wolfgang A. Wall

Chair of Computational Mechanics, TU Munich,
Boltzmannstr. 15, 85747 Garching, Germany
{kuettler, wall}@lnm.mw.tum.de
http://www.lnm.mw.tum.de

Abstract. The simulation of fluid-structure interaction (FSI) problems is a challenge in contemporary science and engineering. This contribution presents an approach to FSI problems with incompressible Newtonian fluids and elastic structures and discusses its realization in a general purpose parallel finite element research code. The resulting algorithm is robust and efficient and scales well on parallel machines. Recent attempts on efficiency improvements are discussed and a numerical example is shown.

1 Statement of Problem

Fluid-structure interaction (FSI) problems are non-overlapping multifield problems coupled at the interface. This contribution is concerned with the coupling of incompressible Newtonian fluids, fluids whose shear stresses depend linearly on the velocity gradient, with nonlinear structures. Both fields are governed by a time dependent nonlinear PDE that reads

$$\rho^S \frac{D^2 \mathbf{d}}{Dt^2} = \boldsymbol{\nabla} \cdot \mathbf{S} + \rho^S \mathbf{f}^S \qquad \text{in } \Omega^S \times (0, T), \tag{1}$$

in the structural field and

$$\left.\frac{\partial \mathbf{u}}{\partial t}\right|_{\chi} + \left(\mathbf{u} - \mathbf{u}^G\right) \cdot \boldsymbol{\nabla}\mathbf{u} - 2\nu\boldsymbol{\nabla}\cdot\boldsymbol{\varepsilon}(\mathbf{u}) + \nabla p = \mathbf{f}^F \quad \text{and} \quad \boldsymbol{\nabla}\cdot\mathbf{u} = 0 \quad \text{in } \Omega^F \times (0, T), \tag{2}$$

in the fluid field. The unknown structural displacements $\mathbf{d}$, fluid velocities $\mathbf{u}$ and fluid pressure p are searched for. At the coupling interface both displacements and forces must balance. The fluid domain deformation is treated by an Arbitrary Lagrangian–Eulerian (ALE) approach.

Please refer to [11] for a profound discussion of the formulation of FSI problems including the required initial and boundary conditions.

The governing equations are discretized in space using finite elements (FE), i.e. the whole domain is covered by an unstructured mesh of elements and nodes. The continuous field variables $\mathbf{d}$, $\mathbf{u}$ and p are replaced by discrete variables

B. Mohr et al. (Eds.): PVM/MPI 2006, LNCS 4192, pp. 366–373, 2006.

at the nodal points and approximations inside the elements. At the element level a sophisticated advection and pressure stabilization respectively a hybrid or mixed element formulation is employed to obtain high quality results. This discretization results in a transient nonlinear system of equations that consists of the structural part and the fluid part

$$\mathbf{M}^S \ddot{\mathbf{d}} + \mathbf{N}^S(\mathbf{d}) = \mathbf{f}^S, \tag{3}$$

$$\mathbf{M}^F \dot{\mathbf{u}} + \mathbf{N}^F(\mathbf{u})\mathbf{u} + \mathbf{K}^F \mathbf{u} + \mathbf{G}^F \mathbf{p} = \mathbf{f}^F, \quad \left(\mathbf{G}^F\right)^T \mathbf{u} = \mathbf{0}, \tag{4}$$

connected by discrete versions of the coupling conditions. These equations are further worked upon through direct time integration. The nonlinearities are treated by a fix-point like or Newton-Raphson scheme.

A detailed discussion of the discrete operators $\mathbf{N}$, $\mathbf{K}$ and $\mathbf{G}$ can be found in [11].

---

for all time steps:
  $i = 1$
  until the nonlinear equations (3) and (4) are satisfied:
    for all elements:
      calculate element matrix $\mathbf{k}_e$ and RHS vector $\mathbf{f}_e$
      assemble $\mathbf{K}_i$ and $\mathbf{f}_i$ from $\mathbf{k}_e$ and $\mathbf{f}_e$
    solve the linearized system of equations

$$\begin{pmatrix} \mathbf{K}_{II,i}^S & \mathbf{K}_{I\Gamma,i}^S & \\ \mathbf{K}_{\Gamma I,i}^S & \left(\mathbf{K}_{\Gamma\Gamma,i}^S + \delta\mathbf{K}_{\Gamma\Gamma,i}^F\right) & \mathbf{K}_{\Gamma I,i}^S \\ & \delta\mathbf{K}_{I\Gamma,i}^F & \mathbf{K}_{II,i}^F \end{pmatrix} \begin{pmatrix} \mathbf{d}_{I,i} \\ \mathbf{d}_{\Gamma,i} \\ \mathbf{u}_{I,i} \end{pmatrix} = \begin{pmatrix} \mathbf{f}_{I,i}^S \\ \mathbf{f}_{\Gamma,i}^S + \mathbf{f}_{\Gamma,i}^F \\ \mathbf{f}_{I,i}^F \end{pmatrix} \tag{5}$$

  $i = i + 1$

Algorithm 1. Basic sketch of monolithic FSI solver

2 Solution of Coupled FSI Problem

2.1 Monolithic Solution Approach

A monolithic solution approach treats all field equations of the FSI problem at the same time [3,4]. That is the algorithm consists in the repeated assembling and solution of a huge linear system of equations with quite diverse entries. In a very compact notation, which abbreviates the left hand side (LHS) of a linearized equation system with $\mathbf{K}$ and denotes by I and Γ the degrees of freedom inside and at the coupling interface of a domain, respectively, algorithm 1 outlines the solving procedure where the δ at the fluid interface contributions in equation (5) accounts for the time discretization of the fluid velocity.

The LHS matrix of equation (5) follows from the discretization of the whole domain by the finite element method and consequently it is banded and very sparse. The direct solution of this system of equations destroys the sparsity pattern and consumes a lot of memory. Because of this memory requirements direct solution approaches are unfeasible for any reasonable problem size. The alternative are iterative solution techniques that do not suffer from excessive memory consumption. The central operation of iterative linear equation solvers is a matrix-vector product. Using an iterative technique the modified solution algorithm is sketched in algorithm 2.

---

for all time steps:
    $i = 1$
    until the nonlinear equations (3) and (4) are satisfied:
        for all elements:
            calculate element matrix $\mathbf{k}_e$ and RHS vector $\mathbf{f}_e$
            assemble $\mathbf{K}_i$ and $\mathbf{f}_i$ from $\mathbf{k}_e$ and $\mathbf{f}_e$
        $j = 1$
        until convergence of the linear system:
            $\mathbf{d}_{j+1} = \mathbf{K}_i \cdot \mathbf{d}_j$
            $j = j + 1$
        $i = i + 1$

Algorithm 2. Sketch of coupled FSI solver with iterative linear equation solver

Another important advantage of iterative solvers for linear systems of equations is better behavior in parallel environments. With a parallel solver for linear systems of equations the above algorithm can easily be run in parallel. This is discussed in more detail in section 3.1.

2.2 Partitioned Solution Approach

Algorithm 2 suffers from various difficulties. On the numerical side it turns out that the linearized system matrix $\mathbf{K}$ is poorly conditioned because of the widely different properties of the participating fields. Very extensive (and expensive) preconditioning is needed to attack this matrix with iterative solvers. From an implementation point of view the derivation of off-diagonal blocks is quite cumbersome [3,2] and also contradicts the aspired software modularity. The later point is particularly important since both field solvers (fluid and structure) solve complex real-world problems and need to be quite sophisticated on their own.

A partitioned approach that avoids these difficulties was presented in [12]. The partitioned approach builds on domain decomposition methods to separate the fluid and the structural domain. In most cases [7,10,1,6] a Dirichlet-Neumann approach is used that prescribes fluid velocities at the coupling interface of the fluid domain and applies the resulting forces to the coupling interface of the structural domain. This approach is natural from an engineering point of view. Furthermore it eases implementation because the field solvers can be handled independently. Available solvers can be integrated easily, which makes the approach most advantageous in industrial applications. The price to pay for these advantages is the additional computation time required by the field iteration.

The FSI algorithm for the relaxed Gauß-Seidel solver as suggested by [12,7] is shown in algorithm 3. The crucial point for efficiency and robustness of the partitioned solution approach is relaxation of interface displacements $\mathbf{d}_\Gamma^S$.

---

for all time steps:
    until convergence of interface displacements $\mathbf{d}_\Gamma^S$
        $i = 1$
        until the nonlinear equations (4) are satisfied:
            for all fluid elements:
                calculate element matrix $\mathbf{k}_e^F$ and RHS vector $\mathbf{f}_e^F$
                assemble $\mathbf{K}_i^F$ and $\mathbf{f}_i^F$ from $\mathbf{k}_e^F$ and $\mathbf{f}_e^F$
            $j = 1$
            until convergence of the linear fluid system:
                $\mathbf{u}_{j+1} = \mathbf{K}_i^F \cdot \mathbf{u}_j$
                $j = j + 1$
        $i = i + 1$
        transfer fluid interface forces $\mathbf{f}_\Gamma^F$ to the structure
        $i = 1$
        until the nonlinear equations (3) are satisfied:
            for all structure elements:
                calculate element matrix $\mathbf{k}_e^S$ and RHS vector $\mathbf{f}_e^S$
                assemble $\mathbf{K}_i^S$ and $\mathbf{f}_i^S$ from $\mathbf{k}_e^S$ and $\mathbf{f}_e^S$
            $j = 1$
            until convergence of the linear structural system:
                $\mathbf{d}_{j+1} = \mathbf{K}_i^S \cdot \mathbf{d}_j$
                $j = j + 1$
        $i = i + 1$
        relax interface displacements $\mathbf{d}_\Gamma^S$
        transfer structural interface displacements $\mathbf{d}_\Gamma^S$ to the fluid

---

Algorithm 3. Sketch of partitioned Gauß-Seidel like FSI solver

3   Parallel Object Based Simulation Code

The above FSI algorithm is implemented in a object based FE code written in C and FORTRAN, where FORTRAN is used for time critical inner loops and all I/O facilities as well as data structures are implemented in C. The choice of language was guided by performance considerations and the availability of development tools on a wide range of platforms. A further point in question was the expected learning curve for casual developers.

The central structures are shown in figure 1. The physical fields are represented by a Field structure that contains any number of discretizations. Each discretization consists of a collection of elements and nodes that make up the FE mesh. For clarity secondary data structure that represent the mesh topology and the boundary conditions are not shown.

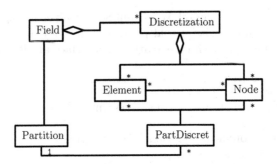

Fig. 1. Core design of the simulation code

3.1 Parallelization

In the lower half of figure 1 there are the data structures needed for parallelization. These data structures mirror the main structures, but depict just a partition
of the entire mesh. The decomposition of the mesh is such that each node belongs
to exactly one processor, consequently the elements at the partitions boundaries
are shared with the adjacent partitions. This way the global systems of equations are naturally distributed among the participating processors because each
equation belongs to exactly one node.

In a parallel environment each processor calculates and assembles just those
global equations that belong to its nodes. This can be parallelized without any
need for communication, it requires to loop all elements adjacent to the nodes
of each processor. The matrix-vector product needed to solve the distributed
system of equations needs some communication of vector elements. However
there are efficient communication patterns and the amount of communication
required depends on the partitioning of the FE mesh. A partitioning with compact submeshes that reduces the required communication can be obtained by
graph partitioning tools such as METIS [5]. Hence a parallel execution of algorithm 3 is feasible without major modifications in the algorithm structure. The
expected speedup warrants the effort.

3.2 Efficiency Improvements

Algorithm 3 is realized in our general purpose FE research software. General
purposeness in this case means a broad range of algorithms with many variants,
easy adoption to new requirements as well as applicability of many different
computing systems. In particular support for cache based machines and vector
machines is needed. This flexibility is of course in conflict with the efficiency of
the code. Our attempts to increase the efficiency include (a) the introduction of
configurable element sets to fill the vector pipes on vector machines and (b) the
change of the sparse matrix format to a node block based format on cache
systems or even a jagged diagonal format on vector machines. Efforts to improve
the linear equation solver are justified by the dominating role of these solvers

in transient FSI simulations. However, considerable time is spend with element calculations, especially in 3d simulations, since a high approximation quality can only be achieved using elaborated element techniques. The effort spend with element formulation pays off, though, by means of a considerable reduction of unknowns in the global system of equations. Results of this efforts can be found in [8,9].

Future tasks in that direction include improved data parallelization of FE meshes and RHS vectors.

4   Numerical Example

As numerical example for the presented algorithm a flexible structure in a 2d channel has been calculated. A snapshot of the simulation is shown in figure 2. This small example consists of 100258 fluid equations, 12400 structural equations and 61360 mesh equations. The unsymmetrical fluid equations are solved with GMRES, the structure and mesh equations are solved with CG. Processor local ILU preconditioning is applied in both cases. The example has been executed on 4, 8 and 16 processors on three hardware platforms.

Xeon   Intel Xeon EM64T 3.2 GHz, Infiniband, 2 processors per node
Opteron   AMD Opteron 850 2.4 GHz, Infiniband, 4 processors per node
Altix   SGI Altix 3700 BX2

Measurements for element matrix calculation, matrix assembly and solution of the global system of equations are shown in figure 3. Interestingly the times required by the Xeon and the Opteron machine for the element calculation and the matrix assembly are practically identical. That is possibly due to identical memory latency. The Altix seems to have some more difficulties with the fine grained work in the element routines and the integer arithmetic of the assembling process. On the other hand does the Altix excel, as expected, during the number crunching of the global linear solver. And this task turns out to dominate the overall execution time.

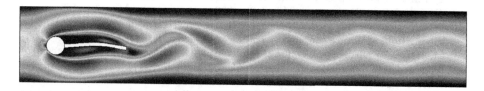

Fig. 2. Velocity |u| of example channel problem with embedded flexible structure

Please note, however, that these are very preliminary numbers. We are currently deploying a new parallel cluster that will enable us to do more profound studies and to solve larger systems. The conference presentation will contain these new cases.

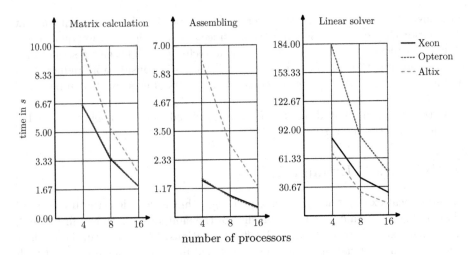

Fig. 3. Execution time in s of two time steps. The element matrix calculation, the matrix assembly and the solution of the global distributed system of equations are measured.

5 Conclusion

This contribution discusses our ongoing effort to establish a FE based FSI solver for general complex real-world problems on parallel machines. The partitioned Gauß-Seidel algorithm with relaxation is presented and its parallelization discussed. The main design points of our general purpose FE research code are sketched including some recent efforts on efficiency improvements. A small example demonstrates the applicability of the proposed algorithm. More examples are to be shown at the conference.

References

1. S. Deparis, M. Discacciati, G. Fourestey, and A. Quarteroni. Fluid-structure algorithms based on Steklov-Poincaré operators. *Comp. Meth. in Appl. Mech. and Engng.*, doi:10.1016/j.cma.2005.09.029, 2006.
2. M.Á. Fernández and M. Moubachir. A Newton method using exact jacobians for solving fluid-structure coupling. *Computers & Structures*, 83(2–3):127–142, 2005.
3. M. Heil. An efficient solver for the fully coupled solution of large-displacement fluid-structure interaction problems. *Comp. Meth. in Appl. Mech. and Engng.*, 193:1–23, 2004.
4. B. Hübner, E. Walhorn, and D. Dinkler. A monolithic approach to fluid-structure interaction using space-time finite elements. *Comp. Meth. in Appl. Mech. and Engng.*, 193:2087–2104, 2004.
5. G. Karypis and V. Kumar. Multilevel k-way partitioning scheme for irregular graphs. *J. Parallel Distrib. Comput.*, 48(1):96–129, 1998.
6. U. Küttler, Ch. Förster, and W.A. Wall. A solution for the incompressibility dilemma in partitioned fluid-structure interaction with pure dirichlet fluid domains. *Comput. Mech.*, doi:10.1007/s00466-006-0066-5, 2006.

7. D.P. Mok and W.A. Wall. Partitioned analysis schemes for the transient interaction of incompressible flows and nonlinear flexible structures. In *Trends in Computational Structural Mechanics, W.A. Wall, K.-U. Bletzinger and K. Schweitzerhof (Eds.)*, 2001.

8. M. Neumann, U. Küttler, S.R. Tiyyagura, W.A. Wall, and E. Ramm. Computational efficiency of parallel unstructured finite element simulations. In *M. Resch, T. Boenisch, K. Benkert, T. Furui, Y. Seo, and W. Bez (Ed.), High Performance Computing on Vector Systems. Proceedings of the High Performance Computing Center Stuttgart, March 2005*. Springer, 2006.

9. M. Neumann, S.R. Tiyyagura, W.A. Wall, and E. Ramm. Robustness and efficiency aspects for computational fluid structure interaction. In *E. Krause, Y.I. Shokin, M. Resch, N. Shokina (eds.), "Computational Science and High Performance Computing II. The 2nd Russian-German Advanced Research Workshop, Stuttgart, Germany, March 14 to 16, 2005"*, volume 91 of *Notes on Numerical Fluid Mechanics and Multidisciplinary Design (NNFM)*. Springer, 2006.

10. T.E. Tezduyar. Finite element methods for fluid dynamics with moving boundaries and interfaces. In E. Stein, R. De Borst, and T.J.R. Hughes, editors, *Encyclopedia of Computational Mechanics*, volume 3, chapter 17. John Wiley & Sons, 2004.

11. W.A. Wall. *Fluid-Struktur-Interaktion mit stabilisierten Finiten Elementen*. PhD thesis, Institut für Baustatik, Universität Stuttgart, 1999.

12. W.A. Wall, D.P. Mok, and E. Ramm. Partitioned analysis approach of the transient coupled response of viscous fluids and flexible structures. In *W. Wunderlich (Ed.), Solids, Structures and Coupled Problems in Engineering, Proceedings of the European Conference on Computational Mechanics ECCM '99, Munich*, 1999.

Optimizing a Conjugate Gradient Solver with Non-Blocking Collective Operations

Torsten Hoefler[1,2], Peter Gottschling[1],
Wolfgang Rehm[2], and Andrew Lumsdaine[1]

[1] Indiana University, Open Systems Lab, Bloomington, IN 47404 USA
{htor, pgottsch, lums}@cs.indiana.edu
[2] Technical University of Chemnitz, Department of Computer Science, 09107
Chemnitz, Germany
{htor, rehm}@cs.tu-chemnitz.de

Abstract. This paper presents a case study about the applicability and usage of non-blocking collective operations. These operations provide the ability to overlap communication with computation and to avoid unnecessary synchronization. We introduce our NBC library, a portable low-overhead implementation of non-blocking collectives on top of MPI-1. We demonstrate the easy usage of the NBC library with the optimization of a conjugate gradient solver with only minor changes to the traditional parallel implementation of the program. The optimized solver runs up to 34% faster and is able to overlap most of the communication. We show that there is, due to the overlap, no performance difference between Gigabit Ethernet and InfiniBand[TM] for our calculation.

1 Introduction

Historically, overlapping communication and computation is the most common approach for scientists to leverage parallelism between processing and communication units [1]. The resulting application is less latency sensitive, and can even, up to a certain extent, run on high latency networks without any change in the parallel speedup. The non-blocking operations allow the applications to ignore process skew or network jitter, which often has negative effects on the running time [2]. Both can be very beneficial on Cluster-Computers (also known as Networks of Workstations, NOW) and on Grid-based systems.

The Message Passing Interface (MPI) standard is currently the de-facto standard for parallel computing and many scientific programs exist which use MPI as their communication layer. MPI-1 offers the possibility to overlap communication and computation and to avoid unnecessary synchronization for point-to-point messages (MPI_ISEND, MPI_IRECV). However, many applications can benefit from using MPI collective communication, which is often optimized for the underlying hardware (e.g., [3,4]) and delivers much better performance than comparable point-to-point communication schemes. Another advantage of collective communication is their abstraction of communication and the resulting

B. Mohr et al. (Eds.): PVM/MPI 2006, LNCS 4192, pp. 374–382, 2006.

ease of use for parallel programs. Gorlatch recently published a good survey of reasons to use collective communication [5].

Especially, applications from scientific computing (SC) are well-suited to benefit from the more abstract parallelization approach of collective communication. Furthermore, many algorithms in SC, e.g., linear solvers, provide a high potential of overlapping communication and computation. In order to combine the advantages of this overlapping and of collective communication, we introduce non-blocking collective operations for the MPI-1 standard and demonstrate their gain in a conjugate gradient solver. An assessment of possible benefits has been presented in [6].

1.1 Related Work

The idea to provide non-blocking collective operations grew out of discussions for the MPI-2 standard. The MPI Forum defined split collectives which were not standardized in MPI-2, but were written down in the MPI-2 Journal of Development (JoD [7]). However, these operations are too limited to be easily usable for scientists. IBM extended the interface and implemented non-blocking collectives as part of their Parallel Environment, but they dropped the support for them in the latest version because they were not part of the MPI standard and were only rarely used by scientists who preferred portability. The upcoming MPI/RT standard [8] defines all operations, including collective operations, in a non-blocking manner. Kale et. al. implemented a non-blocking all-to-all communication as part of the CHARM++ framework [9]. To the best of the authors' knowledge, there are neither explicit studies on performance gain and nor optimized implementations of non-blocking collective operations available.

2 Implementing Non-Blocking Collective Operations

Our implementation aims mainly at portability, low overhead, and ease of use. We built the first prototype library on top of non-blocking point-to-point operations defined in the MPI-1 standard. Therefore, although we cannot leverage special hardware features, the protoype library is portable to all MPI-1 capable parallel computers. Further because we implemented optimized algorithms for all collective operations, we deliver the same performance as the hardware independent blocking collective operations in MPICH2 [10] and Open MPI 1.0 [11].

The interface to the calls is very similar to the blocking MPI collective operations. However, to ensure non-blocking operation, a handler is returned which is comparable to a MPI_REQUEST. The behavior and the application programming interface (API) of those non-blocking collective calls are defined in [12].

The following subsections provide an overview of the implementation of our non-blocking collectives (NBC) library, which offers asynchronous collective support on top of MPI-1. The only difference to the definition in [12] is that all calls and constants are prefixed with NBC_ instead of MPI_ to avoid confusion with MPI standardized operations.

2.1 The Scheduling Engine

To ease implementation, we propose a general framework to support all operations. This framework, our scheduling engine, builds and executes a schedule to perform collective operations. Each collective operation, defined in the MPI standard, can be expressed as a row of sends, receives and operations between ranks of a specific communicator. These functions can be arranged into r communication rounds to build a communicator-specific schedule for each rank. Each round may consist of one or more operations which have to be independent and will be executed simultaneously. Operations in different rounds depend on each other, in a way that operations on round n can only be started after **all** operations in round $n-1$ have finished $\forall\, 0 \le n \le r$.

2.2 Building a Schedule

The schedule defines all required actions to perform the collective operation for a specific rank and a specific communicator. A rank's schedule is specific to each communicator and MPI argument set. It is designed to be reusable if it is saved in association to the communicator and the arguments.

A schedule consists of actions (send, receive, operation) and rounds. It is laid out as a contiguous array in memory to be cache friendly. The memory layout of the simplified example schedule for rank 0, for a MPI_BARRIER implemented with the dissemination principle on a four-node communicator is shown in Fig. 1. This schedule has a send operation to rank 1 and a receive operation from rank 3 in the first round. The round is ended by the **end** flag. The second round issues a send to rank 2 and a receive from rank 2. The dissemination barrier is finished after those operations and NBC_TEST or NBC_WAIT calls return NBC_OK.

| send to 1 | recv from 3 | end | send to 2 | recv from 2 | end |

Fig. 1. Memory Layout of a schedule at rank 0, implementing a Dissemination Barrier between 4 nodes

2.3 Schedule Execution

The schedule array in Fig. 1 consists of four operations in two rounds. The schedule represents the necessary operations to perform a MPI_BARRIER on rank 0 of 4. The non-blocking execution of the schedule begins if the user calls NBC_IBARRIER(comm, handle). The first call to NBC_IBARRIER builds the schedule (if not already done), starts all operations of the first round in a non-blocking manner, initializes the handle, and returns immediately to the user. The user can perform any computation while the operations are processed in the background. The amount of progress made in the background depends on the actual MPI implementation. The current implementation of the NBC library is runnable in environments which offer no thread support. This means that the user should progress the operation manually by calling NBC_TEST(handle). NBC_TEST checks all pending operations for completion and proceeds to the next round if the current round is completed. It returns NBC_OK if the operation (all rounds) is finished, otherwise NBC_CONTINUE to indicate that the operation is still running.

3   Optimization of Linear Solvers

Accelerating parallel applications in scientific computing is a main topic of many research projects. Non-blocking collective communication can be an important contribution to it and we will demonstrate this on a selected case study.

Iterative linear solvers are important components of most applications in SC. They consume, with very few exceptions, a significant part of the overall run-time of typical applications. In many cases, they even dominate the overall execution time of parallel code. Reducing the computational needs of linear solvers will thus be a huge benefit for the whole scientific community.

Despite the very different algorithms and varying implementations of many of them, one common operation is the multiplication of very large and sparse matrices with vectors. Assuming an appropriate distribution of the matrix, large parts of the computation can be realized on local data and the communication of required remote data — also referred to as inner boundaries or halo — can be overlapped with the local part of the matrix vector product.

3.1   Case Study: 3-Dimensional Poisson Equation

For the sake of simplicity, we use the well-known Poisson equation with Dirichlet boundary conditions, e.g., [13]

$$- \Delta u = 0 \quad \text{in } \Omega = (0,1) \times (0,1) \times (0,1), \tag{1}$$
$$u = 1 \quad \text{on } \Gamma. \tag{2}$$

The domain Ω is equidistantly discretized. Each dimension is split into $N+1$ intervals of size $h = 1/(N+1)$. Within Ω one defines $n = N^3$ grid points

$$G = \{(x_1, x_2, x_3) | \forall i,j,k \in \mathbb{N}, 0 < i,j,k \leq N : x_1 = ih, x_2 = jh, x_3 = kh\}.$$

Thus, each point in G can be represented by a triple of indices (i,j,k) and we denote $u(ih, jh, kh)$ as $u_{i,j,k}$. Lexicographical order allows to store the values of the three-dimensional domain into a one-dimensional array. For distinction we use a typewriter font for the memory representation and start indexing from zero as in C/C++

$$u_{i,j,k} \equiv u[(i-1) + (j-1)*N + (k-1)*N^2] \quad \forall 0 < i,j,k \leq N. \tag{3}$$

The differential operator $-\Delta$ is discretized for each $x \in G$ with the standard 7 point stencil represented as a sparse matrix in $\mathbb{R}^{n \times n}$ using the memory layout from (3), confer e.g. [13] for the 2D case.

3.2   Domain Decomposition

The grid G is partitioned into p sub-grids $G_1, \ldots, G_p$ where p is the number of processors. The processors are arranged in a non-periodic Cartesian grid $p_1 \times p_2 \times p_3$ with $p = p_1 \cdot p_2 \cdot p_3$, provided by MPI_DIMS_CREATE. In case that N is divisible by $p_i \forall i$ the local grids on each processor have size $N/p_1 \times N/p_2 \times N/p_3$, otherwise the local grids are such that the whole grid is partitioned and the sizes along each dimension vary at most by one.

Each sub-grid has 3 to 6 adjoint sub-grids if all $p_i > 1$. Two processors P and P' storing adjoint sub-grids are neighbors, written as the relation $Nb(P, P')$. This neighborhood can be characterized by the processors' Cartesian coordinates $P \equiv (P_1, P_2, P_3)$ and $P' \equiv (P_1', P_2', P_3')$

$$Nb(P, P') \quad \text{iff} \quad |P_1 - P_1'| + |P_2 - P_2'| + |P_3 - P_3'| = 1. \tag{4}$$

Fig. 2 shows the partition of G into sub-grids and necessary communication.

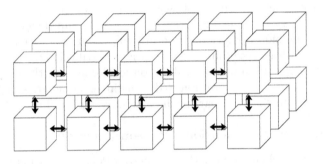

Fig. 2. Processor Grid

3.3   Design and Optimization of the CG Solver

The conjugate gradient method (CG) by Hestenes and Stiefel [14] is a widely used iterative solver for systems of linear equations when the matrix is symmetric and positive definite. To provide a simple base of comparison we restrain from preconditioning [13] and from aggressive performance tuning [15]. However, the local part part of the dot product is unrolled using multiple temporaries, the two vector updates are fused in one loop, and the number of branches is minimized in order to provide a high-performance base case. The parallelization of CG in the form of Listing 1.1 is straight-forward by distributing the matrix and vectors and computing the vector operations and the contained matrix vector product in parallel.

Neglecting the operations outside the iteration, the scalar operations in Listing 1.1 — line 1, 2, 6, 9, and 11 — and part of the vector operations — line 3, 7, and 8 — are completely local. The dot products in line 5 and 10 require communication in order to combine local results with MPI_ALLREDUCE to the global value. Unfortunately, computational dependencies avoid overlapping this

```
1  while (sqrt(gamma) > epsilon * error_0) {
2    if (iteration > 1)
3        q = r + gamma / gamma_old * q;
4    v = A * q;
5    delta = dot(v, q);
6    alpha = delta / gamma;
7    x = x + alpha * q;
8    r = r - alpha * v;
9    gamma_old = gamma;
10   gamma = dot(r, r);
11   iteration = iteration + 1;
12 }
```

Listing 1.1. Pseudo-code for CG method

```
1    fill_buffers(v_in, send_buffers);
2    start_send_boundaries(comm_data);
3    volume_mult(v_in, v_out, comm_data);
4    finish_send_boundaries(comm_data);
5    mult_boundaries(v_out, recv_buffers);
```

Listing 1.2. Pseudo-code for parallel matrix vector product

reductions. Therefore, the whole potential to save communication time in a CG method lies in the matrix vector product — line 4 of Listing 1.1.

3.4   Parallel Matrix Vector Product

Due to the regular shape of the matrix, it is not necessary to store the matrix explicitly. Instead the projection $u \mapsto -\Delta u$ is computed. In the distributed case $p > 1$, values on remote grid points need to be communicated in order to complete the multiplication. In our case study, the data exchange is limited to values on outside planes of the sub-grids in Fig. 2 unless the plane is adjoint to the boundary Γ. Therefore, processors must send and receive up to six messages to their neighbors according to (4) where the size of the message is given by the elements in the corresponding outer plane.

However, most operations can be already executed with locally available data during communication as shown in Listing 1.2. The first command copies the values of v_in needed by other processors into the send buffers. Then an all-to-all communication is launched, which can be blocking using MPI_ALLTOALLV or non-blocking using NBC_IALLTOALLV, which has identical arguments plus a NBC_HANDLE that is used to identify the operation later. The command volume_mult computes the local part of the matrix vector product (MVP) and in case of non-blocking communication, NBC_TEST is called periodically with the handle returned by NBC_IALLTOALLV in order to progress the non-blocking operations, cf. Section 2.1. Before using remote data in mult_boundaries, the completion of NBC_IALLTOALLV is checked in finish_send_boundaries with an NBC_WAIT on the NBC_HANDLE.

3.5   Benchmark Results

We performed a CG calculation on a grid of $800 \times 800 \times 800$ points until the residual was reduced by a factor of 100, which took 218 iterations for each run. This weak termination criterion was chosen to allow more tests on the cluster. We verified on selected tests with much stronger termination criteria that longer executions have the same relative behavior. The studies were conducted on the odin cluster available at the Indiana University which consists of 128 dual 2 GHz Opteron 246 nodes connected with flat InfiniBand[TM] and Gigabit Ethernet networks. Fig. 3 shows the benchmark results up to 96 nodes. We see that the

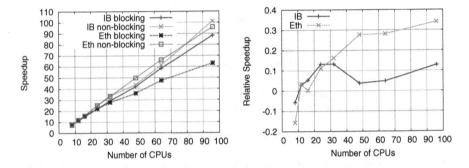

Fig. 3. Parallel Speedup (left) and Relative Performance Gain (right)

usage of our NBC library resulted in a reasonable performance gain for nearly all node counts. The performance loss at 8 processors is caused by relatively high effort to test the progress of communication. Finding simple rules to adapt the testing overhead to communication needs is subject to ongoing research. Due to the implementation design described above, non-blocking point-to-point communication would perform almost equally while requiring the management for multiple communication handlers including the progress enforcement. The overall results show that for both networks, InfiniBand[TM] and Gigabit Ethernet, nearly all communication can be overlapped and the parallel execution times are similar. The factor of 10 in bandwidth and the big difference in the latency of both interconnects does not influence the running time, even if the application has high communication needs. The partially superlinear speedup is due to the calculation of the inner part of the matrix.

3.6   Optimization Impact on Other Linear Solvers

Other Krylov sub-space methods have comparable dependencies on reduction operations which similarly limit the potential of communication overlapping to parts of the execution. Preconditioners of Krylov sub-space methods are often operations similar to MVP, e.g., incomplete LU or Cholesky factorization, and have the potential of overlapping.

Classical iterative solvers, like Gauß-Seidel, only consist of operations comparable with a matrix vector product and, thus, the whole computation is subject to overlapping. Due to very slow convergence, their importance as iterative solvers is limited. However, these methods are very important components of multigrid methods (MG) [16]. Other operations in MG, which project values between two grids, have a high potential to overlap communication, too. The computation on on the small grids introduces severe communication bottlenecks where non-blocking communication can provide significant improvements. As multigrid methods are solvers with minimal complexity, they are extremely important in SC and we will investigate them in detail in future work.

4 Conclusions and Future Work

We demonstrated the easy use of the NBC library and the principle of non-blocking collectives for a class of application kernels. We were able to improve the parallel application running time by up to 34% with minor changes to the application. The CG solver source code and the NBC library are available at: http://www.unixer.de/NBC/.

Future work includes an optimized MPI-2 implementation of the NBC library, hardware optimized non-blocking collective operations, and the analysis of more applications. The possibility of asynchronous progress, which removes the need for testing, with a separate thread will also be investigated. However, this may have other implications because the user can not control when the library gets called and possibly wipes out the CPU cache.

Acknowledgments

The authors want to thank Jeff Squyres, George Bosilca, Graham Fagg and Edgar Gabriel for helpful discussions. This work was supported by a grant from the Lilly Endowment and National Science Foundation grant EIA-0202048.

References

1. Liu, G., Abdelrahman, T.: Computation-communication overlap on network-of-workstation multiprocessors. In: Proc. of the Int'l Conference on Parallel and Distributed Processing Techniques and Applications. (1998) 1635–1642
2. Petrini, F., Kerbyson, D.J., Pakin, S.: The case of the missing supercomputer performance: Achieving optimal performance on the 8, 192 processors of asci q. In: Proceedings of the ACM/IEEE SC2003 Conference on High Performance Networking and Computing, 15-21 November 2003, Phoenix, AZ, USA, CD-Rom, ACM (2003) 55
3. Hoefler, T., Mehlan, T., Mietke, F., Rehm, W.: Adding Low-Cost Hardware Barrier Support to Small Commodity Clusters. In: 19th International Conference on Architecture and Computing Systems - ARCS'06. (2006) 343–350

4. Liu, J., Mamidala, A., Panda, D.: Fast and scalable mpi-level broadcast using infiniband's hardware multicast support (2003)
5. Gorlatch, S.: Send-receive considered harmful: Myths and realities of message passing. ACM Trans. Program. Lang. Syst. **26**(1) (2004) 47–56
6. Hoefler, T., Squyres, J., Rehm, W., Lumsdaine, A.: A Case for non Blocking Collective Operations (2006) submitted to ISPA - preprint available at: http://www.unixer.de/sec/nbcoll.pdf.
7. Message Passing Interface Forum: MPI-2 Journal of Development (1997)
8. Kanevsky, A., Skjellum, A., Rounbehler, A.: MPI/RT - an emerging standard for high-performance real-time systems. In: HICSS (3). (1998) 157–166
9. Kale, L.V., Kumar, S., Vardarajan, K.: A Framework for Collective Personalized Communication. In: Proceedings of IPDPS'03, Nice, France (2003)
10. MPICH2 Developers: http://www-unix.mcs.anl.gov/mpi/mpich2/ (2006)
11. Gabriel, E., Fagg, G.E., Bosilca, G., Angskun, T., Dongarra, J.J., Squyres, J.M., Sahay, V., Kambadur, P., Barrett, B., Lumsdaine, A., Castain, R.H., Daniel, D.J., Graham, R.L., Woodall, T.S.: Open MPI: Goals, Concept, and Design of a Next Generation MPI Implementation. In: Proceedings, 11th European PVM/MPI Users' Group Meeting, Budapest, Hungary (2004)
12. Hoefler, T., Squyres, J.M., Bosilca, G., Fagg, G.: Non Blocking Collective Operations for MPI-2 (2006) preprint available at: http://www.unixer.de/sec/standard_nbcoll.pdf.
13. Hackbusch, W.: Iterative solultion of large sparse systems of equations. Springer (1994)
14. Hestenes, M., Stiefel, E.: Methods of conjugate gradients for solving linear systems. J. Res. Natl. Bur. Stand. **49** (1952) 409–436
15. Gottschling, P., Nagel, W.E.: An efficient parallel linear solver with a cascadic conjugate gradient method. In: EuroPar 2000. Number 1900 in LNCS (2000)
16. Trottenberg, U., Oosterlee, C., Schüller, A.: Multigrid. Academic Press (2000)

Parallel DSMC Gasflow Simulation of an In-Line Coater for Reactive Sputtering

A. Pflug, M. Siemers, and B. Szyszka

Fraunhofer Institute for Surface Engineering and Thin Films IST,
Bienroder Weg 54e, 38108 Braunschweig, Germany
andreas.pflug@ist.fraunhofer.de

Abstract. There is an increasing demand for high precision coatings on large areas via in-line reactive sputtering, which requires advanced process control techniques. Thus, an improved theoretical understanding of the reactive sputtering process kinetics is mandatory for further technical improvement. We present a detailed Direct Simulation Monte Carlo (DSMC) gas flow model of an in-line sputtering coater for large area architectural glazing. With this model, the pressure fluctuations caused by a moving substrate are calculated in comparison with the experiment. The model reveals a significant phase shift in the pressure fluctuations between the areas above the center and the edges of the substrate. This is a geometric effect and is e. g. independent of the substrate travelling direction. Consequently, a long sputtering source will observe pressure fluctuations at its center and edges, which are out of phase.

For a heuristic model of the reactive sputtering process, we show that in certain cases a two-dimensional model treatment is sufficient for predicting the film thickness distribution on the moving substrate. In other cases, a strong phase shift between averaged pressure fluctuations and reactive sputtering process response is observed indicating that a three-dimensional model treatment is required for a realistic simulation of the in-line deposition process.

1 Introduction

Reactive sputtering is a key technology for a large variety of technical applications such as thin film photovoltaics, displays, architectural energy-saving and automotive coatings. With increasing size of coated substrates and with increased performance of the coating devices, advanced process control techniques e. g. in order to maintain a precise film homogeneity or to obtain a specific crystalline phase of the coated film become important. To improve the theoretical understanding of the reactive sputtering process kinetics, heuristic models as originally introduced by Berg et al. [Ber87, Ber05] enabling a qualitative understanding are investigated by several groups. In this "Berg model", the reactive sputtering process is treated within a volume with homogeneous partial pressure by simplified balance equations between the gas phase and the oxidation degrees of target and substrate surfaces.

In order to obtain a heuristic sputtering simulation for more realistic recipient geometries, an approach of coupling multiple Berg models via flow conductances

B. Mohr et al. (Eds.): PVM/MPI 2006, LNCS 4192, pp. 383–390, 2006.

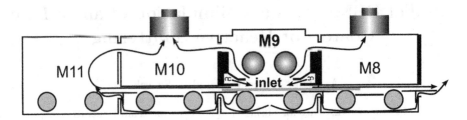

Fig. 1. Two dimensional cross section of modules M8–M11 of the "BigMag" in-line coater. Module M9 contains a double rotatable sputtering target, while in the upper regions of M8 and M10 three turbo molecular pump with a pumping speed of approx. $1.4 \, m^3/2$ are located, respectively. The load lock is located in module M1 (not shown), additional pumping is installed in modules M5 and M3.

and effective pumping speeds was introduced in [Pfl03]. For typical pressures in the range of $10^{-2} \ldots 10^{-4}$ mbar, the Knudsen number of a sputtering system is in the order of $0.1 \ldots 10$. Thus we need to consider rarified gas flow conditions, which can be simulated by a "Direct Simulation Monte Carlo" (DSMC) approach as given in [Bir94].

In a cooperation with Applied Films & Co. KG we performed a DSMC simulation of a so-called "BigMag" in-line coater with a moving glass substrate which causes pressure fluctuations in the sputtering compartment in the order of 5% [Pfl04]. The coater setup around the sputtering compartment under investigation is shown in Fig. 1 With a subsequently developed, twodimensional heuristic model, where the pressure fluctuations are averaged along the sputter source direction, i. e. perpendicular to the transport direction, the resulting film thickness profile on the substrate could be correctly predicted in the case of a reactive ZnO deposition. However other experiments reveal that in many cases the fluctuations of e. g. the sputter target voltage are not in phase with the averaged pressure fluctuations. In these cases the simulated film thickness profile strongly deviates from the experiment.

A detailed analysis of the simulated threedimensional pressure distribution, as presented in this paper, shows that this is due to a strong phase shift in the pressure fluctuations between the area at the center of the glass substrate and the edges, which occurs even without any plasma discharge.

2   DSMC Simulation of the "BigMag" In-Line Coater

For the DSMC method based on Ref. [Bir94] a parallel code has been implemented at Fraunhofer IST in C++ using g++ 3.3.5 and the parallel environment PVM3[1]. The simulation runs are carried out on a Linux cluster with ten AMD/Opteron-250 processors and a GBit ethernet network.

In this DSMC system the worker processes are pure C++ programs which are capable of handling a set of simple geometric units such as rectangular boxes

---

[1] PVM stands for "parallel virtual machine", see http://www.csm.ornl.gov/pvm/

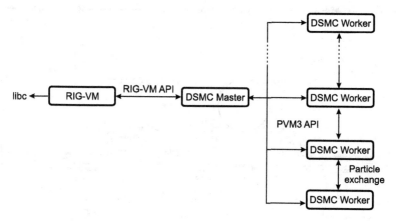

Fig. 2. Software framework of the DSMC simulation system developed at Fraunhofer IST

or cylinders. Each volume contains a distribution of super particles, whereof each represents a number N_R of gas molecules. Typically, N_R is in the order of $10^{10} \ldots 10^{14}$. It is possible to define rectangular or circular connecting surfaces between adjacent volume units in order to exchange super particles. Depending on whether these volume units are hosted by the same worker process or by two different processes the particle transfer is either performed internally or via exchanging PVM messages.

The master process is also written in C++ as a sub-class of a scripting language "RIG-VM" developed at Fraunhofer IST. RIG-VM[2] has a C-like syntax and is designed for simplifying communication between different embedded numerical algorithms. In this case, we use RIG-VM for definition of the recipient geometry – i. e. the set of volume units, their connections and assignments to the worker tasks – as well as the control of the overall calculation schedule. As shown in Fig. 2 the master process can be controlled by the scripting language via the RIG-VM API, while the communication with the worker tasks uses PVM.

The chamber geometry is decomposed into many rectangular volume units, as shown for M8-M10 in Fig. 3. The whole DSMC model comprises modules M5-M11, i. e. a total volume of approx. 7.5 m^3, and consists of 1005 rectangular units. The substrate size is 3.21×1.0 m^2, i. e. the substrate length in move direction is 1.0 m. For simulation of substrate movement the area traversed by the substrate is divided into stripes sized $3.21 \times 0.05 \times 0.004$ m^3. Each stripe can be either connected to the surrounding gas volume or – if the substrate is present – disconnected. By subsequently disconnecting / connecting stripes at the front / rear side of the substrate, the movement can be resembled at a resolution of 5 cm in transport direction.

The DSMC simulation was carried out for an Argon inflow of 600 sccm distributed homogeneously below the two shieldings beneath the cylindrical

---

[2] For a documentation of "RIG-VM" see http://www.simkopp.de/rvm/

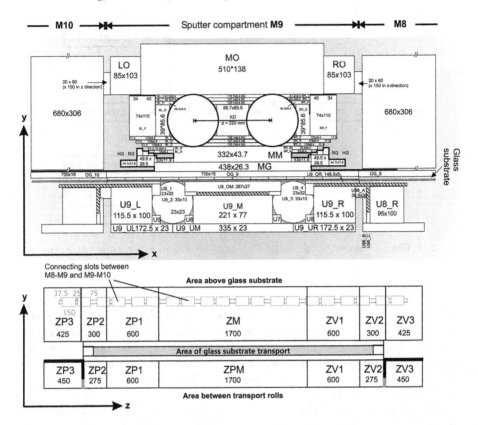

Fig. 3. Decomposition of the "BigMag" volume into rectangular boxes. The upper graph shows the decomposition of module M9 in the x-y plane, while the lower graph contains the extrusion of the geometry along the sputter target direction, i. e. the z-axis. The whole DSMC model comprises seven sputtering compartments M5-M11 and consists of 1005 rectangular boxes in total.

targets. For the initial substrate position, where the left glass edge is located at the connection between M7 and M8, 150000 DSMC time cycles at a time step of $\delta t = 2.5 \times 10^{-5}$ s are performed, while the total number of simulation particles stabilizes in the order of 5×10^6. Additional 10000 steps of time-averaging are performed thereafter. For each subsequent position of the glass substrate, which is travelling to the left direction, the positions and velocities of all particles are stored in temporary files which allows to continue the calculation after application of the geometry modification. 25000 time steps are performed at each position, which corresponds to a substrate travelling speed of 4.8 m/s. For 57 substrate positions in total, this takes a calculation time of approx. five days on a Linux cluster with 10 Opteron processors at 2.4 GHz.

For parallelization, N volume units have to be assigned to the M worker processes. The *individual load* L_i for each volume unit i can be either estimated by its actual number of particles or measured with the clock() function during

Table 1. Load balancing results for 1005 volume units on a 10-CPU Opteron-250 cluster. Simulation runs comprise approx. 10^6 super particles at an average Ar-pressure of 100 mPa.

M	Network connections		Time for 1000 cycles [s]			
	Particles	CPU-load	Particles		CPU-load	
1	0	0	438		438	
2	202	80	297	(74%)	272	(81%)
3	286	83	255	(57%)	219	(67%)
4	366	119	241	(45%)	170	(64%)
5	366	158	213	(41%)	150	(58%)
6	406	162	169	(43%)	141	(52%)
7	446	185	148	(42%)	122	(51%)
8	489	209	152	(36%)	114	(49%)
9	498	250	127	(38%)	122	(40%)
10	474	247	145	(30%)	118	(37%)

a couple of test cycles. Additionally, a $N \times N$ matrix C is constructed during setup of the geometry, whereof $C_{ik}(i < k)$ is the number of connections between volume elements i and k. If $P_i \in [1 \ldots M]$ is the process, where volume element i is assigned to, the quantity E to be minimized by the load balancer is

$$
E = \gamma_1 \underbrace{\left(\sum_{i<k \,\wedge\, P_i \neq P_k} C_{ik} \right)^2}_{\text{network traffic}} + \gamma_2 \underbrace{\sum_{i=1,M} \left(\overline{L} - \sum_{k=1,N;\; P_k=i} L_k \right)^2}_{\text{load fluctuations}}, \qquad (1)
$$

whereof $\overline{L}$ is the average load per process, and γ_1, γ_2 are weighting factors. The minimization is carried out with the "Simulated Annealing" [Kir83] method by randomly changing the assignments between volume units and processes. With L_k given in [μs], a ratio of $\gamma_1/\gamma_2 = 100 \ldots 1000$ has been found to yield a good compromise between minimization of load fluctuations and network traffic.

The load balance results for 1005 volumes and approx. 10^6 simulation particles are summarized in Tab. 1, with L_k either given by the number of particles or by the measured CPU-load of volume unit k. The latter method yields better results, since the computational effort of the collision calculation is strongly nonlinear in pressure. With a random initial process assignment vector P, the initial number of communication paths going over the network is in the range of 2000, for $M = 2$ it is still in the range of 1400. As shown in Tab. 1 this quantity is reduced by about 90% after load balancing. For $M > 1$ the relative speedup compared to $M = 1$ is given in percent, while for $M \geq 7$ the communication overhead obviously becomes dominant. Better speedup efficiencies are obtained at higher gas pressures with higher ratio between CPU-load and communication overhead. Further improvements might be achieved by using a low-latency network such as Myrinet and by using MPI instead of PVM.

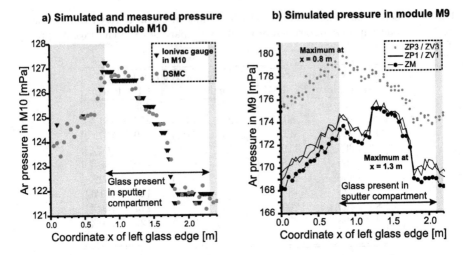

Fig. 4. (a) Simulated and measured pressure at a Ionivac pressure gauge located in segment "ZP3" of module M10 and (b) simulated pressure in module M9 at the center (segments "ZM" and "ZP1/ZV1") as well as at the boundaries (segments "ZP3/ZV3"). All pressure values are plotted against the substrate coordinate.

3   Results and Discussion

By comparing pressure measurements of an Ionivac pressure gauge located at the coign of module M10 with DSMC simulations, a very good agreement between measurement and simulation is obtained, as shown in graph (a) of Fig. 4. It shall be noted in this context, that the measured pressure had to be rescaled by a common factor since the absolute value of the Ionivac gauge was decalibrated.

In this picture, a monotonically decreasing pressure is found for the time period, during which the travelling glass substrate is being deposited in chamber M9. Thus, if a two-dimensional model of the reactive sputtering process in the x-y plane is an appropriate description and if the sputtering process follows immediately the pressure fluctuations, a monotonic shape of the resulting film thickness profile on the substrate is expected.

However, a closer look at the simulated pressure fluctuations in the sputtering module M9 – as shown in graph (b) of Fig. 4 – reveals that the shape of the pressure fluctuations is different for the central region (segments "ZM" and "ZP1/ZV1", see Fig. 3) of M9 in comparison to its boundary regions (segments "ZP3/ZV3"): While in the boundary regions a pressure maximum occurs at a substrate coordinate of $x = 0.8$ m, in the center a maximal pressure is obtained for $x = 1.3$ m. For the latter case of the pressure fluctuations, a "u" or "n" shaped film thickness profile on the glass substrate is expected rather than a monotonic profile. It shall be noted that – in the x-y plane – all simulation data is taken from the center of volume unit "MM" (see upper graph in Fig. 3).

With a two-dimensional heuristic model of the reactive in-line sputtering process in the "BigMag" coater, the process response and film thickness profile

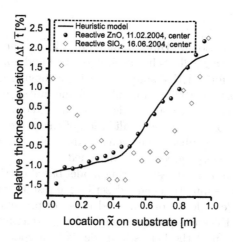

Fig. 5. Simulated and measured relative film thickness profile on a glass substrate sized 3.21×1.0 m^2 after a one-pass dynamic deposition in the "BigMag" in-line coater. In case of a ZnO deposition the prediction of the two-dimensional heuristic model of reactive sputtering is in good agreement with the experiment while it fails for other experiments, e. g. a reactive SiO$_2$ deposition, where the process conditions do not allow a simplified treatment of the system in two dimensions.

due to the DSMC simulated pressure fluctuations were calculated as reported in [Pfl04]. In this case the pressure field was averaged over the z-direction of the geometry. As shown in Fig. 5, the two-dimensional treatment could be appropriate under certain conditions as shown in case of a measured ZnO-thickness profile. This profile has been obtained by sputtering in oxide mode at high total pressure. However, there are also experiments – such as the measured SiO$_2$ deposition profile – showing that a simplified two-dimensional treatment within the heuristic sputtering simulation.

The reason for these deviations is obtained by a detailed look on the pure DSMC gas flow simulation data: During substrate movement a long sputtering source is confrontated with differently shaped pressure fluctuations with respect to the location at the target surface. For a strong coupling via the electron drift current it may be appropriate to assume a homogeneous ion generation probability in front of the target surface which allows a two-dimensional treatment of the system. However for a weaker coupling along the target race track, a three-dimensional treatment, i. e. modeling of a segmented target with inhomogeneously distributed oxidation degrees is necessary.

The nature of the coupling via the ring currents depends on many details. In addition to usual process parameters such as discharge power or total pressure it may also depend on whether a DC or a pulsed power is applied to the targets. Also the presence of shieldings close to the target surfaces could have an impact on the coupling behaviour. It will be subject of further investigations to develop a heuristic model of an in-line sputtering process based on the detailed information obtained from threedimensional DSMC simulations.

4 Conclusion

A parallel, threedimensional "Direct Simulation Monte Carlo" (DSMC) simulation system for rarified gas flows has been developed and applied on the problem of an in-line coater for reactive sputtering with a moving glass substrate. The DSMC model consists of 1005 rectangular volume units, comprises seven modules of an in-line coater as usually designed for architectural glazing which is a total volume of approx. 7.5 m^3. For validation of the model, the measured pressure fluctuations during substrate movement are compared with the simulation; as a result, a very precise agreement is obtained.

A two-dimensional heuristic sputtering model based on the gas flow data obtained by the DSMC simulation could describe the resulting process fluctuations and thickness profile at least for a limited range of process conditions. In case of a measured film thickness profile obtained from a reactive sputtering process of ZnO the prediction of the heuristic model is in a good agreement with the experiment. However, in other cases a strong disagreement is found between the two-dimensional simulation and the experiment as in the case of a reactive SiO$_2$ deposition experiment.

A detailed look on the DSMC simulation data reveals that the reason for this disagreement is most probably a phase shift of the pressure fluctuations at different positions above the substrate in the sputtering compartment. This phase shift even occurs under pure gas flow conditions, i. e. with no discharge plasma. The development of a heuristic model in three dimensions with a segmented target will be an issue for further investigations.

Acknowledgment

The authors gratefully acknowledge the financial contribution to parts of this work from the BMBF under contract No. 02PP2001 and the VolkswagenStiftung under contract No. I/79 263.

References

[Ber05]   S. Berg, T. Nyberg, *Fundamental understanding and modeling of reactive sputtering processes* Thin Solid Films **476** (2005) 215-230.

[Ber87]   S. Berg, H. O. Blohm, T. Larsson, C. Nender, *Modeling of reactive sputtering of compound materials*, J. Vac. Sci. Technol. **A5** (1987) 202-7.

[Bir94]   G. A. Bird, *Molecular gas dynamics and the direct simulation of gas flows*, Oxford Engineering Science Series 42 (1994).

[Kir83]   S. Kirkpatrick, C. D. Gelatt, M. P. Vecci, *Optimization by Simulated Annealing*, Science **220** (1983) 671-680.

[Pfl04]   A. Pflug, B. Szyszka, M. Geisler, A. Kastner, C. Braatz, U. Schreiber, J. Bruch, *Modeling of the film thickness distribution along transport direction in in-line coaters for reactive sputtering*, Proc. 47th SVC Tech. Conf. (2004) 155-160.

[Pfl03]   A. Pflug, B. Szyszka, V. Sittinger, J. Niemann, *Process Simulation for Advanced Large Area Optical Coatings*, Proc. **46**th SVC Tech. Conf. (2003) 241-247.

Parallel Simulation of T-M Processes in Underground Repository of Spent Nuclear Fuel

Jiří Starý, Radim Blaheta, Ondřej Jakl, and Roman Kohut

Institute of Geonics, Academy of Sciences of the Czech Republic
stary@ugn.cas.cz, blaheta@ugn.cas.cz, jakl@ugn.cas.cz, kohut@ugn.cas.cz

1 Introduction

In the background of our interest in the modelling of thermo-mechanical phenomena is its relevancy to the assessment of underground repositories of nuclear waste - a highly urgent topic worldwide, with great impact on the future of nuclear power utilization. In this context, one of the most internationally recognised project is the Äspö Prototype Repository in Sweden, which is a full-scale experimental realisation of the KBS-3 concept of spent nuclear fuel repository [3], where modelling of phenomena such as heat transfer, moisture migration, solute transport and stress/strain development can be verified.

This paper deals with mathematical simulation of the KBS prototype nuclear waste repository in a simplified form. We consider the finite element solution of thermo-elasticity problems, which are one-sidedly coupled. Thus, we can divide the problem into two parts. Firstly, we determine the temperature distribution by solving a nonstationary heat equation. Secondly, we solve a linear elasticity problem at required time levels.

The numerical solution of both problems leads to the repeated solution of large linear systems. For this purpose, we developed iterative solvers based on the conjugate gradient method with Schwarz-type preconditioners. As we shall demonstrate, their parallelization greatly improves the efficiency of the solution.

2 From Thermo-Elasticity to Linear Equations

The thermo-elasticity problem is formulated to find the temperature $\tau = \tau(x,t)$ and the displacement $u = u(x,t)$,

$$\tau \colon \ \Omega \times (0,T) \to R \,, \qquad u \colon \ \Omega \times (0,T) \to R^3 \,,$$

that fulfill the following equations

$$\kappa \rho \frac{\partial \tau}{\partial t} = k \sum_i \frac{\partial^2 \tau}{\partial x_i{}^2} + q(t) \qquad\qquad \text{in} \quad \Omega \times (0,T) \,,$$

$$-\sum_j \frac{\partial \sigma_{ij}}{\partial x_j} = f_i \quad (i = 1,\ldots,3) \qquad\qquad \text{in} \quad \Omega \times (0,T) \,,$$

B. Mohr et al. (Eds.): PVM/MPI 2006, LNCS 4192, pp. 391–399, 2006.

$$\sigma_{ij} = \sum_{kl} c_{ijkl} \left[\varepsilon_{kl}(u) - \alpha_{kl}(\tau - \tau_0) \right] \quad \text{in} \quad \Omega \times (0, T),$$

$$\varepsilon_{kl}(u) = \frac{1}{2} \left(\frac{\partial u_k}{\partial x_l} + \frac{\partial u_l}{\partial x_k} \right) \quad \text{in} \quad \Omega \times (0, T)$$

together with the corresponding boundary and initial conditions. Above, κ is the specific heat, ρ is the density of material, k is coefficient of the heat conductivity, q is the density of the heat source, f is the density of the gravitational forces, c_{ijkl} are components of the elasticity tensor, α_{kl} are the coefficients of the heat expansion and τ_0 is the reference (initial) temperature. The values of the material constants, used in the solved problem, can be found in [1].

After the variational formulation, the whole thermo-elasticity problem is discretized by the finite elements in space and the finite differences in time. Linear finite elements and the simplest time discretization lead to the solution of linear equations for vectors τ^j, u^j of nodal temperatures and displacements at the time levels t_j $(j = 1, \ldots, N)$ with the time steps $\Delta t_j = t_j - t_{j-1}$. It gives the time stepping algorithm presented in Figure 1.

> **find** τ^0: $M_h \tau^0 = \tau_0$
> **find** u^0: $A_h u^0 = b^0 = b_h(\tau^0)$
> **for** $j = 1, \ldots, N$:
>
>   **compute** $d^j = \left(M_h - (1-\vartheta)\Delta t_j K_h \right) \tau^{j-1}$
> $+ \vartheta q_h^j + (1 - \vartheta) q_h^{j-1}$
>
>   **find** τ^j: $(M_h + \vartheta \Delta t_j K_h)\tau^j = d^j$
>   **find** u^j: $A_h u^j = b^j = b_h(\tau^j)$
>
> **end for**

Fig. 1. The time stepping algorithm for thermo-elasticity problems. In practice, we prefer the backward Euler time steps given by $\vartheta = 1$.

Here, M_h is the capacitance matrix, K_h is the conductivity matrix, A_h is the stiffness matrix, q_h comes from the heat sources, b_h represents volume and surface forces including a thermal expansion term and $\vartheta \in \langle 0, 1 \rangle$ is a parameter.

We aim at the development of robust, stable methods and therefore we restrict our attention to implicit methods with $\vartheta \in \langle \frac{1}{2}, 1 \rangle$. Particularly, we shall consider two cases, with $\vartheta = \frac{1}{2}$ and $\vartheta = 1$, which correspond to the Crank-Nicolson (CN) and backward Euler (BE) method, respectively.

To optimize the solution, we use adaptive time steps. Very roughly, it means that we test the time change of the solution and change the time step size if the variation is too small or large. Practically, the testing is based on a local comparison of the BE and CN steps [1].

3   Solution of Large Linear Systems

Most of the computational work is concentrated in the repeated numerical solution of two large systems of linear equations. For each time step, we must solve the linear system for the heat conduction,

$$(M + \Delta t K)\tau = d\,.$$

Thereafter, but only at given time levels, we also must solve the linear system for the elasticity,

$$Au = b\,.$$

In the existing in-house finite element software, we use an iterative solution of both systems based on the well proven preconditioned conjugate gradient (PCG) method. Whereas in the sequential case the preconditioning is based on the incomplete factorization, parallel solvers take advantage of the additive Schwarz method [2] for the preconditioning step. It means, that the PCG search directions are constructed from pseudoresiduals rather than from residuals. To compute the pseudoresidual from residual, the domain and the correspondingly space of solution vectors are decomposed into subdomains Ω_k and corresponding subspaces. Then the residual is restricted to subspaces, local contributions from subspaces are computed by solving restricted problems and results are summed up to the pseudoresiduals. The subspace computations can be done in parallel.

For elliptic elasticity problems, it is important to add a subspace corresponding to a global coarse grid to improve the preconditioner performance and ensure numerical scalability. In this case, we speak about two-level Schwarz method. In another papers [4], we describe how this coarse grid space can be built algebraically and investigate its properties. Furthermore, in a recent paper [5], we show that the Schwarz preconditioner without any coarse grid space (one-level Schwarz method) is sufficiently efficient for parabolic problems like the considered time dependent heat conduction.

4   Parallel Implementation

We conceived the realisation also as an opportunity to make a practical comparison between the two main standards in parallel programming, message passing and shared memory, and its main representatives, MPI and OpenMP standards. That is why we implemented the parallel solver in two variants.

Note that OpenMP requires shared-memory parallel hardware and allows bottom-up directive-based parallelization, as a rule focusing on the most time-consuming loops, whereas message passing of MPI is supported and generally available on all parallel architectures including distributed-memory systems, and may require fundamental restructuring of the original sequential code.

In our case, both solvers, written in Fortran, follow the same algorithm and apply the same parallel decomposition, thus being directly comparable. In this decomposition, the k-th of m parallel processes corresponds to the subproblem Ω_k and works with its particular portion of data, including the matrices M_k, K_k and the vectors τ_k, q_k, for example, and follows the time stepping algorithm presented in Figure 1.

Due to a special one-dimensional domain decomposition, the communication requirements are fairly small in this approach. In the iterative phase, the k-th process communicates just locally with its neighbours, i.e. the $(k+1)$-th and $(k-1)$-th processes, mainly when the matrix-by-vector multiplication and the preconditioning are performed. Moreover, the amount of data transferred is small, proportional to the overlapped region. Thus, even the MPI-based parallel solver has potential to be efficient and scalable.

Having some background in MPI and PVM parallel programming, it was quite easy for us to switch to OpenMP and to mimic the MPI parallelization using the parallel region constructs. Merely addressing of subdomains in the global (shared) data structures was a little bit tricky.

5   Parallel Computing of the KBS Model

The Äspö prototype repository consists of a 65 m long tunnel, which lies 450 m below the ground surface within crystaline lithology. It is divided into two separate sections with four and two identical deposition holes, respectively. In these 1.75 m diameter and 8 m deep holes, heater canisters simulating the heating from the radioactive waste are emplaced. The engineered barriers are created by canisters surrounded by bentonite, transport tunnel closed by a backfill material and the natural rock massif.

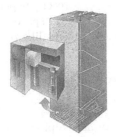
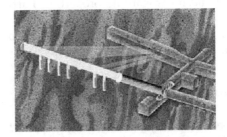

Fig. 2. KBS-3 concept of the SNF prototype repository

A constructed 3D model of the prototype repository, shortly named KBS, considers a coupled thermo-mechanical problem in the computational domain having dimensions $158 \times 57 \times 115$ m. The thermal source, decayed exponentially in time, is given by the radioactive waste. The rock is isotropic and its mechanical properties do not change with the temperature variations. We assume that the heat is transferred only by conduction.

The boundary conditions for the mechanical part consist of the weight of the overburden at the upper face of the model and zero normal displacements and zero shear stresses on the other faces. For the thermal part, we assume zero heat flux on the face corresponding to the plane of the symmetry and the original rock temperature $\tau = 10\,^{\circ}$C on the rest of the model boundary. This temperature also gives the initial condition.

The model is discretized by linear tetrahedral finite elements with $2\,586\,465$ DOF for the heat transfer and $7\,759\,395$ DOF for the elasticity computations. The time interval is to be 100 years, the adaptive time stepping begins with the time step 10^{-4} and requires 47 time steps in total. The development of temperature and stress fields is monitored in selected time levels of 1, 4, 10, 19, 50, 75 and 100 years.

5.1   Finite Element Software GEM

The described KBS model is implemented within the in-house FEM software named GEM. The discretization is based on structured meshes, which can be viewed as an adaption of uniform, reference mesh to the geometry of the solved problem. The hexahedra are divided into tetrahedra and the flexibility of the regular meshes is enhanced by the fact that some tetrahedra can remain void.

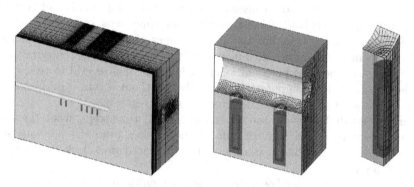

Fig. 3. FE mesh for the KBS model

The pre-processor strategy supports starting from an initial very coarse approximation of the situation and gradual refinement of this initial approximation with the aid of interpolation of nodal coordinates and material distribution. During this refinement, some details can be modified by supplying the data corresponding to the smaller and smaller details of the constructed model.

5.2   Results of Computations

First, let us consider the computation of the nonstationary heat conduction. In Table 1, we can observe the dependence of the number of PCG (preconditioned conjugate gradient) iterations on the time step size Δt and various number of subproblems/processors #P. To show this dependence, just one time step, which started from the initial zero guess and continued up to the relative residual accuracy 10^{-6}, is considered. The behaviour of the solvers in the other time steps is similar.

The results demonstrate the numerical stability of the parallel solvers based on additive Schwarz domain decomposition without a coarse grid, i.e. the number

Table 1. The dependence of the number of PCG iterations on the time step size Δt (in years) and various number of subproblems #P

#P	Δt								
	0.0001	0.001	0.01	0.1	1.0	5.0	10.0	100.0	1000.0
	Without coarse grid								
1	11	11	16	26	38	46	60	109	193
2	12	12	16	26	38	49	64	118	222
4	12	12	16	26	38	49	64	125	238
8	14	16	20	26	39	50	68	146	281
12	14	16	20	25	42	54	78	183	328
16	14	16	20	26	42	56	84	212	395
	With coarse grid								
4	18	17	17	27	41	50	53	83	142

of iterations remains almost constant with the increasing number of subproblems/processors. This holds for sufficiently small time steps, say $\Delta t \leq 5$, acceptable for most applications. For the given number of subproblems, the number of iterations naturally grows with increasing time step. This fact encourages the idea to employ one-level preconditioner without the coarse grid instead of the two-level one, cf. the computations for #P=4 without a coarse grid and with the coarse grid of $60 \times 10 \times 17$ nodes created by aggregation.

Further, we shall consider the full sequence of 47 time steps, when the linear system is always solved with the initial guess taken from the previous step. Table 2 shows the number of iterations, the measured wall-clock time and the relative speedup.

The tests were performed on a shared memory multiprocessor Sun Fire E15000 with the theoretical peak performance 86 GFlops. In total, it consists of 48 UltraSPARC-III/900 processors, 48 GB of shared memory, Sun Fireplane system interconnect with data transfer capacity up to 9.6 GB/s and 3.4 TB disk storage. The system is divided into 4 virtual servers and we used the largest one with 36 CPUs and 36 GB of memory assigned.

Table 2. Parallel computations on Sun SMP. The total number of iterations # It and the computation time T in dependence on the number of subproblems # P. The relative speed-up S of the parallel solver is related to the sequential run of the same code.

#P	OpenMP			MPI		
	# It	T [s]	S	# It	T [s]	S
1	1341	6292		1344	5931	
2	1421	4101	1.63	1424	3169	1.87
4	1425	2082	3.44	1428	1577	3.76
8	1514	1120	6.34	1514	833	7.12
12	1578	872	8.48	1581	596	9.95
16	1614	751	10.09	1618	483	12.28

Both the parallel solvers show a good scalability up to 16 processors. The MPI code is roughly 36% faster than its equivalent OpenMP counterpart. This suggests further investigation and optimization of the OpenMP code.

To give an idea about the whole modelling procedure, we present also the computations of one elasticity problem in Table 3. The efficiency of the parallel MPI solver based on the conjugate gradient method and both one-level Schwarz preconditioner without a coarse problem and two-level Schwarz preconditioner with auxiliary global problem created by regular 6×6×6 aggregation was tested using zero initial approximation.

The tests were performed on a Beowulf cluster. This distributed memory system consists of 8 computing nodes, each equipped by AMD Athlon/1400 processor, 1.5 GB of memory and 2 FastEthernet interfaces. The system includes also one interactive node identical to the computing nodes and the fileserver with two AMD Athlon MP/1900 processors, 1 GB of memory and 100 GB of disk space.

Table 3. Parallel computations on Beowulf cluster. The total number of iterations # It, the computation time T and the computation time per one iteration T_1 in dependence on the number of subproblems # P. The coarse grid is created by aggregation of 6×6×6 mesh nodes.

# P	Without coarse grid			With coarse grid		
	# It	T [s]	T_1 [s]	# It	T [s]	T_1 [s]
1	332	6264	18.87			
2	395	3347	8.47	144	1273	8.84
4	491	2190	4.46	160	752	4.70
7	546	1494	2.74	170	534	3.14

5.3 From the Engineering Point of View

The temperature reaches maximum values in the time level 19 years after the installation of the cannisters, see Figure 4. The temperatures depend on the

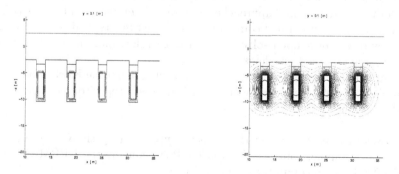

Fig. 4. The temperature field after 1 year (left) and 19 years (right). Range 10 - 85 °C.

initial temperature of rocks and the heating power of the canisters, which is determined by the time of cooling the spent nuclear fuel in an interim repository. The knowledge of the temperature distribution is necessary for computation of the stresses in rock and assessment of the stability issues.

The computation of stresses in rocks is a more complicated task. These stresses are developed at least in three subsequent phases: (1) phase of virgin rocks and the initial in-situ stress, (2) phase of excavation and stresses induced by it, (3) phase after installation of the canisters and filling the deposition holes and the transport tunnel with stresses induced by the thermal load. Here, we present the results from very simplified modelling, which putted all the phases together and assumed the initial in-situ stress to be caused only by the weights of rocks. The corresponding values of hydrostatic pressure (range 0 - 7.5 MPa) and shear stress intensity (range 0 - 5.5 MPa)are then shown in Figure 5.

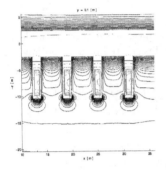 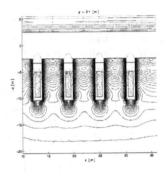

Fig. 5. Hydrostatic preassure (left) and shear stress intensity (right) after 19 years

6   Conclusion

The paper deals with the parallel simulation of thermo-mechanical processes in the underground SNF repository. We briefly describe the numerical methods used for the modelling and approaches to parallel computing on both distributed and shared memory parallel computers. The experiments confirms a very good efficiency of the developed solvers and their usefulness for the solution of such kind of large practical problems.

Note that the analysis of repository constructed along the KBS-3 concept can be applied also to the assessment of nuclear waste repository projects in the Czech Republic.

Acknowledgement. This work has been supported by the Grant Agency of the Czech Republic, contract No. 105/04/P036, and by the UPPMAX project P2004009 "Parallel computing in Geosciences".

References

1. R. Blaheta, P. Byczanski, R. Kohut, A. Kolcun, R. Šňupárek: *Large-Scale Modelling of T-M Phenomena from Underground Reposition of the Spent Nuclear Fuel.* In: P. Konečný et al (eds.): EUROCK 2005. A.A.Balkema, Leiden, 2005, pp. 49–55.
2. B. Smith, P. Bjørstad, W. Gropp: *Domain decomposition. Parallel multilevel methods for Elliptic Partial Differential Equations.* Cambridge University Press, New York, 1996.
3. C. Svemar, R. Pusch: *Prototype Repository - Project description.* IPR-00-30, SKB, Stockholm, 2000.
4. R. Blaheta, J. Nedoma eds.: *Numerical Models in Geomechanics and Geodynamice.* Special issue of Future Generation Computer Systems, volume 22, issue 4 Elsevier, 2006 pp. 447–448.
5. R. Blaheta, R. Kohut, M. Neytcheva, J. Starý: *Schwarz Methods for Discrete Elliptic and Parabolic problems with an Application to Nuclear Waste Repository Modelling,* submitted to Mathematics and Computers in Simulation, IMACS/Elsevier, special issue Modelling 2005.

On the Usability of High-Level Parallel IO in Unstructured Grid Simulations

Dries Kimpe[1,2], Stefan Vandewalle[1], and Stefaan Poedts[2]

[1] Technisch-Wetenschappelijk Rekenen, K.U.Leuven,
Celestijnenlaan 200A, 3001 Leuven, België
{Dries.Kimpe, Stefan.Vandewalle}@cs.kuleuven.be
[2] Centrum voor Plasma-Astrofysica, K.U.Leuven,
Celestijnenlaan 200B, 3001 Leuven, België
Stefaan.Poedts@wis.kuleuven.be

Abstract. For this poster, the usability of the two most common IO libraries for parallel IO was evaluated, and compared against a pure MPI-IO implementation. Instead of solely focusing on the raw transfer bandwidth achieved, API issues such as data preparation and call overhead were also taken into consideration. The access pattern resulting from parallel IO in unstructured grid applications, which is also one of the hardest patterns to optimize, was examined.

Keywords: MPI, parallel IO, HDF5, parallel netcdf.

Parallel IO is of vital importance in large scale parallel applications. MPI offers excellent support for parallel IO since version 2, particularly because of its fundamental and complete support for user defined data types. However, as demonstrated by the popularity of netcdf[5] and HDF5[3], applications are in need of a higher level API that enables them to deal with data more naturally. Parallel netcdf[4] and the implementation of MPI-IO support in HDF5 fulfill this need for MPI applications. As the software stack for this kind of storage can be quite complex, performance is easily lost if the coupling between the layers is not done carefully.

The motivation for this work originates in the investigation of a performance problem in a parallel unstructured grid code relying on HDF5 for file storage. In this code, after partitioning, all CPUs need to read the coordinates and values of all grid points that were assigned to them by the mesh partitioner. This results in an almost random access pattern consisting of collective read operations. While eventually the total dataset is read, a non-contiguous subset is accessed during every read operation.

The authors did everything possible to assure efficient IO, for example by utilizing collective data transfers with complete HDF5 type descriptions and a continuous storage layout. Still, the code performed poorly when scaling to larger CPU counts. Examination of the HDF5 source code revealed that no parallel IO was supported for point selections in a dataset.[1] This resulted in

---

[1] This holds for both the latest stable release (1.6.5) and the current alpha release (1.8.0).

B. Mohr et al. (Eds.): PVM/MPI 2006, LNCS 4192, pp. 400–401, 2006.
© Springer-Verlag Berlin Heidelberg 2006

every CPU executing an independent read request for every accessed element of the dataset, leading to seriously degraded IO performance.

HDF5 has extensive support for partial dataset selection and another method to access the same subset was found. While this method did support parallel IO, it suffers from another kind of problem. Although the final read operation itself takes advantage of parallel IO and custom data types, the API needed to setup this selection requires repeatedly calling a function with time complexity $O($`number_of_currently_selected_elements`$)$. For a random selection of n points, this leads to $n!$ operations. Searching for alternatives, parallel netcdf was tested as well, but was also shown to have issues preventing efficient data access (for the described access pattern).

For this poster, an effort was made to describe best practices to achieve high performance with the discussed storage libraries, evaluated in the context of unstructured grid applications. Problems affecting performance, in both the API and internal implementation, are highlighted. Actual performance measurements, demonstrating achievable bandwidth, were made in combination with true parallel filesystems such as lustre[1] and PVFS2[2] and more traditional ones such as NFS. All tests were performed on an opteron based cluster situated at K.U.Leuven.

As a preliminary conclusion, application writers in need of directly available performance are better off directly using MPI-IO whenever possible. This is particularly true for the class of irregular access patterns considered in our study. Relying on a storage library that fails to utilize the flexibility and power that MPI-IO offers results in a significant loss of performance. In principle, nothing prevents high level IO libraries from achieving the same performance as raw MPI-IO. However, at this moment their implementations need to mature somewhat more before this becomes true.

References

1. Lustre: A Scalable, High-Performance File System, white paper, November 2002, http://www.lustre.org/docs/whitepaper.pdf.
2. Rob Latham, Neil Miller, Robert Ross and Phil Carns: A Next-Generation Parallel File System for Linux Clusters, LinuxWorld, Vol. 2, January 2004.
3. HDF5: http://hdf.ncsa.uiuc.edu/HDF5/.
4. Li, J., Liao, W., Choudhary, A., Ross, R., Thakur, R., Gropp, W., Latham, R., Siegel, A., Gallagher, B., and Zingale, M. 2003. Parallel netCDF: A High-Performance Scientific I/O Interface. In Proceedings of the 2003 ACM/IEEE Conference on Supercomputing (November 15 - 21, 2003). Conference on High Performance Networking and Computing. IEEE Computer Society, Washington, DC, 39. Jianwei Li, Wei-keng.
5. Rew, R., Davis, G., and Emmerson, S., "NetCDF User's Guide, An Interface for Data Access Version 2.3," available at ftp.unidata.ucar.edu, April 1993.

Automated Performance Comparison

Joachim Worringen

C&C Research Laboratories, NEC Europe Ltd.
http://www.ccrl-nece.de

Keywords: benchmark, performance comparison, perfbase, test automation.

1 Motivation

Comparing the performance of different HPC platforms with different hardware, MPI libraries, compilers or sets of runtime options is a frequent task for implementors and users as well. Comparisons based on just a few numbers gained from a single execution of one benchmark or application are of very limited value as soon as the system is to run not only this software in exactly this configuration. However, the amount of data produced for thorough comparisons across a multi-dimensional parameter space quickly becomes hard to manage, and the relevant performance differences hard to locate. We deployed *perfbase* [3] as a system to perform performance comparisons based on a large number of test results yet being able to immediately recognize relevant performance differences.

2 Automation of Performance Comparison

perfbase is a toolkit that allows to import, manage, process, analyze and visualize arbitrary benchmark or application output for performance or correctness analysis. It uses a SQL database for data storage and a set of Python command line tools to interact with the user. Its concept is to define an *experiment* with *parameter and result values*, import data for different *runs* of the experiment from arbitrarily formatted text files, and perform queries to process, analyze and visualize the data. The presented framework for automated performance comparison is a set of shell scripts and *perfbase* XML files. With this framework, only four simple steps are required to produce a thorough comparison:

1. Define the range of parameters for execution (i.e. number or nodes or processes) in the *job creation script*.
2. Execute the *job creation script*, then the *job submission script*. Wait for completion of the jobs.
3. Run the *import script* which uses *perfbase* to extract relevant data from the result files and store it in the *perfbase* experiment.
4. Run the *analysis script* which issues *perfbase* queries to produce the performance comparison. Changing parameters in the analysis script allows to modify the comparison result.

Two examples will illustrate the application of this framework to a single micro-benchmark (*Intel MPI Benchmark*) or a suite of application kernel benchmarks (*NAS Parallel Benchmarks*).

B. Mohr et al. (Eds.): PVM/MPI 2006, LNCS 4192, pp. 402–403, 2006.

2.1   Intel MPI Benchmark

The *Intel MPI Benchmark* [2] is a well-known and widely used MPI micro bench-mark which measures the performance of individual MPI point-to-point commu-nication patterns and collective communication operations. A single run of this benchmark with 64 processes will perform 80 tests with 24 data sizes each. For each data size, between 1 and 3 latencies are reported, resulting in more than 5000 data points. This amount of data can hardly be analyzed manually. In-stead, we define a threshold for results being considered as differing. Only for these cases, we report a single line with the key information like percentage of data points being different, the average difference and the standard deviation. The full range plots showing absolute and relative performance is generated as well and can be analyzed based upon the summary report.

2.2   NAS Parallel Benchmarks

The *NAS Parallel Benchmarks* [1] are an established set of application kernels often used for performance evaluation. The execution of the NPB can be varied across the kernel type, data size and number of processes. Together with the vari-ation of the component to be evaluated and recommended multiple executions, a large number of result data (performance in MFLOPS) is generated. From this data, we generate a report consisting of a table for each kernel with rows like C 64 4 6.78. In this case, the 64 process, 4 processes per node execution of the corresponding kernel for data size C delivered 6.79% more performance with variant A than with variant B. The data presented in the tables is also visualized using bar charts.

3   Conclusion

The application of the *perfbase* toolkit allows to thoroughly but still conveniently compare benchmark runs performed in two different environments. The impor-tant features are the management of a large number of test runs combined with the filtering of non-relevant differences. This allows to actually do in-depth com-parisons based on a large variety of tests. The framework can easily be applied to other benchmarks. The *perfbase* toolkit is open-source software available at http://perfbase.tigris.org and includes the scripts and experiments described in this paper.

References

1. D. H. Bailey et al. The nas parallel benchmarks. *The International Journal of Supercomputer Applications*, 5(3):63–73, Fall 1991.
2. N.N. *Intel MPI Benchmarks: Users Guide and Methodolgy Description*. Intel GmbH, http://www.intel.com, 2004.
3. J. Worringen. Experiment management and analysis with perfbase. In *IEEE Cluster 2005*, http://www.ccrl-nece.de (Publication Database), 2005. IEEE Computer Society.

Improved GROMACS Scaling on Ethernet Switched Clusters

Carsten Kutzner[1], David van der Spoel[2], Martin Fechner[1], Erik Lindahl[3],
Udo W. Schmitt[1], Bert L. de Groot[1], and Helmut Grubmüller[1]

[1] Department of Theoretical and Computational Biophysics, Max-Planck-Institute of
Biophysical Chemistry, Am Fassberg 11, 37077 Göttingen, Germany
[2] Department of Cell and Molecular Biology, Uppsala University, Husargatan 3,
S-75124 Uppsala, Sweden
[3] Stockholm Bioinformatics Center, SCFAB, Stockholm University, SE-10691,
Stockholm, Sweden

Abstract. We investigated the prerequisites for decent scaling of the
GROMACS 3.3 molecular dynamics (MD) code [1] on Ethernet Beowulf
clusters. The code uses the MPI standard for communication between
the processors and scales well on shared memory supercomputers like the
IBM p690 (Regatta) and on Linux clusters with a high-bandwidth/low
latency network. On Ethernet switched clusters, however, the scaling
typically breaks down as soon as more than two computational nodes are
involved. For an 80k atom MD test system, exemplary speedups Sp_N on
N CPUs are $Sp_8 = 6.2$, $Sp_{16} = 10$ on a Myrinet dual-CPU 3 GHz Xeon
cluster, $Sp_{16} = 11$ on an Infiniband dual-CPU 2.2 GHz Opteron cluster,
and $Sp_{32} = 21$ on one Regatta node. However, the maximum speedup
we could initially reach on our Gbit Ethernet 2 GHz Opteron cluster was
$Sp_4 = 3$ using two dual-CPU nodes. Employing more CPUs only led to
slower execution (Table 1).

When using the LAM MPI implementation [2], we identified the all-
to-all communication required every time step as the main bottleneck.
In this case, a huge amount of simultaneous and therefore colliding messages "floods" the network, resulting in frequent TCP packet loss and
time consuming re-trials. Activating Ethernet flow control prevents such
network congestion and therefore leads to substantial scaling improvements for up to 16 computer nodes. With flow control we reach $Sp_8 = 5.3$,
$Sp_{16} = 7.8$ on dual-CPU nodes, and $Sp_{16} = 8.6$ on single-CPU nodes.

For more nodes this mechanism still fails. In this case, as well as
for switches that do not support flow control, further measures have to
be taken. Following Ref. [3] we group the communication between M
nodes into $M - 1$ phases. During phase $i = 1 \ldots M - 1$ each node sends
clockwise to (and receives counterclockwise from) its ith neighbouring
node. For large messages, a barrier between the phases ensures that the
communication between the individual CPUs on sender and receiver node
is completed before the next phase is entered. Thus each full-duplex link
is used for one communication stream in each direction at a time.

We then systematically measured the throughput of the ordered all-to-
all and of the standard MPI_Alltoall on $4 - 32$ single and dual-CPU nodes,
both for LAM 7.1.1 and for MPICH-2 1.0.3 [4], with flow control and without. The throughput of the ordered all-to-all is the same with and without

B. Mohr et al. (Eds.): PVM/MPI 2006, LNCS 4192, pp. 404–405, 2006.
© Springer-Verlag Berlin Heidelberg 2006

flow control. The lengths of the individual messages that have to be transferred during an all-to-all fell within the range of 3 000 . . . 175 000 bytes for our 80k atom test system when run on 4 − 32 processors. In this range the ordered all-to-all often outperforms the standard MPI_Alltoall. The performance difference is most pronounced in the LAM case since MPICH already makes use of optimized all-to-all algorithms [5].

By incorporating the ordered all-to-all into GROMACS, packet loss can be avoided for any number of (identical) multi-CPU nodes. Thus the GROMACS scaling on Ethernet improves significantly, even for switches that lack flow control.

In addition, for the common HP ProCurve 2848 switch we find that for optimum all-to-all performance it is essential how the nodes are connected to the ports of the switch. The HP 2848 is constructed from four 12-port BroadCom BCM5690 subswitches that are connected to a BCM5670 switch fabric. The links between the fabric and subswitches have a capacity of 10 Gbit/s. That implies that each subgroup of 12 ports that is connected to the fabric can at most transfer 10 Gbit/s to the remaining ports. With the ordered all-to-all we found that a maximum of 9 ports per subswitch can be used without losing packets in the switch. This is also demonstrated in the example of the Car-Parinello [6] MD code. The newer HP 3500yl switch does not suffer from this limitation.

Table 1. GROMACS 3.3 on top of LAM 7.1.1. Speedups of the 80k atom test system for standard Ethernet settings (Sp), with activated flow control (Sp_{fc}), and with the ordered all-to-all (Sp_{ord}).

CPUs	single-CPU nodes						dual-CPU nodes				
	1	2	4	8	16	32	2	4	8	16	32
Sp	1.00	1.82	2.24	1.88	1.78	1.73	1.94	3.01	1.93	2.59	3.65
Sp_{fc}	1.00	1.82	3.17	5.47	8.56	1.82	1.94	3.01	5.29	7.84	7.97
Sp_{ord}	1.00	1.78	3.13	5.50	8.22	8.64	1.93	2.90	5.23	7.56	6.85

References

1. van der Spoel, D., Lindahl, E., Hess, B., Groenhof, G., Mark, A.E., Berendsen, H.J.C.: GROMACS: Fast, Flexible, and Free. J. Comput. Chem. 26 (2005) 1701–1718
2. The LAM-MPI Team. http://www.lam-mpi.org/
3. Karwande, A., Yuan, X., Lowenthal, D.K.: An MPI prototype for compiled communication on Ethernet switched clusters. J. Parallel Distrib. Comput. 65 (2005) 1123–1133
4. MPICH-2. http://www-unix.mcs.anl.gov/mpi/mpich/
5. Thakur, R., Rabenseifner, R., Gropp, W.: Optimization of collective communication operations in MPICH. Int. J. High Perform. Comput. Appl. 19 (2005) 49–66
6. Hutter, J., Curioni, A.: Car-Parrinello molecular dynamics on massively parallel computers. ChemPhysChem 6 (2005) 1788–1793

Asynchrony in Collective Operation Implementation

Alexandr Konovalov[1], Alexandr Kurylev[2],
Anton Pegushin[1], and Sergey Scharf[3]

[1] Intel Corporation, Nizhny Novgorod Lab, Russia
[2] University of Nizhni Novgorod, Russia
[3] Institute of Mathematics and Mechanics Ural Branch RAS, Ekaterinburg, Russia

Abstract. Special attention is being paid to the phenomenon of divergence between synchronous collective operations and parallel program load balancing. A general way to increase collective operations performance while keeping their standard MPI semantics suggested. A discussion is addressed to internals of MPICH2, but approach is quite common and can be applied to MPICH and LAM MPI as well.

Collective operations significantly increase both programmer performance and expressibility of message-passing programs. But increase in expressiveness level should either be supplemented by a good load balance between interacting processes or a program performance can suffer from not ready yet processes waiting overhead. Looking from the scalability perspective, it becomes more and more problematic to guarantee satisfactory load balancing, but most of the papers on collective operation implementation lack estimations of poor load balancing influence on collectives' performance: it's assumed that computations are well-balanced.

Mentioned problem can definitely be ignored in case of using asynchronous collective operations. But asynchronous collectives haven't become a part of the MPI-2 standard. It can be speculated that authors of MPI-2 standart assumed that asynchronous collectives could be supported through generic requests. Our belief is that trying to implement all this logic on the user-level (outside of MPI library) is more or less a reinventing the wheel. Progress Engine is a well thought-out, effective mechanism which only partially suffers from the lack of user interoperability.

Poor load balance between collective operation participants can influence algorithmic part of a collective operation. For example, an optimal broadcasting algorithm in the worst "total unreadiness" case would send data from a root node to everyone, not by using binomial tree. Optimal solution for "real-world tasks" is somewhere in between, supposedly, in the field of highly branched trees. This paper addresses the problems of unbalancing on the implementation level.

Despite the collective operation algorithms diversity [1], they have a common feature: they use some nodes for message transit, i.e. for retranslation of received information further. But current MPI Progress Engines do not support transit and retranslation, so collective operations have to be implemented via "common"

B. Mohr et al. (Eds.): PVM/MPI 2006, LNCS 4192, pp. 406–407, 2006.

point-to-point operations. As a result, performance may suffer significantly if a transit node is busy with computing and is not ready to participate in a collective operation yet. It's clear that the subset of nodes, which will suffer from performance degradation depends on the collective operation algorithm. If we do a broadcasting using binomial tree, lagging node descendants suffer. If there is ring-based gather, lower nodes of lagging node are affected.

We came up with a prototype implementation of active broadcasting for eager (i.e., "quite small") messages in the MPICH2-1.0.x environment. It works in the following way. Message re-sending starts right after broadcast message came to a transit node. The algorithms of original Bcast (for example, binomial tree) is used for retranslation, but message sending performed in an asynchronous manner. Thus, useful time from receiving message to Bcast function call is utilized by background transmissions. In the best case our background broadcasting finalizes before actual user's Bcast call and as a result node's siblings complete their broadcastings earlier (maybe before the beginning of Bcast call on a parent transit node at all).

Next parameters were added to MPICH2's packet for proposed optimization:

1. target communicator;
2. collective operation details;
3. additional tag.

The first two items are used to determine important Bcast parameters without actual user's Bcast call. Communicator was added because it is impossible to send the message without it. As a result, we have to know all communicators on each processes, so exchanges were added into all communicator management operations (communicator creation never turned out to be a performance-critical operation). "Operation details" include only root's rank for Bcast. Additional tag was added to escape the problems of not-in-time receiving like in point-to-point case, because collective operations are now internally asynchronous.

Idea behind async Bcast can be used for optimization of other collectives as well. Let's draw a quick sketch of a possible Gather implementation. It's possible to send transit packets right after arriving supplementing them with a transit node info, if Gather is already called on the node. Key point for the performance is a good lagged-packets-operating strategy: "Should they be send via binomial tree or directly to a root process?"

According to preliminary results, some variant of GAMESS demonstrates 2.5% computation speedup. It's quite significant, because Bcast takes only 6.8% of run time. Additional details including project sources can be found at http://parallel-debugger.itlab.unn.ru/en/optimization.html. Work by one of the co-authors, Sergey Scharf, was done with the financial support of Russian Fund of Basic Research (#04-07-90138-b).

References

1. Thakur R., Gropp W.: Improving the Performance of Collective Operations in MPICH // Proc. of the Euro PVM/MPI 2003, LNCS **2840** (2003) 257–267.

PARUS: A Parallel Programming Framework for Heterogeneous Multiprocessor Systems

Alexey N. Salnikov

Moscow State University Faculty of Computational Mathematics and Cybernetics,
119992 Moscow, Leninskie Gory, MSU, 2-nd educational building,
VMK Faculty, Russia
salnikov@cs.msu.su

Abstract. PARUS is a parallel programing framework that allows building parallel programs in data flow graph notation. The data flow graph is created by developer either manually or automatically with the help of a script. The graph is then converted to C++/MPI source code and linked with the PARUS runtime system. The next step is the parallel program execution on a cluster or multiprocessor system. PARUS also implements some approaches for load balancing on heterogeneous multiprocessor system. There is a set of MPI tests that allow developer to estimate the information about communications in a multiprocessor or cluster.

Most commonly, parallel programs are created with the libraries that generate parallel executable code, such as MPI for cluster and distributed memory architectures, and OpenMP for shared memory systems. The tendency to make parallel coding more convenient led to the creation of software front-ends for MPI and OpenMP. These packages are intended to get rid the user from the part of problems related to parallel programming. Several examples of such front-ends are DVM [2], Cilk [3], PETSc [4], and PARUS [5]. The latter is being written by the group of developers headed by the author.

PARUS is intended for writing the program as a data-dependency graph. The data-dependency graph representation gives the programmer several advantages. In the case of splitting the program into very large parallel executed blocks, it is convenient to declare the connections between the block and then execute each block on its own group of processors in the multi-processor system. The algorithm is represented as a directed graph where vertices are marked up with series of instructions and edges correspond to data dependencies. Each edge is directed from the vertex where the data are sent from to the vertex that receives the data. Afterwards, the vertex processes the data and collects the data in memory for delivering to other vertices of the graph. Thereby, the program may be represented as a network that has source vertices (they usually serve for reading input files), internal vertices (where the data are processed), and drain vertices, where the data are saved to the output files and the execution terminates. Then, the graph is translated into a C++ program that uses the MPI library. The resulting program automatically tries to minimize processors load imbalance and data trasmission overhead.

B. Mohr et al. (Eds.): PVM/MPI 2006, LNCS 4192, pp. 408–409, 2006.

One of the targets of this research was to investigate how the data-dependency graph approach to writing parallel programs can be applied to the following examples: 1) a distributed operation over a large array of data, 2) an artificial neural network (perceptron), 3) multiple alignment problem. In order to evaluate the performance of PARUS, we designed the following tests.

The first test uses a recursive algorithm that computes the result of an associative operation to all elements of an array. Two examples of such operations are summation and maximization. Every block is treated by its own processor and the transmission delays are ignored. The algorithm requires $O(log_m(n))$ operations, where m is the parameter of the algorithm that corresponds to number of array elements per processor. Value of m is set to cover data transmission overhead. Testing this implementation on MVS-1000M with 100 processors revealed a 40 times speedup on an array sized 10^9.

Second, PARUS was used to simulate a three layer perceptron with a maximum of 18,500 neurons in each layer. The maximum acceleration that was achieved was over 7 times.

Third, an algorithm of multiple sequence alignment was implemented. The problem is important in molecular biology. The parallel implementation was based on the MUSCLE package (http://www.drive5.com/muscle). The procedure of construction of an alignment was parallelized. The parallelism is based on the alignment profiles and evolutional tree (cluster sequence tree). We perform align of profiles concerned with each level of tree in parallel. The speedup of parallel program in comparison with original MUSCLE depends on the degree of cluster tree balance. Well balanced tree will provide high perfomance of parallel program. The program was used to align all human-specific LTR class 5 in the EMBL data bank [1]. The test has demonstrated a 2.4 times speedup on 12 processors on a Prime Power850 machine. This work was a part of the project supported by CRDF grant No. RB01227-MO-2 and by RFBR grant No. 05-07-90238.

PARUS has been installed and tested on the following multiprocessors: MVS-1000M http://www.top500.org/system/5871,
http://www.jscc.ru/cgi-bin/show.cgi?path=/hard/mvs1000m.html&type=3 (cluster of 768 Alpha processors), IBM pSeries690 (SMP 16 processors Power4+), Sun Fujitsu PRIMEPOWER 850 Server (SMP 12 processors SPARC64-V).

References

1. Alexeevski A.V., Lukina E.N., Salnikov A.N., Spirin S.A. Database of long terminal repeats in human genome: structure and synchronization with main genome archives //Proceedings of the fourth international conference on bioinformatics of genome regulation and structure, Volume 1. BGRS 2004, pp 28-29 Novosibirsk.
2. The DVM system: http://www.keldysh.ru/dvm/
3. The Cilk language: http://supertech.csail.mit.edu/cilk/
4. The PETSc library: http://www-unix.mcs.anl.gov/petsc/petsc-as/
5. The PARUS system: http://parus.sf.net/

Application of PVM to Protein Homology Search

Mitsuo Murata

Tohoku Bunka Gakuen College,
6-45-16 Kunimi, Aoba-ku, Sendai 981-8552, Japan

Although there are many computer programmes currently available for searching homologous proteins in large databases, none is considered satisfactory for both speed and sensitivity at the same time. It has been known that a very sensitive programme could be written using the algorithm of Needleman and Wunsch [1]. This algorithm first calculates the maximum match score of two protein sequences on a two-dimensional array, MAT(m,n), where m and n are the lengths of the two sequences (the average length is 364 amino acids in the Swiss-Prot database [2]). The similarity or homology between the two sequences is then assessed statistically by comparing the score from the real sequences and the mean score from a large number (>200) of pairs of random sequences that are produced by scrambling each of the original sequences. Homology search using this algorithm means that this statistical analysis must be carried out between the query sequence and every sequence in the database sequentially. Consequently, as the size of database increases – the well-known TrEMBL database now contains over 2,500,000 protein sequences (about 962 Mbytes), homology search by this method becomes very time consuming.

A new programme that was named SEARCH, written in C and based on the Needleman-Wunsch algorithm was created for homology search. A large amount of CPU time required by the straightforward implementation of the algorithm was reduced to a practical level by improving the algorithm and by optimising the programme, i.e. full statistical analyses were not carried out on *a priori* non-homologous pairs and the most CPU intensive parts of the programme were written in assembly language. SEARCH was run to find sequences homologous to cucumber basic protein (CBP, 96 amino acids) [3] in the Swiss-Prot database, which contained 204,086 sequences. The search was completed in 5 min 32 sec on a Pentium 4 2.8 MHz computer. There were 159 homologous proteins, which included 20 plastocyanins: plastocyanin is a photosynthetic electron transport protein, and which has been known to be homologous to CBP from physicochemical characteristics. When the same search was carried out, for comparative purposes, using BLAST [4] and FASTA [5], which are the two most frequently used programmes (run at http://www.expasy.ch/tools/), however, these programmes found no plastocyanin. This seems to indicate that SEARCH is a more sensitive programme.

When Swiss-Prot and TrEMBL were combined to include 2,710,972 sequences and used as the database, the search time was 1 hr 15 min 16 sec. To improve the search time, the PVM system was employed: PVM 3.4.5 was installed in 41 Pentium 4 2.8 GHz computers, consisting of 1 master and 40 slaves, and running under Linux. A small C programme was first written and used to divide the database file into 40 smaller files containing an equal number of sequences (except the last one), which

B. Mohr et al. (Eds.): PVM/MPI 2006, LNCS 4192, pp. 410–411, 2006.
© Springer-Verlag Berlin Heidelberg 2006

were then distributed to the slaves. The file size varied from 17 to 35 Mbytes (median 26 Mbytes) depending on the sizes of the proteins therein. The schedule of programme execution is as follows: The master initiates SEARCH in the slaves by sending out the query sequence. Each slave carries out the search and, whenever homology is found, it sends back the name and score of the homologous sequence to the master. The master sorts the reported sequences according to score, and when the search is completed in all slaves, it produces a result file which contains the names and scores of homologous proteins. This type of PVM application, data parallelism, seems particularly suited in this application, in which a large database is divided into smaller parts in the slaves. This is allowed as the statistical analysis of the Needleman-Wunsch algorithm is, unlike with some other search programmes, carried out only between the query sequence and one sequence in the database at a time.

When SEARCH was run under this system using the same query sequence and databases as above, the search was completed in 2 min 6 sec, improving the search time about 36-fold. Considering the communication overhead inherent in the PVM system and the fact that the time spent on statistical analysis is not uniform among the slaves – it takes longer if the proteins are larger and also if there are more potentially homologous proteins in the database, the 36-fold improvement using 40 computers seems reasonable. Furthermore, the names and scores of homologous proteins listed were the same as the ones obtained in the single computer system. Therefore, it was concluded that no data were lost while being sent from the slaves to the master. In a separate experiment, database files in the slaves were made to contain not the same number of proteins but a similar amount of data i.e. a similar number of amino acids (about 26 Mbytes/slave). When SEARCH was run with this database system, however, search was slower by about 14% (2 min 24 sec). Similarly sized databases may have contributed to lowering the efficiency of communication between the master and slaves.

That the task of each slave is completely independent of those of other slaves and rather infrequent communication using small amounts of data seem to make the PVM system very effective in the sort of application described here.

References

1. Needleman, S., Wunsch, C.: A General Method Applicable to the Search for Similarities in the Amino Acid Sequence of Two Proteins. J. Mol. Biol. 48 (1970) 443-453
2. Web site for Swiss-Prot and TrEMBL: http://www.expasy.ch/sprot/sprot-top.html
3. Murata, M., Begg, G.S., Lambrou, F., Leslie, B., Simpson, R.J., Freeman, H.C., Morgan, F.J.: Amino Acid Sequence of a Basic Blue Protein from Cucumber Seedlings. Proc. Natl. Acad. Sci. USA 79 (1982) 6434-6437
4. Altschul, S.F., Gish, W., Miller, W., Myers, E.W., Lipman, D.J.: Basic Local Alignment Search Tool. J. Mol. Biol. 215(3) (1990) 403-410
5. Pearson, W.R., Lipman, D.J.: Improved Tools for Biological Sequence Analysis. Proc. Natl. Acad. Sci. USA 85 (1988) 2444-2448

Author Index

Lecture Notes in Computer Science

For information about Vols. 1–4065

please contact your bookseller or Springer

Vol. 4116: R. De Prisco, M. Yung (Eds.), Security and Cryptography for Networks. XI, 366 pages. 2006.

Vol. 4115: D.-S. Huang, K. Li, G.W. Irwin (Eds.), Computational Intelligence and Bioinformatics, Part III. XXI, 803 pages. 2006. (Sublibrary LNBI).

Vol. 4114: D.-S. Huang, K. Li, G.W. Irwin (Eds.), Computational Intelligence, Part II. XXVII, 1337 pages. 2006. (Sublibrary LNAI).

Vol. 4113: D.-S. Huang, K. Li, G.W. Irwin (Eds.), Intelligent Computing, Part I. XXVII, 1331 pages. 2006.

Vol. 4112: D.Z. Chen, D. T. Lee (Eds.), Computing and Combinatorics. XIV, 528 pages. 2006.

Vol. 4111: F.S. de Boer, M.M. Bonsangue, S. Graf, W.-P. de Roever (Eds.), Formal Methods for Components and Objects. VIII, 447 pages. 2006.

Vol. 4110: J. Díaz, K. Jansen, J.D.P. Rolim, U. Zwick (Eds.), Approximation, Randomization, and Combinatorial Optimization. XII, 522 pages. 2006.

Vol. 4109: D.-Y. Yeung, J.T. Kwok, A. Fred, F. Roli, D. de Ridder (Eds.), Structural, Syntactic, and Statistical Pattern Recognition. XXI, 939 pages. 2006.

Vol. 4108: J.M. Borwein, W.M. Farmer (Eds.), Mathematical Knowledge Management. VIII, 295 pages. 2006. (Sublibrary LNAI).

Vol. 4106: T.R. Roth-Berghofer, M.H. Göker, H. A. Güvenir (Eds.), Advances in Case-Based Reasoning. XIV, 566 pages. 2006. (Sublibrary LNAI).

Vol. 4104: T. Kunz, S.S. Ravi (Eds.), Ad-Hoc, Mobile, and Wireless Networks. XII, 474 pages. 2006.

Vol. 4102: S. Dustdar, J.L. Fiadeiro, A. Sheth (Eds.), Business Process Management. XV, 486 pages. 2006.

Vol. 4099: Q. Yang, G. Webb (Eds.), PRICAI 2006: Trends in Artificial Intelligence. XXVIII, 1263 pages. 2006. (Sublibrary LNAI).

Vol. 4098: F. Pfenning (Ed.), Term Rewriting and Applications. XIII, 415 pages. 2006.

Vol. 4097: X. Zhou, O. Sokolsky, L. Yan, E.-S. Jung, Z. Shao, Y. Mu, D.C. Lee, D. Kim, Y.-S. Jeong, C.-Z. Xu (Eds.), Emerging Directions in Embedded and Ubiquitous Computing. XXVII, 1034 pages. 2006.

Vol. 4096: E. Sha, S.-K. Han, C.-Z. Xu, M.H. Kim, L.T. Yang, B. Xiao (Eds.), Embedded and Ubiquitous Computing. XXIV, 1170 pages. 2006.

Vol. 4095: S. Nolfi, G. Baldassare, R. Calabretta, D. Marocco, D. Parisi, J.C. T. Hallam, O. Miglino, J.-A. Meyer (Eds.), From Animals to Animats 9. XV, 869 pages. 2006. (Sublibrary LNAI).

Vol. 4094: O. H. Ibarra, H.-C. Yen (Eds.), Implementation and Application of Automata. XIII, 291 pages. 2006.

Vol. 4093: X. Li, O.R. Zaïane, Z. Li (Eds.), Advanced Data Mining and Applications. XXI, 1110 pages. 2006. (Sublibrary LNAI).

Vol. 4092: J. Lang, F. Lin, J. Wang (Eds.), Knowledge Science, Engineering and Management. XV, 664 pages. 2006. (Sublibrary LNAI).

Vol. 4091: G.-Z. Yang, T. Jiang, D. Shen, L. Gu, J. Yang (Eds.), Medical Imaging and Augmented Reality. XIII, 399 pages. 2006.

Vol. 4090: S. Spaccapietra, K. Aberer, P. Cudré-Mauroux (Eds.), Journal on Data Semantics VI. XI, 211 pages. 2006.

Vol. 4089: W. Löwe, M. Südholt (Eds.), Software Composition. X, 339 pages. 2006.

Vol. 4088: Z.-Z. Shi, R. Sadananda (Eds.), Agent Computing and Multi-Agent Systems. XVII, 827 pages. 2006. (Sublibrary LNAI).

Vol. 4087: F. Schwenker, S. Marinai (Eds.), Artificial Neural Networks in Pattern Recognition. IX, 299 pages. 2006. (Sublibrary LNAI).

Vol. 4085: J. Misra, T. Nipkow, E. Sekerinski (Eds.), FM 2006: Formal Methods. XV, 620 pages. 2006.

Vol. 4084: M.A. Wimmer, H.J. Scholl, Å. Grönlund, K.V. Andersen (Eds.), Electronic Government. XV, 353 pages. 2006.

Vol. 4083: S. Fischer-Hübner, S. Furnell, C. Lambrinoudakis (Eds.), Trust and Privacy in Digital Business. XIII, 243 pages. 2006.

Vol. 4082: K. Bauknecht, B. Pröll, H. Werthner (Eds.), E-Commerce and Web Technologies. XIII, 243 pages. 2006.

Vol. 4081: A. M. Tjoa, J. Trujillo (Eds.), Data Warehousing and Knowledge Discovery. XVII, 578 pages. 2006.

Vol. 4080: S. Bressan, J. Küng, R. Wagner (Eds.), Database and Expert Systems Applications. XXI, 959 pages. 2006.

Vol. 4079: S. Etalle, M. Truszczyński (Eds.), Logic Programming. XIV, 474 pages. 2006.

Vol. 4077: M.-S. Kim, K. Shimada (Eds.), Geometric Modeling and Processing - GMP 2006. XVI, 696 pages. 2006.

Vol. 4076: F. Hess, S. Pauli, M. Pohst (Eds.), Algorithmic Number Theory. X, 599 pages. 2006.

Vol. 4075: U. Leser, F. Naumann, B. Eckman (Eds.), Data Integration in the Life Sciences. XI, 298 pages. 2006. (Sublibrary LNBI).

Vol. 4074: M. Burmester, A. Yasinsac (Eds.), Secure Mobile Ad-hoc Networks and Sensors. X, 193 pages. 2006.

Vol. 4073: A. Butz, B. Fisher, A. Krüger, P. Olivier (Eds.), Smart Graphics. XI, 263 pages. 2006.

Vol. 4072: M. Harders, G. Székely (Eds.), Biomedical Simulation. XI, 216 pages. 2006.

Vol. 4071: H. Sundaram, M. Naphade, J.R. Smith, Y. Rui (Eds.), Image and Video Retrieval. XII, 547 pages. 2006.

Vol. 4070: C. Priami, X. Hu, Y. Pan, T.Y. Lin (Eds.), Transactions on Computational Systems Biology V. IX, 129 pages. 2006. (Sublibrary LNBI).

Vol. 4069: F.J. Perales, R.B. Fisher (Eds.), Articulated Motion and Deformable Objects. XV, 526 pages. 2006.

Vol. 4068: H. Schärfe, P. Hitzler, P. Øhrstrøm (Eds.), Conceptual Structures: Inspiration and Application. XI, 455 pages. 2006. (Sublibrary LNAI).

Vol. 4067: D. Thomas (Ed.), ECOOP 2006 – Object-Oriented Programming. XIV, 527 pages. 2006.

Vol. 4066: A. Rensink, J. Warmer (Eds.), Model Driven Architecture – Foundations and Applications. XII, 392 pages. 2006.